Missions for Science

Missions for Science

U.S. TECHNOLOGY AND MEDICINE IN AMERICA'S AFRICAN WORLD

DAVID MCBRIDE

Rutgers University Press

New Brunswick, New Jersey, and London

Library of Congress Cataloging-in-Publication Data

McBride, David
    Missions for science : U.S. technology and medicine in America's African world / David McBride.
        p. cm.
    Includes bibliographical references and index.
    ISBN 0-8135-3067-9 (cloth)
    1. Technology transfer—Southern States. 2. Technology transfer—North America. 3. Technology transfer—Liberia. 4. Medical innovations—Southern States. 5. Medical innovations—North America. 6. Medical innovations—Liberia. I. Title

T174.3 .M38 2002
338.96'06—dc21

                                                                    2001048790

British Cataloging-in-Publication information is available from the British Library.

Manufactured in the United States of America

CONTENTS

ACKNOWLEDGMENTS

Throughout the modern era global society has become infused with technological advances honed in the United States and its industrial partner-nations. This transfer of technologies from the United States has been conducted by a variety of American institutions—from philanthropies setting up medical projects and businesses seeking raw materials and markets to the military and media institutions establishing outlets around the world. The international impact of U.S. technologies has been so rapid and multidimensional that historians and cultural analysts have barely been able to provide chronicles and critiques of the roots and effects of this technological revolution. It was my long-standing recognition of the need to further explore America's technological influence that shaped my desire to write this book. Now that the book is complete, I must acknowledge that it is the product of the support I received along the way from numerous colleagues, institutions, friends, and family members.

Fellow investigators across many disciplines and professions as well as my students at Pennsylvania State University have helped me sharpen many points throughout the book. Among my colleagues, Monroe H. Little has been the most inspiring. He is one of our nation's most learned thinkers, with a truly unique grasp of the place of science and education in America and the black community. Other colleagues who were most helpful over the years in spurring me through the writing of this book include Darryl Thomas, Toni Miles, Christine Ahmed-Saidi, and Jan Guan.

I also must thank several associations and centers for national conferences and forums they organized. Over the years at these meetings I was able to present several papers that developed into aspects of this book. These organizations include the American Association for the History of Medicine; the National Library of Medicine Division of History; the Institute for Global Cultural Studies (SUNY—Binghamton); the Center for Multidisciplinary Applied Research in Urban Issues (State College of Buffalo); the John B. Slaughter Annual Conference on Science, Technology, and the Black Community, at the University of Maryland—College Park; the Latin American and Caribbean Studies Program at the University of Michigan—Ann Arbor; and the New York Academy of Science—History and Philosophy Section. I also benefited from a visiting professorship at the Rollins School of Public Health at Emory University and Morehouse Medical College. The editors and staff at Rutgers University Press, especially senior editor Helen Hsu, Jill Stuart, and Karen Johnson were indispensable in helping my book finally see the light of day.

Last, I must thank my friends and family. Lorraine Brown, Ida Belle Minnie, Simone Summers, Yasmin Villuendas, and Marida and Nuria Padro as well as the entire McBride-Jordan clan have been most encouraging. Thanks to Julian, Patrice, and TeeJay, in particular, and all my brothers and sisters, nieces and nephews, and our spiritual leader, Ruth McBride Jordan. These kinfolk have been relentless in confronting me at every family gathering in recent years, each demanding, "Let me have that copy of your new book now!" When I explained that I was still working on it, each of them backed away, the adults forgivingly, the teens and children suspiciously. Any shortcomings or errors in this book are, of course, mine alone. Any positive influences that grow from this book, like the positive changes technologies often bring, have only been made possible because of the support of genuinely good people and institutions like all those mentioned above.

Missions for Science

Introduction

*A*ll agree that science and technology have been central to America's rise to global prominence. Religious and cultural expansion, markets and military control, foreign allies, and scientific exploits—all inspired the European and American empires built across the Atlantic Ocean since the Age of Discovery. But what sets the ascent of the United States to world power apart from earlier civilizations are its technological advances. The description of biological and physical properties in natural phenomena and materials and the mechanical application of this knowledge were integral to the earliest stages of the North American colonies. This scientific knowledge led to exceptionally productive innovations in agriculture, manufacturing, and energy resources. As the United States passed through the nineteenth century, both its economy and government became even more wedded to nurturing large-scale technologies. Eventually, technologies in its military and industrial sectors empowered the United States to expand across the continent and beyond its Atlantic and Pacific coasts. Vital to the triumph of the capitalist economy in the United States, today technology and science hold the highest place in America's collective social values.[1]

In this book we explore how the pursuit of the scientific ideal and the technical and medical outgrowths of this pursuit have shaped African Diaspora populations in the United States and the larger Atlantic world. Our focus is on the societies in four tropical or semitropical Atlantic regions: the southern United States, the Panama Canal Zone, Haiti, and Liberia. This exploration begins with the pre–Civil War period of U.S. history but focuses primarily on the late nineteenth and twentieth centuries. The four Atlantic regions chosen for our study's geographic focus were selected because each grew into an important site for U.S. technical, military, and medical activities and each has a large African Diaspora population. They shared in common rural geography, tropical climate, and endemic diseases. Thus, these regions posed similar challenges in the transfer of the most advanced scientific, medical, and military resources. Moreover, as largely the creation of U.S. foreign politics, Liberia and the Canal Zone have a unique place in America's political and moral heritage as well as collective world view. Along with Haiti, each of these foreign regions has been an ongoing target for the science-based projects of U.S. philanthropy, government, or business. And all of these projects were themselves offshoots of the spirit or mission of science, a mission that captured the national imagination of the emerging modern United States.

As the United States industrialized, engaged their foreign territorial interests throughout the Atlantic world, and rose to the center of Western science and technology, it was helped forward by millions of people of African and indigenous American descent. Indeed, people from the Atlantic side of the African continent have been one of the longest-standing building blocks for this nation's astonishing agricultural and industrial growth. Ironically, the scholarship and popular writings about the development of the United States into the world's leading technological nation have been largely one-sided. These works focus most on how American science and technology drew from European intellectual, cultural, and economic connections. But the technological links across the Atlantic Ocean *after* slavery, links that bridged the two massive continents— America (North and Central) and Africa—remain largely overlooked.

Once slavery collapsed in the United States, a wide variety of administrative, commercial, and medical professionals with scientific skills transferred technology to the four regions we cover. They included religious workers, engineers, military officials, business managers, public health experts, physicians, and philanthropists. By tracing these activities, this book establishes the timing, nature, and geographic patterns in the movement of technology from the United States to the larger Atlantic world. This study looks at U.S. technical expansion from both sides of the Atlantic Ocean. On one hand, it covers the changing stages of technology and scientific specialism emerging in the "donor," urban industrial United States. On the other hand, it examines the effects of U.S. technical missions on the "host" U.S. rural South and three African-background Atlantic societies. These effects in the southeastern cotton states, Panama, Haiti, and Liberia include changes in traditional economies, disease patterns, health and educational institutions, public health resources, and politics of development. Also, nationalist leaders in these areas attempt to bend the power or charisma of science toward their individual political ends.

This story begins in the pre–Civil War period, when the industrial and slavery sectors battled to dominate America's economy and national politics. It ends with the 1970s, when the United States had become the leading power in the international political system. By this time agencies, military forces, corporations, and technological institutions of the United States were enmeshed throughout not only the Atlantic world, but a shrinking globe of six billion people.[2] The U.S. technical initiatives in the Panama Canal Zone, Haiti, and Liberia—and these initiatives' links to the U.S. Deep South—are particularly important yet little-studied chapters in the emergence of modern American technology.[3] These regions have long been popular in U.S. society's historic imagination. As such, they have been among the first in the evolving Third World to receive direct transfers of U.S. technology.

Although there has been virtually no historical writing on U.S. technology and post-slavery black Atlantic regions, scholarship on other aspects of the African Diaspora in the New World and the modern United States has grown. The focus of this growing historical and sociological scholarship has been on comparative slavery, the demography and disease patterns among Africans and Europeans in the New World, and the politics of post-emancipation race relations.[4] There has also been an expanding body of studies examining the influence of African American politics, the civil rights movement, and black religious activism on U.S. foreign affairs and African nations.[5] These works have been valuable in a number of ways. First, they have illuminated much about the slave trade and the institutionalization of slavery. These studies reveal valuable insights into the different socio-demographic and biological experiences of blacks throughout the Americas during the slavery era. Finally, these works have covered the subsequent political and social movements, involving legal discrimination, citizenship struggles, and race relations, that emerged throughout the post-emancipation Americas.

But to date there is no scholarship explaining how modern industrial and scientific advances—changes that came after emancipation and legal race segregation had long passed—shaped black Atlantic population centers. Moreover, there is little historical analysis of how shifting environmental factors and disease-control resources affected the social life and collective development of these regional populations and societies. Finally, we have little scholarship that describes how independent black Atlantic republics with close historical links to the United States themselves envisioned and attempted to use science and technology to build their nations.

The absence of scholarship on the influence of U.S. technology and science on post-emancipation black societies of North America, the Caribbean, and West Africa is ironic. It would seem logical that tracing U.S. technological development should include the labor, skills, and ingenuity contributed by the black Atlantic populations to this process. Enslaved Africans were the first people in our society who were transplanted to America en masse specifically so that they could function as a productive tool. Yet a look for scholarship on the connections between black Atlantic nations and population centers and the United States' modern scientific and technological development yields almost nothing. Why?

There have been two scholarly approaches especially popular in the history of American science, medicine, and technology. Each has contributed to the blind spot in current scholarship concerning the relationship between large-scale technologies and black population regions of the Atlantic community. First, many studies of America's technological past (including its medical advances) focus on small clusters of leading inventors, scientists, and industrial magnates. Mostly biographies and political and entrepreneurial histories, these conventional studies depict ingenious inventors or crafty business figures fostering industries for a young nation destined for world eminence. These entrepreneurial dynamos, according to these writings, propel America to international commercial and political leadership. It is America's great inventors and business speculators who pragmatically capitalized on new scientific discoveries and managerial know-how hatched in the firms of small-town or cosmopolitan America.

In recent decades a second cluster of scholars who use a "social constructionist" paradigm has emerged to explain America's technological past. These scholars, primarily sociologists and social and medical historians, have emphasized sociological approaches. According to this school of intellectuals, scientists are imbedded in clusters of competing professional peer groups—for example, engineers, physician-researchers, production managers, and academic scientists. These scientists mobilize knowledge, techniques, and research resources in ways that suit their own intellectual styles, professional stature, and (indirectly) society's nationalism.[6] Furthermore, scholars of the social constructionist paradigm take a systems approach to explain technological growth. They have argued that effective large-scale technological projects require the interrelation of social, economic, and political factors *in addition to* scientific experimentation and entrepreneurial drive.[7]

Both the biographical-political and social constructionist approaches describe American technology's development and transfer as a passionate process. These scholars emphasize that individual and group efforts making our society's complex, technological "black boxes" result from competition among scientists, new styles of thought, and even ideological, social, or political class interests. In the end the process contributes or yields to the universal and neutral workings of scientific principles, experimentation, and engineering—the three anchors of modern Western science. But the science historians have tended to neglect another key dimension. They have failed to adequately consider ethnic culture and gender patterns. More specifically, they have not weighed how racial group encounters and the institutions established to govern race relations can influence the making and maintenance of technologies. In the United States these race relations in the social-historical context crystallized into legal institutions such as slavery and Jim Crow segregation. These relations also take less direct forms such as racial friction between personnel in military institutions or specific industries, or practices in public education and housing that segregate racial group members.

Another factor explains why black societies of the Atlantic world have been overlooked in the history of U.S. technological modernization. In addition to blind spots in the key historical and sociological approaches to U.S. technology, there has been the influence of a widespread public perception or cultural assumption. This perception is that the technological environment is a fixed and unalterable backdrop to our lives and social concerns. This popular idea that technology is an autonomous pregiven, a neutral and inevitable force, is deeply rooted in middle-class American culture, past and present.[8] From the events leading to the Civil War to our current times, policy makers and educators involved in America's racial problem have tended to ignore technological values and institutions. Many assume that technology, especially engineered technology involving the national infrastructure, is mere "natural" background woven around our communities and cities and not especially relevant to the burning political issues of the times involving ethnic and race relations. Others believe that newer, revolutionary technology in, say, computers and telecommunications automatically works for the general good and, therefore, will somehow automatically lessen racial problems.

The blindness of much of the nation's educated to the pervasive strength of technological development is a major reason why the general lay public assumes that African Americans have never played an important role in the evolution of science and technology.[9] According to this popular view, blacks have only been simple consumers of modern commodities (TVs, cars, and the like). As for the role of blacks in the history of science and technology, this view is captivated exclusively with profiling "the-first-black-to" inventors.[10] Unfortunately, these intellectual and popular tendencies also divert the common citizenry from understanding that there may be technological causes and solutions for many of the most critical problems facing black America and poor countries such as Haiti and Liberia. These problems include large underclasses, political instability, high levels of preventable diseases, and educational institutions and economic systems that function inadequately if at all.

Missions for Science uses primarily the historical approach as well as aspects of the sociological approaches. However, this book broadens these approaches to take into account additional factors. These factors include cultural and social ideas and goals, relating to race, ethnicity, and gender, that are involved in the planning and execution of technical projects. Also covered is the influence of government- or state-sanctioned racial practices throughout U.S. history. These practices influence the racial division of labor and participation in technologies in those Atlantic regions hosting U.S. technical initiatives. It is only by using such a broad perspective that we establish the central thesis of this study.

This study argues that the deepening specialization and rationalization of American technical experts, medical professionals, and foreign aid authorities— accepting the assumption that social utopia will inevitably evolve from science— cause them to overestimate the impact of their technical projects on the national development of these societies. U.S. experts abroad and their sponsors at home become most concerned with scientific specialization and professionalism. Their highest priority becomes understanding scientific processes that are not seeable by the naked eye. With knowledge derived from the molecular world, these experts believe their professional specialties will yield great gains for the public good. However, these priorities have overshadowed available knowledge and policies that could have been used to help the populations of these regions experience less disease and increase public and higher education and overall living standards. Following World War II, we further argue, leaders in Haiti and Liberia pursued their own brand of scientific modernism. They undertook campaigns for national development and politics ostensibly modeled on U.S. scientific values and technological institutions. While these political leaders became riveted to importing the scientific mission and expertise of the United States, actual benefits from modern technology and biomedicine for their majority national populations never came to pass. As a result, these societies remain among the poorest in the Atlantic world, their political and educational institutions literally collapsing numerous times in recent decades.

The specific questions this book centers on are: What specific technologies and medical resources were transferred by U.S. government, philanthropy, and industries to these black population centers and why? How did the professed aims of U.S. technical projects, public health, and military activities differ from their actual effects and unforeseeable consequences in the health and social lives of populations in these four Atlantic regions? Were U.S. technical experts and programs in the black Atlantic societies able to sustain their advances for use by the broader, indigenous populations and institutions of these regions? Or, did the U.S. technical transfer amount to a form of hegemony—one in which regional leaders endorse the technology to reinforce a disguised dependency and anti-democratic rule? Finally, what lessons may we learn from the history of technology and medicine in these four geographic regions of the Atlantic world as we open a new century?

Chapter 1 covers the influence of technology on black Americans from the era of slavery to the post-Reconstruction South. Following the Civil War and Reconstruction, black Americans remained demographically and economically concentrated in the Deep South under the weight of the revived plantation system. The technology of the plantation shaped the economy and communities of southern black Americans into a region known as the Black Belt. Plantation agriculture became the central factor underlying the racial caste, poverty, and disease problems that vexed the South's poor, but most persistently the South's Black Belt.

Chapter 2 focuses on the first Atlantic region abroad in which U.S. technology had a long-term presence and social impact: the Panama Canal Zone and the surrounding Central American and Caribbean region. Beginning with the Spanish-American War and opening decades of the twentieth century, the United States forged its role as a new power among the industrial Western nations. By the opening decade of the twentieth century, the "unregenerate" South—its racial politics, planter class, and black subsociety—had drifted off the national political agenda. Europe had already colonized much of the tropical world. The Isthmus of Panama became the first stop abroad for large-scale technical projects administered directly by an expanding U.S. commercial and military establishment in a frenzy to catch up after the Spanish-American War.

Chapter 3 focuses on the United States and pre–World War II Haiti. The United States found it necessary to try to administer the technological and medical development of this nation through direct occupation. Once the United States became more involved in political and military friction with an aggressive Germany, later joined by Italy and Japan, Liberia become the fourth region in which U.S. technical projects concentrated. The gradual build up of interest in Liberia on the part of medical researchers, educational philanthropists, and U.S. businesses is the subject of chapter 4.

Chapter 5 covers the effects of early farm mechanization, mixed industry, and modern public health on the Black Belt, the core region of the plantation South. From the interwar decades through the 1940s American technology and medicine advanced greatly. A growing legion of scientists developed inside U.S. government, corporations, and academic institutions. In philanthropy the Rockefeller Foundation's International Health Board took the lead in research and in spurring public health programs to eliminate diseases such as hookworm and malaria. The Rockefeller Foundation first initiated medical and public health programs in the rural South, including the Black Belt, and then abroad throughout the tropical world. The United States' new network of scientists and technicians in industry and medicine built technologies that complemented the global spread of American military and economic influence. Products, machines, and techniques derived from experimentation in basic physics, chemistry, and biology began to lead technological innovation in the United States. But also, this technological modernization changed permanently the Black Belt South. The new industrialization and the expanding federal agencies and infrastructure of public works increased urban growth throughout the nation. Mechanization in agriculture slowly displaced manual production on cotton plantations. Public health programs and sanitation drastically reduced infectious diseases such as malaria, venereal diseases, and tuberculosis. The new urban growth, technological changes, and public health campaigns improved the South but ironically eroded and depopulated the South's traditional Black Belt plantations and community.

In chapters 6 and 7 we see that following World War II the United States' new, highly specialized scientific and technical institutions were reflected in the technical, medical, and public health resources that the United States sent to Haiti and Liberia. Both of these black republics attempted to incorporate the ideals and technical products of the modern U.S. scientific community. However, neither the U.S. technical initiatives nor those undertaken by Liberia's and Haiti's leaders themselves significantly furthered the political and economic autonomy of these black nations. The mission for science that propelled the United States to the top of global political leadership contributed to the failures of the Atlantic world's two oldest black republics.

PART I

Enduring the Technology Take-Off

*Today we have discovered a powerful new way to understand
the universe, a method called science.*
—Carl Sagan

CHAPTER 1

Machines and Plantations

BIRTH OF THE BLACK BELT

The Science Mission Unfolds

The coming of Africans to the western side of the Atlantic Ocean and the making of the many societies and subsocieties of African Diaspora populations throughout the Americas began over four centuries ago. Since the discovery of the New World, Africans have journeyed with European explorers and settlers. Together the newcomers opened paths and outposts for flows of settlers and slaves. Searches for natural riches, exotic goods, or fertile soil were interwoven with quests for religious and political freedom. The New World environment offered territory for the settler communities to develop polities free from the restrictions of Europe's monarchies and orthodox churches. Gradually farms, skilled trades, plantations, and manufacture emerged. As this loose network of mechanics, traders, and technical activities grew, it laced together the early economy of colonial North America.

During the colonial period, the European settlers used a mosaic of tools, farming styles, and plantation methods, as well as artisan skills and weaponry. Some of these techniques and tools were transplanted from the Old World. Others were adopted from Native Americans and/or enslaved Africans, or invented in the colonial settlements. Only with these tools and techniques could the settlers convert the new wilderness into economic use. Mechanical manufacturing, mathematical landscaping, and new armaments from Europe's artisans and scientists were also incorporated into this early regime of technical activity. As the wave of technical activities and instruments grew, the colonies needed more energy resources that, in turn, made necessary large energy technologies. By the time the North American colonies were a young republic, water wheels and steam engines, mills, factories, coach routes, and rail lines were radiating west throughout the interior of the New World along the Atlantic coast.

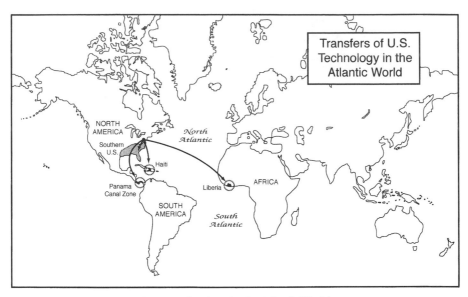

Map 1.1. Transfers of U.S. Technology in the Atlantic World.

The spirit or mission of science galvanized this expansion in mechanics, manufacture, and technical farming. This pursuit of using practical rationalism and mechanics to shape economic and social progress was rooted in the Enlightenment, that period that exalted logic, mathematics, and experimentation.[1] Now, as the North American colonies came into being, the mission to expand scientific knowledge became interwoven with the settlers' political and economic aspirations to build prosperous, unified colonies. From the colonial period, the political and commercial leaders in North America envisioned a Heaven-like, scientific universe as the overarching ideal that would hold the colonies under one government. Organized religion would be put outside of government, and the government of the new nation would be formed inside the realm of science.

Thomas Paine was an exemplary promoter of this idea. In his classic revolutionary tract, *The Age of Reason* (1794), he denounced the Christian church and "all other schemes of invented systems of religion" for holding humanity "in ignorance." To Paine organized Christianity had been "founded in nothing; it rests on no principles; it proceeds by no authorities; it has no data; it can demonstrate nothing; and it admits of no conclusion." Thus, Paine railed, "The study of theology, as it stands in Christian churches, is the study of nothing."[2] Paine believed that humanity could only obtain political democracy and harmony by releasing itself from the strictures of the Judeo-Christian religion, pursuing instead the understanding of the universe and the building of a national community through science: "All the knowledge . . . man has of science and of machinery, by the aid of which his existence is rendered comfortable upon earth, and without which he would be scarcely distinguishable in appearance and condition from a common animal, comes from the great machine and structure of the universe . . . not Moses and the prophets, nor Jesus Christ, nor his apostles. . . . The Almighty is the great mechanic of the creation; . . . the original teacher of all science. Let us then learn to reverence our master."[3]

The founding fathers passionately embraced ideals of science and the so-called mechanical arts. Thomas Jefferson followed a number of scientific interests, especially natural history. Benjamin Franklin was a prominent experimental scientist. He studied everything from magnetism, electricity, and geology to the physiology of humans and animals. Alexander Hamilton was trained in medicine. James Madison envisioned the Constitution as a Newtonian concept, an assortment of devices that would balance and maintain equilibrium between opposing groups and movements. When Madison and Jefferson planned the University of Virginia, the institution was to function as a "temple dedicated to science and liberty." Indeed, Jefferson believed that freedom itself was "the first-born daughter of science."[4]

The nation's political founders thought that by combining the new revelations of sciences with government America had gallantly and irreversibly broken from the monarchical European past. With science as its guidepost, they expected the American States would become a center for the highest levels of human intelligence, political democracy, and economy. Jefferson confided to John Adams in an 1813 letter that America's and Europe's new science was opening boundless opportunities for young democracies against "the artificial aristocracy, founded on wealth and birth." "Science had liberated the ideas of those who read and reflect," he wrote. Moreover, "the invention of gunpowder has armed the weak as well as the strong with [the power to inflict] missile death," making enlightened government even more necessary. Jefferson exhorted that "the American example had kindled feelings of right in the people. An insurrection has consequently begun, of science, talents, and courage, against rank and birth, which have fallen into contempt. . . . Science is progressive."[5]

The fixation on science—whether mathematical formulations, mechanical contraptions, rudimentary engineering, new uses for machinery and weaponry—gained a pivotal place in the political fabric of the United States. Indeed, by the mid-twentieth century, the belief that the nation's very existence depended on science's progress was a virtual absolute principle in American political and intellectual circles. It had been American scientists who provided the leadership for the defeat of Nazi Germany. The ability of U.S. scientists and weapon specialists in the Manhattan Project to "split the atom" and deliver atomic bombs against foes like Japan raised to a fever pitch postwar America's political faith in the power of science. Interest spiraled in two directions: the physical sciences leading to innumerable chemical products and nuclear energy; and the life sciences, from molecular biology to genetics, reshaping medicine, food processing, and agriculture.

Today, the ideal that science is opening a limitless frontier for progress is our nation's most cherished social value. The late astronomer Carl Sagan pleaded with his fellow Americans to stay true to the scientific ideal. Sagan called the world "our beleaguered little planet [with its] widespread torture, famine, and governmental criminal irresponsibility." To him, America's, indeed humanity's, only hope rested on our continued deep belief in what Jefferson called the "light of science."[6]

The Diaspora Unfolds

During the centuries between the colonial period and the Civil War, the scientific ideal became woven deep inside America's political, economic, and cultural institutions. In the meantime, the nation's working classes—manual laborers, factory workers, initially including women and children, slaves and ex-slaves, immigrant workers—plodded into the Industrial Age. The first great mass of African Americans merging into North America's early front line of mechanics and industry were the slaves concentrated in the colonial South. Also, a small segment of freed black people spread throughout the continent. During the nineteenth century, after the legal abolition of slavery up and down the colonies of the Americas, other people of the African Diaspora, as well as original populations of Central America and the West Indies, were brought within U.S. industry's sweep.

Throughout these regions in which black masses and America's scientific, mechanical, and industrial activities converged, the place of blacks in labor, agriculture, education, and public health took on patterns that differed from their European counterparts. Patterns also varied between the African Diaspora populations themselves. These variations depended on the geographies and economic systems where they were placed, born, or freely settled. Before emancipation the great mass of blacks throughout the Americas had been ensnared in dissimilar slave systems. The plantations for sugar and cotton production, for instance, tended to be larger and had harsher work environments compared to those plantations involving tobacco. Likewise, following the Civil War and Reconstruction, the role of blacks in the new large-scale manufacturing, agricultural, transportation, and public health institutions varied from nation to nation. Each society differed by the type and intensity of industrialization. Also, they varied by the extent that racial thought influenced the political and economic segments that nurtured or sponsored technologies.

In Haiti and Liberia, a political and ideological nexus with the United States was first laid down. Later, U.S. interests would set in place technical projects involving the agricultural and environmental characteristics unique to each nation. Before the Civil War both Liberia and Haiti held a special place in the politics of slavery and international relations of the United States. Both Haiti and Liberia were born out of the lingering battle to end slavery in the Americas. In 1806, after armed revolution by its African slaves against the French, Haiti was founded. It was hailed by radical abolitionists in both the United States and Latin America as a symbol of popular democracy's triumph over slavery, but dreaded by the more moderate or gradual abolitionists, colonizationists, and certainly the planter classes. To these factions, the Haitian Revolution was a monstrosity. Since they considered black masses as either biologically incapable or socially unready for civil government and economy, these conservatives believed the Haitian debacle could unduly influence slaves in other nations of the Americas to take up arms. A more fitting response to slavery, America's gradualists believed, was Liberia. This nation was established in 1821 by the American Colonization Society. An organization of conservative abolitionists, the society favored resettling the nation's slaves abroad as the means to end slavery. The society established this West African settlement of ex-slaves and freed blacks. These settlers, in turn, were to build a miniature, constitutional black republic ruled by educated and enlightened African Americans, or Americo-Liberians.

Despite their importance to slavery as a domestic political issue in the United States, Haiti and Liberia were not yet of serious concern to America's military strategists, scientists, or industrialists. Throughout the antebellum period, owing to the political influence of slave-holding states, U.S. political links with the two black republics hardly existed. This changed following the outbreak of the Civil War. On December 3, 1861, President Abraham Lincoln gave his first annual message to Congress. He took up the matter of official recognition of Haiti and Liberia. "If any good reason exists why we should persevere in withholding our recognition of the independence and sovereignty of Haiti and Liberia, I am unable to discern it," he remarked. Lincoln reasoned that this diplomatic link was a matter of "expediency" because "important commercial advantages might be secured by favorable commercial treaties with them."[7]

The economic ties that President Lincoln envisioned would unfold ever so painfully in the following century. Indeed, only vacillating links between the United States and the two black nations emerged throughout the nineteenth and early twentieth centuries. These links had roots in the mission of U.S. political and business leaders to expand industrial-military technologies as well as scientific investigation in these foreign lands. It was this pursuit of nurturing engineering and science within the economy, the military, and medicine that grew and intensified as the twentieth century unfolded. In the United States, waterpower, railroads, factories, and sanitary engineering first appeared primarily in the Northeast; then later and only faintly in the slave South's Black Belt. But during the late nineteenth century, U.S. industrial and military might was exhibited forcefully in the defeat of the Confederate South and the opening of the Panama Canal Zone. After the United States was well on its way toward becoming a commercial and military world power, Haiti and Liberia were brought into the widening scope of U.S. scientific and industrial expansionism.

The technology exported by the United States for commerce, military operations, and public health reached Haiti and Liberia primarily after the two world wars. Sanitary engineering, medicines for disease treatment, and education assistance stressing vocationalism made up a good part of U.S. technical missions in these nations. However, the U.S. technical transfers and health work with these nations throughout the twentieth century did not result in great leaps of national development for either Haiti or Liberia. Instead, the "autonomous," authoritarian government leaders of Haiti and Liberia were the prisms through which these transfers had to pass. As such, the Haitian and Liberian presidents largely commandeered U.S. scientific influences and programs into their own versions of scientific progressivism.

In post-emancipation North and South America, societies and subsocieties with heavy concentrations of African-descent people differed widely. Each region varied by degree of industrial and urban development and sociopolitical fabric. What most black masses in the post-emancipation period had in common was that as workers they were sifted mostly into the lower rungs of their societies' respective industrial and plantation institutions. Alongside other labor masses— Native Americans and Asian and European immigrant workers—the Industrial Age Negro helped push these institutions forward into the twentieth century.

The Technical Impulse

The technical factor initially most influential in shaping the African Diaspora subsocieties was neither mechanical devices nor the factory. It was the plantation. In North America, the labor and cropping methods of plantation agriculture created and held together the largest black population center in the United States, known as the Black Belt. Stretching across the South from Virginia to Texas, this land was known for its black soil, its domination by cotton crops, and its heavy concentration of black American slaves, later sharecroppers. Through the 1930s, social scientists, politicians, and newspapers used the term "Black Belt" in two ways. They usually meant the term to refer to the entire black population who resided in the Deep South states. By World War I this region entailed over two hundred counties; and each county had mostly black residents. However, the term "Black Belt" also was sometimes used to denote sections, inside individual southern states, that were contiguous clusters of mostly black counties.[8]

Despite whether the onlooker's lens was focused on only one state in the Deep South or the region as a whole, the plantation in the service of "King Cotton" was the heart of the Black Belt. Describing rural southern black life in his 1934 sociological study, *Shadow of the Plantation,* Charles S. Johnson emphasized that in the South the "whole of life is bound up with the slow and tedious decadence of the plantation system." Recalling the work of the southern historian U. B. Phillips, Johnson stated that the plantation "formed the industrial and social frame of government in the Black Belt counties, while slavery [had] provided merely the code of laws for the perpetuation of the system."[9]

But neither Phillips nor Johnson recognized, in their perspectives on the southern plantation, that the forces that shaped the South's Black Belt were not just the lingering shadows of the Old South's backward slave culture. Instead, the plantation was a dynamic institution that evolved from and remained energized by the mission or impulse toward technical efficiency. This technical instinct involves knitting together isolated simple skills and rituals or techniques into large-scale technical systems. Archaeologists and historians trace this process in human history as far back as the ancient civilizations of Egypt, Nubia, and the West and later societies such as Iran, Iraq, North and South China, India, and Southeast Asia.[10]

More specifically, the South's plantations were the products of the technical dynamic flowing from two centers in the New World. Initially, the technical drive came from the northeastern colonies of North America with their use of mechanical power and manufacturing. Later, the technical activism swelled from within the South itself in the form of the plantation worked by regimented unskilled and skilled labor. In pre–Revolutionary War America, energy from natural resources and laborers was captured, combined, and applied in combination with mechanical inventions and tools. This mechanization of labor and natural energy then engulfed the colonies—enabling settlers to clear, farm, and mine the new lands. Moreover, mechanized weaponry transported from European homelands tipped battle after battle in favor of the English settlers of the American colonies. In turn, the colonies' European populations increased, while Native American populations and land control declined.

The prominent modern geographer A. G. Price recognized that technical superiority was the insurmountable advantage Europeans brought into the New World. In colonial America, he wrote, "[t]he dominating factors were not the differing cultures of the native Indian peoples but the greater material development of the whites. These latter possessed ships, horses, wheeled vehicles, firearms and manufactured goods."[11] The greater effectiveness of European tools for travel, production, and weaponry was central to the gradual sweeping aside of native societies. The prominent scholar-philanthropist Edwin Embree believed the beginning point of America's racial disharmony was when the Native Americans were conquered as the continent was initially colonized. Each local victory over Native American tribes fed a growing sense among Euro-Americans that their colonies reigned because of their technical superiority. To Price and Embree, this superiority was indeed the key reality in the settler-and-Native-American encounter: "However long and bitter the struggles, the end was always the same: defeat for the Indians, the loss of more and more land to the white hordes which came pressing onward in ever greater numbers. . . . By 1750, the population of the English colonies had grown to one million—probably more than the total native population had ever been throughout the whole area of the present United States. And tens of thousands of white settlers, seeking new lands to farm, were beginning to push westward from the seaboard into the interior."[12] In addition to military defeats and growing white settlements, the dissemination of whiskey and diseases common throughout Europe for which Native Americans had little tolerance also lowered their physical durability to compete with the new colonies from Europe. The destruction of Native American's ecological and food systems also weakened their societies.[13]

By the mid-nineteenth century, America's manufacturing and financial sectors were concentrated overwhelmingly in the Northeast.[14] Water wheels and roads, factories with masses of wageworkers, and steam engines and railroads, all had proliferated throughout the Northeast. As industrial historian Roger Burlingham has written, the "nation which was in the making in 1850, though a democracy in constitutional form and spirit [had] much of the look of an empire." The South and the West provided raw materials and food supplies to the Northeast, and were in effect "colonies of the Northeast."[15] The growing productivity of the North's energy and transportation resources and factories was only one dimension of the region's political and economic superiority over the South's slave economy. The second technological factor was the North's military institutions. In the end, the North's industrialized military unraveled the South's paramilitary slave patrols and Confederate army altogether, paralyzing its plantation system. It is the head-on military collision between the nation's two divergent economic regions that is the subject of W.E.B. Du Bois's classic *Black Reconstruction.*[16]

During the early nineteenth century, the same missionary drive to mechanize the natural energy, manufacture, and labor resources of early settler colonies evolved into the large-scale technologies later in the century known as the Western nations' "second industrial revolution." These technical developments— the water wheel, factory manufacture, the steam engine, and the cotton gin— also were shaping the economic role of the South's enslaved Africans as well as its large and small planters. Today, the historical stereotype of the pre–World War II South as a slow-moving rural society of backward white farmers and indolent, raggedly clad black farm workers still recurs.[17] The historical idea that complements this stereotype is that in the early nineteenth century the South was merely a vast patch quilt of loosely laid out plantations leisurely attended to by black slaves. Yet undercurrents and jolts of advancing technical and managerial organization had occurred even inside the antebellum South's rural plantation regions.

The Technology of the Plantation

During slavery the impulse toward efficiency took hold of the large landowners. They strove to make their plantations function as self-sufficient units. Each plantation developed a cadre of skilled workers who could make and repair farm equipment, houses, wagons, clothing, and other manufacture. In the eighteenth century, water transportation had been the major mode for transporting the produce of the grand "slave society."[18] But by the mid-nineteenth century, railroads were knitting the South's port cities to the region's smaller hinterland towns and large cities. These railway lines stretched into the edges of plantation regions.

Many planters with large holdings of slaves also favored public projects such as roads, canals, bridges, and the like because these structures enabled produce from their plantations to reach railway and shipping centers.[19] Once railroad construction and operations gained a foothold in the South, small industrial sites—railroad yards, packaging plants, loading sites—gradually expanded throughout the slave South's towns and cities. Skilled blacks (slave and free) and white laborers could be found in these shops and forges, scattered offshoots of the machines and large-scale industrialization growing throughout the northern region of the nation. The string of cotton plantations, woven together by rudimentary modern transportation, throughout the South both before and after the demise of legal slavery, would become the Black Belt or Cotton Belt of the early twentieth century.

In the Old South, plantation owners and managers worked to establish a machine-like organization of their black slaves, the South's human tools. The plantation managers arranged slave work to mimic the assembly-line factory. Fields became covered with groups of farmhands working assembly-line fashion but without factory walls surrounding them. Slave drivers pitted hoe gangs against plow gangs. In the cotton fields, hoe gangs were pushed forward to work faster and faster by oncoming plows. On tobacco plantations in Virginia, "the Negroes were worked in both gang and task systems," writes the slave specialist Herbert Klein. "For preparing the fields and planting," he continues, "the Negroes usually worked in gangs of ten or more with a driver who set the pace." Owners of the large, most productive plantations usually developed specific, printed manuals and rules. "A plantation might be considered as a piece of machinery," one slave owner named B. H. Barrow wrote in his plantation rules, and "to operate successfully, all its parts should be uniform and exact, and its impelling force regular and steady."[20]

Economic historians have revealed even more clearly the mechanized layout of southern plantations. In 1910 U. B. Phillips described the factory-like quality of plantations: "Its concentration of labor under skilled management made the plantation system with its overseers, foremen, blacksmiths, carpenters, hostlers, cooks, nurses, plow-hands, and hoe-hands practically the factory system applied to agriculture."[21] In 1935 Du Bois perceptively demonstrated that a rationalized system of utilizing downtrodden labor had an even broader, indeed national scale of operation and influence in pre-Reconstruction America. In the South, the slave system of production positioned black and white labor in competition with each other. But the machine-like organization of industrial work in the North also pitted whites against black labor. In the industrial North, white workers struggled against their freed-black competitors to maintain the highest skills and wages. Modern economic historians such as Fogel and Engerman have reemphasized the productive power of the slave labor arrangement: "This feature of plantation life—the organization of slaves into highly disciplined, interdependent teams capable of maintaining a steady and intense rhythm of work—appears to be the crux of the superior efficiency of large-scale operations on plantations."[22]

The labor needs of the plantation system generated an increase in slave purchases and persistent use of physical coercion against slaves. Soil exhaustion caused by the growing demand for land for cotton cultivation caused the institution's southwestward spread. The geographic and socioeconomic embryo of the Black Belt grew from 1810 to 1860. During this period the population of the Alabama-Mississippi region increased from 40,000 to 1.66 million, about one-half of whom were slaves.[23] The threat and use of punitive violence was at the heart of the labor "incentive" for slaves. But still, it was the plantation organization or system that fashioned black slaves into the efficient cogs governing other black slaves. About seven-tenths of the overseers on plantations were black, and an estimated one-third to over one-half of adult slaves worked in gang labor units.[24]

In addition to the methods of the large plantations that regimented field slaves, other industrial techniques and management styles—worked by black hands and white hands—were pushing up throughout the Old South. Small industrial work sites and clusters of skilled slaves began to appear throughout the plantation and urban South. The slaveholder segment owned most of the South's factories, machine shops, sawmills, flour mills, gin houses, and threshing machines. In addition, these slaveholders also held the workshops on which railroads depended. Gradually, steam- and coal-powered railroad lines were constructed, weaving throughout the South's thick forests and long stretches of farmland. The slaves from Africa appeared to be immune to the insect-borne diseases of the South's coastal and hinterland wilderness—malaria, yellow fever, and typhus. Thus, from the Carolinas to Maryland, slaves provided the labor that was pivotal to the building and start-up of the South's railroads.[25]

On plantations and inside towns and villages, the South's large planters relied heavily on the artisanship and skills of black labor. Slaves and freedpersons, these African Americans worked as carpenters, blacksmiths, shoemakers, draymen, hackmen, and operators of livery stables, shops, and restaurants.[26] In the South's cotton factories, mills, and foundries, slaves worked in both skilled and unskilled jobs. For factory work, frequently slave owners combined slave and free labor. The Tredegar Iron Works in Richmond, Virginia, was the largest iron plant in the antebellum South and the fourth largest in the United States. Slaves comprised one-half of Tredegar's work force.[27] Other masters derived steady income from hiring out their slaves to live in cities and work in the skilled trades that were accumulating in the urban setting.[28]

In the South's agrarian sector, free blacks and slaves did skilled jobs that a few decades after emancipation were incorporated into service establishments, plants, railroads, and shipping lines. As mentioned earlier, plantations were organized as self-sufficient units and slaves were central in all aspects of plantation maintenance. According to slavery historian J. B. Sellers, "records show that slaves developed the skills needed [that] kept the equipment and buildings of the plantation in repair, often constructing the homes in which their masters lived." Slaves, especially the women, also performed home manufacture "such as the weaving of cloth and the making of shoes."[29] Overseers had to supply nurseries for the slaves' infants and children as well as isolation buildings or hospitals for ill slaves. Frequently, conjurers, "doctoresses," and other slaves skilled in infant and sick care staffed these nurseries and buildings.[30]

In addition to plying individual skills in large towns, cities, and railroad sites, slaves and free blacks became skilled workers or tight-knit work crews vital for internal water and sea transportation. On rivers and bays, blacks worked as boatmen, fishermen, and oystermen. A large workforce of black laborers worked on the forty or so bateaux operating on the Appomattox River. In Deep South states, like Alabama, slaves were widely employed on steamboats. Divided into stevedores and deckhands, these slave gangs worked the cotton bales onto and off the steamboats. According to Sellers, "[i]f the boat belonged to an individual, the deck hands were his slaves; if the boat was owned by the company, they were often the property of the company."[31]

Modern War and the End of Legal Slavery

The Civil War further industrialized the black slave population. This conflict unleashed the aspect of technology that became the ultimate source for America's power in the twentieth century: the capacity to combine the most advanced industrial planning, management, and engineering into the weapon delivery systems for its military. Conventional histories of the Civil War focus on sectional politics and personal dramas, across the social class spectrum, that attended the rise of the Republican Party, the Union army's triumphs, and Reconstruction. Also important in this standard view of the Civil War and Reconstruction is the emphasis on the politics that led to the emancipation of the slaves and constitutional rights for former slaves.

But most political leaders, political-minded lay activists, educators, and journalists, then as now, failed to realize that the Civil War turned not on the political idealism of President Lincoln and an amorphous North advocating a republic free of slavery. Instead, the Civil War was won by the ability of the North's military leaders to accomplish its strategic ends by using its armed masses in combination with large-scale industrial production, transportation, and communications. After being bloodied by the initial assaults of the Confederate army, the chief aim of the Union military became to literally smash both the South's military units and its civilian supply areas. And this, by the war's end, the Union army did supremely well.

We have seen that on the brink of the Civil War, industrialization was subtly drifting into the South's rural black and white societies. By contrast, in the North the impulse toward scientific efficiency became central to its manufacturing sector. Indeed, firms steadily refined their factories and blended in traditional skilled artisans to the extent that commodity production soared, as did the making, maintenance, and expansion of railroads. Owing to this early adeptness at improving factory techniques, machinery, and railroads, the arms industry flourished in the Northeast. In New England special-purpose machine tools were designed, enabling factories to rapidly produce thousands of rifles, each the exact replica of the other.[32] It was this manufacturing-railroad complex throughout the northern states that equipped and transported to the South an essentially self-contained Union army—the fighting force that would eventually shatter the Confederate forces.

The Civil War has been called the world's first modern war. In no other previous war had so many technical developments been brought into the service of war. According to leading military historians, the American Civil War was the first time "the technological resources of a whole nation were ultimately mobilized to overwhelm an opponent. There was mass production of weapons and ammunition, of uniforms and boots. Canned food was supplied to armies, transported for the first time by rail."[33] Just three months after Lincoln's call for volunteer militia, U.S. military personnel increased by 2,700 percent. As one leading military historian stated recently, this was "the biggest mobilization in terms of time and scale the country would ever manage [and] more impressive in some ways than any mobilization since."[34]

Add to the northern forces and their technical advantages the minds and muscles of the African American free and slave population. Along with mechanized mass production and modern weaponry, Union generals mustered almost two hundred thousand black soldiers as well as tapped the South's rebellious and fleeing slaves for support labor. As slaves escaped to Union-held areas, they were deliberately withdrawing their hands from the farm tools and labor gangs of their masters and adding them to the Union effort.

The Confederacy had much less success utilizing slaves for their war operations. But they managed to incorporate some slave workers. As white workers joined the Confederate ranks, southern military officials mobilized slaves into their engineering and medical units. Black laborers made up nearly one-half of the work force in Georgia's nineteen military hospitals. They also worked in ordnance plants and as fortification builders in key strategic sites throughout the South. The Confederate military paid masters for the use of these workers and often paid the slave workers cash as an extra incentive at the work sites as well. How was it possible that black labor nursed confederate soldiers and fortified the very plants and armies that were in combat to keep them enslaved? Two factors in particular came into play: frequently, slaves lacked awareness of the war; and, second, the Confederate employers used black slaves to work only in brief, specific work roles. Black military workers for the Confederacy were used on sites far away from northern armies. Moreover, impressed black laborers were often brought in from rural plantations, worked for short intervals, then were rotated out, and, hence, may have not known about the war at all, or the seriousness of the war.[35]

The mobilization of black slaves during the war also occurred in the South's industrial sector. A Chicago stationary engineer, trained in the South, recalled that throughout the shops and work sites, "the white laborer and mechanic had been supplanted almost entirely by the slave mechanics at the time of the breaking out of the Civil War. Many of the railroads in the South had their entire train crews, except the conductors, made up of the slaves—including engineers and firemen."[36]

But the Union military by far harnessed the greatest technical resources and mass of workers—both northern labor and southern black peasant—for its cause. With this combination of labor power, Union generals commanded an unstoppable attack force. General William Sherman, for the first time in the modern history of warfare, exploited the opportunity to combine the advantages of railroads, telegraphs, and steamships into one military attack-machine. Using this powerful network of technologies, Sherman and other Union military leaders were able to weld together white and black soldiers, supplies, weapons, and explosives and land the decisive blows against the Confederacy. When General Sherman destroyed not only the Confederate army, but also the Confederate army's civilian supply centers, modern American warfare had been born. The ultimate instrument produced by and in the service of the U.S. mission for scientific progress had been unveiled. Less than a century later, this technical weapon would take the form of a vast network of nuclear warhead missiles.

The Black Belt South

Manufacturing, engineering works, and mass planning were implanting industry and the modern military primarily in the U.S. North. In turn, these institutions destroyed the South's political and military rebellion in the Civil War. Following Reconstruction, the nation's march toward industrialization reinvigorated the South's plantation economy and triggered an extensive southern textile industry. The raw materials and produce of the South were supplied to northern factories and consumer markets that were expanding inside an urbanizing America. As for the black American's role in the post-Reconstruction national economy, it remained primarily in the South's plantation regions.

When America opened the doors to the twentieth century, the demographic and social center for the nation's black population was the Black Belt of the Deep South. While a small number of black southerners were finding industrial or service work in southern cities or by trekking to the North or West, the vast majority remained in or near the Black Belt. In 1918, when the U.S. Census Bureau officials published a special report on black Americans, they found that "as regards the Negro population, the proportion resident in the South has not varied greatly from census to census." In 1790, 91.1 percent of the nation's blacks resided in the South, and still 89 percent as of 1910. The three states with the largest black populations were Georgia, Mississippi, and Alabama—with 1.2 million, 1 million, and 900,000 black residents, respectively.[37]

The South's Black Belt at the opening of the twentieth century comprised the land running from eastern Virginia and North Carolina and South Carolina through central Georgia and Alabama and a separated area of the lower Mississippi River Valley. Within this region there were 286 counties with black

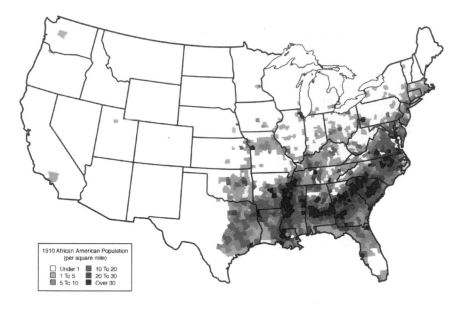

MAP 1.2.   The Black Belt: African Americans in the Southeastern United States, 1910. Reprinted from U.S. Bureau of the Census, *Negro Population, 1790–1915* (Washington, D.C.: GPO, 1918), 110.

populations of more than 50 percent. Traced by U.S. census officials since 1860, the Black Belt experienced a population increase from that year to 1900, when its total black population reached slightly over 4 million (see map 1.2).[38]

The central economic and social institution of the Black Belt was the plantation. After slavery plantations remained large units of land farmed by anywhere from a score of tenants or sharecroppers to several hundred of such farm workers. In the Caribbean, emancipation did little to alter the economic organization or social importance of the plantation. In the West Indies, as Gordon Lewis has emphasized, the formal abolition of slavery in the nineteenth century did not "release the Caribbean Negro masses from servitude in any real sense. It merely replaced the whip of slavery with the prison of low-cost agricultural 'free' labor." In the decades following abolition of slavery, large firms from both western European nations and the United States owned and extended plantations throughout the Americas. Commercial plantations also held their own in Africa and Asia. Throughout the Western and Eastern worlds, the plantations were producing crops such as rubber, cotton, spices, and sugar for the West's growing appetite for these raw materials and delicacies.[39]

Several basic organizational and financial features were common to the modern (that is, post–1900) plantations and gave them durability throughout the twentieth century. Modern plantations are usually large farming estates or settlements. These land holdings have a central hierarchical administration, access to a permanent as opposed to a migratory cheap labor force, and strong distant or international markets for their crops. Unlike the urban factory in which workers can commute to work, plantations often provide housing for workers or generate adjacent squatter communities that over time become permanent worker communities. As one recent authority has written, plantations utilize "a core of local labor or those housed on the plantations usually subject to a discipline more stringent than that applied to industrial wage labor."[40]

In the U.S. South, the operations of the large plantation took on a form of rural "industrialization." This linking of labor into organized groups to tend the land provided an efficient and profitable way for owners to delegate supervision to the on-site managers. Writing about the cotton industry of the South in the 1920s, one of the region's leading economists, Rupert P. Vance, emphasized the absolute staying power of the plantation method. He described the plantation in the U.S. South as the primary "system under which the labor force was apportioned to production of southern agricultural staples."[41] A southern plantation required four main elements. First, it had to contain a large acreage of "fertile soil, cheap, level or rolling, and to some extent homogenous in texture and topography." Second, the plantation necessitated a "labor supply of low social status, docile and comparatively cheap." Third, a form of total supervision of the work tasks and group life of this labor force had to be developed. As Vance emphasized, "the management required is social as well as economic supervision." Finally, the type of crop had to be one capable of both mass cultivation and attracting consistently strong prices in the market. "The products must be staples," according to Vance, "routine crops easily cultivated by set rules; cash crops for which no problem of marketing exists."[42]

The Black Belt was one of the world's most important cotton-producing regions, but at the same time a region of extreme poverty for its tenant farmer population. Throughout the early decades of the twentieth century, sociologists trying to unravel the roots of southern black poverty pointed their fingers at the share-cropping plantations. Their surveys established clearly the triangular link between plantations, southern poverty, and a caste system of race relations. Most notable among these sociologists and economists specializing on the South and race relations were Rupert Vance, the Fisk scholar Charles Johnson, and T. J. Woofter. In a 1936 survey of the South's cotton region for the Works Progress Administration, Woofter gave a blunt description of plantation life inside the Black Belt: "[This area is] characterized by a high percentage of tenants, a high degree of concentration of land ownership, a heavy proportion of Negroes, a very mobile population, per capita farm incomes . . . lower than those in other farming sections of the country, small proportions of urban and village dwellers, scarcity of industries, large families, poor school facilities, especially for Negroes, and utter subjection to King Cotton: boom when the King is prosperous and gloom when the King is sick."[43]

When radical Reconstruction ended, northern political and industrial leaders took no substantial interest in developing the South. Northern industrialists were preoccupied with their investments in the western territories and in new foreign regions such as Central America and the Far East. Furthermore, the North had its own public health problems. Northern cities were facing housing, education, and public health pressures associated with new influxes of immigrants from southern and eastern Europe.[44] The massive physical disruption experienced by the South in its loss in the Civil War, as well as its sparse industrial development, made this region uninviting to northern capitalists. By contrast, investors swooned over the tremendous flood of new markets that would be produced by a canal in Panama.

Southern financial leaders also lacked political or economic incentives to organize on a mass scale public health resources for protecting workers' health. The agrarian South had an abundant supply of cheap labor and a large poor population. The tenant workforce, supplemented with prison labor, was large enough to provide quick replacements for the disabled among the tenant workers. With so much cheap labor available, the large southern planters did not even fear epidemics as potentially disruptive to productivity and the flow of profits. In this context, the South's financial elite did not feel compelled to sink their limited capital into expanding the region's municipal and state public health and welfare services.

Alfred Stone, a southern planter and economic advocate widely published in national political magazines, wrote about the dismal situation facing the region's blacks in 1904. Following Reconstruction, the South entered "a period of rapid, almost revolutionary, industrial change." Superficially, this change did not seem to affect the large plantations on which many southern blacks resided and worked. But now, Stone pointed out, the South was becoming permeated with "industrial competition." Antiblack sentiment associated with the South's defeat and radical Reconstruction politics was widespread in the region. Consequently, southern whites were hiring whites for work traditionally performed by black folk. With the exception of the "two so-called learned professions of teachers and clergymen," the black in the Deep South "has lost ground in the whole in the following skilled occupations: carpenter, barber, tobacco and cigar factory operative, fisherman, engineer or fireman (not locomotive) . . . blacksmith . . . laundry work, hackman or teamster, [and] steam railroad employee."[45]

In addition to sharecroppers, plantations used other forms of cheap labor. In the post-Reconstruction South peonage and the use of convicts on labor gangs were frequently encouraged by local politicians and large landowners. In town, county, and state jurisdictions, criminal and property laws were enforced to confine black workers to share-cropping or to labor on roads and other amenities that benefited large landholders.[46]

Imprisonment of blacks swelled throughout the southern states. The convict-lease system developed, forcing black prisoners—including older boys along with the men—to do exhausting and dangerous work on levees, highways, and plantations.[47] Black convicts in the South became used widely as labor in mines and on farms operated by both public and private interests. Criticism from abroad began to emerge concerning this practice. In 1903 the English political economist J. A. Hobson made a survey of the conditions of blacks throughout the South. He found "terrible revelations" of "the cruelties inflicted upon coloured men and women both under the convict gang system and under [the] practice of 'peonage.'" Hobson emphasized that "taken together they constitute a very real and considerable recrudescence of slavery."[48] A recent critically acclaimed study of prisons in Mississippi at the turn of the century has borne Hobson out. It emphasizes that living and work conditions for the black convict population were "worse than slavery."[49]

Incarceration was the blunt means to block the possible drain of black peasant labor into socially protected segments such as the physically or mentally disabled. Still, local planters and law officials had to relinquish substantial numbers of blacks to asylums. Incarceration of mental defectives became widespread throughout the post-Reconstruction South. However, local authorities could not regiment, through prison life or forced labor, the seriously deranged and dysfunctional. Thus, the psychotics of the black poor were useless economically and eventually placed in asylums. It is likely that the neglect of health problems such as infectious diseases in local black populations increased the numbers of blacks with psychoses. In Georgia, for example, in 1860 it was estimated that only 1 out of every 10,584 blacks was a patient in a ward for the insane. However, by 1870 this figure was 1 out of 1,764; and in 1890, 1 out of 943.[50]

Disease, Race, and the Plantation Heritage

Interwoven into the plantation economy and the Black Belt was the residential segregation of tenant-class blacks from their white counterparts and both these segments from the South's white middle and upper classes. Sanitary housing, piped water systems, and safe maternity and childhood social conditions all were more frequently established for white communities than black ones. Sharecroppers, black and white, fared worst in work and living conditions compared to the South's population as a whole. Furthermore, black sharecroppers usually had to face hazardous and unhygienic conditions the longest and, more frequently, across generations. The belief that blacks were more disease-prone or disease-proof discouraged local and regional authorities from including them in public health measures. Thus, improvements by local health authorities in physical conditions to prevent insect-, animal-, or water-borne diseases tended to reach the Black Belt communities last.

The heavy toll that preventable diseases took on lives inside the Black Belt was the product of long, tortuous exposure to disease and unhealthy environmental factors that its residents experienced. The South's warm climate was not the only persistent ecological factor conducive to high rates of infectious "tropical" diseases in the region. The popular tendency has been to view mosquito-borne diseases such as malaria and yellow fever as found only in regions like the semitropical South or tropical Third World countries. However, prior to the early twentieth century, conditions such as swampy and stagnant water supplies, unquarantined ports, and congested housing caused these diseases to flourish throughout the colder regions of the United States as well. As living and health standards improved, both infectious and deficiency diseases, such as typhoid fever, malaria, yellow fever, and pellagra, steadily disappeared in the industrializing North; but they remained embedded in the largely semitropical, impoverished South.

From 1693 to 1870, eighty yellow fever outbreaks alone occurred in eight states north of Virginia.[51] The early European immigrants had brought malaria to North America during the colonial era from regions where it was widespread: England, France, and Spain. African slaves from West Africa also brought malaria into the American colonies. By the mid-nineteenth century, this disease was entrenched in virtually all settlements from New England to south Florida and westward to the Columbia River Valley and the California hinterlands.[52]

In the North, population distribution and decreases in deadly mosquito-borne diseases had followed the geography of this section's urban and industrial growth. In the South, a similar pattern unfolded, but much more slowly. By the start of the twentieth century, the North had made progress against these diseases, while the South had not. By the late nineteenth century, improvements in social and work conditions in particular caused disease rates in the North to drop. The standard of living rose and municipalities implemented public health measures, subtly reducing the size of populations who were sick in the clinical sense. The improvements included better housing, more plentiful and cleaner food and water supplies, expansion of sewage disposal, and isolation of persons known to be infectious. To be sure, city medical authorities often prejudged and publicly harangued the new immigrants from Ireland, Germany, and Italy as unhealthy, uncooperative hordes.[53] But still, the combination of environmental and municipal health improvements, frequently championed by progressive doctors, drove diseases like malaria, typhoid fever, pellagra, and rickets down to negligible levels.

During the Civil War the breakup of plantation patriarchy caused by the military calamity created waves of refugees from among the former slaves. The physical conditions of settled life on the plantations as well as the folk institutions that slaves had nurtured to protect their health and morale were either destroyed or weakened. Where the black refugees massed—from the District of Columbia to Florida to Mississippi—famine, exposure, and disease were common. Throughout the Civil War one-quarter of all hospital admissions were due to malaria fever.[54] The District of Columbia experienced massive influxes of freedpersons lacking adequate clothing, food, or shelter. When in the 1870s thirty to forty thousand black refugees crowded into the city, tuberculosis became epidemic and death rates from this disease climbed until the 1890s. In Florida the ex-slaves suffered smallpox and cholera outbreaks that killed local whites as well.[55]

Overall, diseases such as malaria and hookworm remained endemic to the South due to the combination of both ecological and socioeconomic factors. The South's year-round warm temperatures encased its many valleys and slow-moving bodies of water as well as its abundant cotton plantations. Thus, the typical southern plantation, small farm, and town areas were stable breeding environments for the mosquitoes that carry malaria parasites as well as the nematodes that produce hookworm. According to J. A. LePrince, a key public health administrator during the early twentieth century, "in military camps in the Southern states in the Spanish-American War period so many soldiers were infected by malaria before leaving for Cuba that only a small part of some regiments was fit for service." Widespread poverty also resulted in high levels of nutritional problems such as pellagra and rickets.[56]

By the early decades of the twentieth century, disease burdens became concentrated most heavily in the South's black populations and to a lesser extent among whites living in similar poverty. In some outbreaks, such as of yellow fever and malaria, blacks had a higher survival rate than whites in some localities, probably because of immunities that these black populations had acquired over generations to local strains of these diseases. In New Orleans, for example, when a major yellow fever epidemic broke out in 1905, blacks comprised only 19 of the 434 fatalities that occurred.[57] However, as the twentieth century progressed and southern health departments began to record mortality rates, blacks were found dying from malaria, tuberculosis, and venereal disease at a rate that was much higher than that of the region's whites. If black slave populations had brought with them from Africa unique racial immunities, the hands of time were removing these protections with each passing generation.

The medical researcher Charles Wardell Stiles summarized the southern health predicament in a 1909 study, "The Industrial Conditions of the Tenant Class (White and Black) As Influenced by the Medical Conditions." An official with the U.S. Public Health and Marine Hospital Service, Stiles was the nation's leading researcher on the South's hookworm parasite. He wrote that "with the insanitary [sic] conditions so prevalent in the South, it need not be surprising that the tenant labor is inefficient and the death rate high. As the sanitary conditions under which the negroes are living are in general much worse than those for the whites, and as the nourishment of the blacks is irregular and poor, the present much higher death rate and the low degree of efficiency among the negroes need cause no surprise."[58]

Ironically, in the decades immediately following Reconstruction, the mere fact that diseases such as malaria and venereal diseases were highly prevalent inside the Black Belt only reinforced deficient public health resources and segregationist beliefs. Southern health officials often assumed blacks were biologically a dying or sickly race who would not benefit from public health assistance anyway. Thus, hazardous living and work conditions, on one hand, and defeatist notions among health officials, on the other, tended to perpetuate unwillingness of local authorities to improve public health conditions in the black South.[59]

Black Labor and the New Industrial Order

As the twentieth century dawned, the grinding work on the fields of the plantations bound together the South's black farmer communities. At the same time, the rows of machinery clattering in the textile mills in adjacent villages and towns bound together the region's white working class. The plantation enterprise best succeeds when there is an abundant supply of laborers. In the U.S. South of the early twentieth century, cash-strapped farm workers were so plentiful that plantation owners hired them cheap and then let them go with each harvest since new tenants were almost always waiting nearby.[60]

In remote rural and paternalistic political settings, such as the plantation communities of the U.S. South—and, as we see in later chapters, in the rubber-growing region of Liberia—investors were able to acquire cheaply large tracts of land with little government regulation or taxation. The commercial owners of the South's big plantations also benefited from public works provided by the local or state governments. Roads, for instance, frequently were built by convict labor. Another political factor encouraged the southern economy to concentrate on plantation agriculture in the early twentieth century. Southern plantations were reinforced by the racial discrimination interlaced in local and state government and law. This Jim Crow political and legal system solidified the division of blacks from whites in all major social institutions, such as housing, education, and recreational life. In this way, racial caste became a pervasive force that, on one hand, exacerbated economic class divisions but, on the other, spread the social idea of immutable racial categories.[61]

In the 1890s most of the South's blacks remained tied to the plantation, especially in the region's cotton counties, which relied heavily on black male labor. Black women of the plantation South worked the fields as well, although more often as part of a share-cropping family. As labor historian Jacqueline Jones has written, among the rural black families at the turn of the century "even the poorest families . . . sought to preserve a division of labor between the sexes." The families did this in ways such that "that fathers assumed primary responsibility for the financial affairs of the household . . . and mothers oversaw domestic chores first and labored as field hands or wage earners when necessary."[62]

In the meantime, mills became the provinces of the South's white working class. The managers of these mills and large farms restricted black labor to the lower-paying operative jobs outside, not inside, the mills.[63] While most southern black laborers were still in rural agriculture (about 1.75 million as of 1890), some were absorbed into domestic and personal service work (slightly less than 1 million) located in town and country regions. Finally, a small percentage remained or became skilled workers or professionals, especially ministers and teachers, rapidly moving into urban areas.[64]

In rural societies undergoing modernization or rural-to-urban transitions, population growth can be accelerated by just modest public health improvements, such as making childbirth safer or bettering nutrition and sanitation.[65] Reconstruction policies had done much to improve the standard of living and health environment throughout the South. Consequently, the region's population became one of the fastest growing in the nation. In the North, even with the influx of European immigrants, the ratio of children under five years of age to white women fifteen to forty-four years of age was 470 to 1,000. By contrast, this ratio for children to white women in the South was 633 to 1,000, and for black women 621 to 1,000.[66]

The growth in the South's population due to its rising birthrates foreshadowed a new era of protracted competition for jobs throughout this region. Alfred Stone, a leading southern economist, observed in 1904 that blacks in the South were keeping pace with the region's whites, at least in terms of physical survival. However, the numerical growth of the black southern population was "the net results of . . . various and complex industrial changes [that can] best be measured by [such] vital statistics of the race." Stone emphasized that "the future of the Negro race in the United States seems to be essentially an industrial and economic question, turning upon their efficiency in comparison with classes of the population who compete with them in their staple occupations."[67]

As the industrial organization of the nation's economy spread, it was difficult for blacks to obtain skilled or professional positions. It was even more difficult for blacks to obtain capital ownership and investments in the new technical systems and their products. Railroads, shipping, electrification, communications, and iron, steel, and oil production were all out of the reach of black capital and mass labor. The growth of industries and urban markets thirsting for plantation produce gave the South's plantations the basis for their continuous, profitable existence. At the same time, the new industrialization of post-Reconstruction pushed northern black labor—both intellectual and skilled—outside of large industry and capital. Skilled jobs, management, and ownership in new industry all became beyond the grasp of blacks.

Before the Civil War free blacks in the North and Midwest had obtained a measure of economic security, even relative prosperity as small business owners and artisans. However, by the beginning of the twentieth century these economic gains were eroding. New types of managerial organization washed asunder small-scale proprietors. Also, hiring and training practices involving the new managerial revolution racially segregated the skilled workforce. The new managers simply did not hire or apprentice blacks for these increasingly sophisticated office, factory, and distribution work sites.

By the 1890s, white entrepreneurs, many of whom were first-generation European immigrants, were using much more effective management practices than the traditional approaches of artisans and small family businesses. White firms were simply outproducing and underpricing the older black businesses. At the 1906 convention of the National Negro Business League in New York, the successful merchandiser John Wanamaker pointed out that at one time Philadelphia's black businesses were the equal or better of all the local competitors. However, by the new century, these very same black businesses had disappeared. "As an old business man I am speaking the fact," Wanamaker told the audience, "they lost their business because the Swiss, the Germans, and others who were American white men did that same business better than they did it. Their color had not the least thing to do with it."[68]

Back in the South, while blacks were concentrated in sharecropping, the region's whites moved into towns and small cities with textile mills and other processing sites for agricultural produce. As industrial analyst Herbert Northrup has written, by the early twentieth century cotton textiles in the South had become an "economic restoration mechanism and a source of employment for poor whites."[69] Between 1880 and 1907 the number of textile mills in the South grew from 161 mills with 17,000 workers to 600 mills with 125,000 workers. In this new regime of millwork, black workers were limited mostly to operatives' jobs or field, yard, or janitorial work.[70]

Booker Washington's Romantic Land

Leaders throughout the black South, especially Booker T. Washington, fundamentally misperceived the nation's industrial revolution and the growing influence of the experimental, applied sciences within this economic transformation. Washington and his ideologues did not fathom the immense distance that U.S. economic and science profession leaders and their political sponsors had traveled and dreamed to travel in their mission to intensely use scientific technology in industry and government. Nor could Washington grasp the unbending power that resultant large-scale industries and managerial bureaucracies were gaining over workers and farmers. As labor and farm leaders from Samuel Gompers and Tom Watson to A. Philip Randolph realized, this power could only be counterbalanced by mass unions and farmer organizations or active government regulation. Instead, Washingtonians accepted the hegemonic idea that modern scientific developments inevitably yield social efficiency and progress and that blacks, therefore, need not concern themselves with the highest reaches of the laboratory and clinical sciences.

Invisible to Washington and his network of educators, ministers, and newspaper editors was the growing power of technologies and scientific management throughout the nation's black communities in the North and South. These black leaders did not understand that large companies and government entities were refining engineering techniques and building science research laboratories to improve and control industry, transportation, sanitation, and cities. Beginning with the Progressive Era, scientists were becoming increasingly professionalized as they fanned out into academe, business, and government.[71] The nation's business elite in particular began to invest in large-scale technologies and scientific innovation, including nurturing science professionals and laboratories. In turn, the United States was propelling itself into leadership of the second industrial revolution throughout the Western hemisphere, while bringing tropical societies into its commercial and, later, military sphere.

In the meantime Booker T. Washington misguidedly pleaded that now, more than ever, black Americans had to hold onto farming and vocational interests in the South. Scientific innovation was indeed advancing throughout American society, Washington acknowledged, but black Americans needed to concentrate on moral character and efficiency as workers, not lofty science. In 1901 he wrote in Tuskegee Institute's annual report, "More and more each year, I feel that . . . the salvation of my race will largely rest upon its ability and willingness to secure and cultivate properly the soil."[72] His educational campaign stressed the need for "special education" for the South's blacks—an education in hands-on skills. In one of his popular inspirational books, Washington wrote: "At Tuskegee . . . when a student is trained to the point of efficiency where he can construct a first-class wagon, we do not keep him there to build more vehicles, but send him out into the world to exert his trained influence and capabilities in lifting others to his level, and we begin our work with the raw material all over again."[73]

With heavy industry and scientific management burgeoning throughout much of America, Washington's emphasis on manual trades and farm skills for cultivating small parcels of land was out of step. Nor was ownership of small farms or businesses the potential basis for national social and economic power for African Americans facing a society of science-driven industry and government. By encouraging his students and patrons to stand aside from the groundswell of science developing inside industry and academe, Washington in effect consented to a hegemonic role for the management and ownership of technologies in the Black Belt South. Scientists and their economic backers would expand and refine technologies while black Americans stood outside this circle of innovators, investors, and managers.

In fact, black laborers in southern agriculture—either farm workers, sharecroppers, or small-farm owners—were no longer absolutely indispensable to the South as Washington and his followers had assumed. The southern economy was shifting to less labor-intensive crops and rural-based industry—for example, lumber and furniture manufacturing—as well as mechanized farming. Unskilled or semiskilled farm workers from white communities were in constant supply and in general amenable to the low wage scales.

Like Booker T. Washington, the scientist George Washington Carver believed strongly in the importance of basic agricultural training for the South's black youth. But Carver was more enthralled with the new science of what he called "analytical" and "creative" chemistry. Carver came to Tuskegee in 1896 to direct the school's experimental farm. He went on to acquire legendary status throughout the South and the nation as America's Negro "plant wizard." His reputation stemmed from his development of numerous products and practical techniques involving the cultivation of peanuts, sweet potatoes, and cotton and his travels throughout the South teaching this new knowledge to small and commercial farmers.

Yet under Booker Washington's administration at Tuskegee, Carver suffered mightily. Washington repeatedly excoriated Carver for spending too much time in his laboratory doing experiments. Washington criticized Carver for neglecting to teach more classes at Tuskegee on basic farming as well as for not doing more public instruction and exhibits for blacks who owned small farms. As for Carver's botanical and chemical experiments, Washington once wrote to Carver, "When it comes to the matter of practical farm managing which will secure definite, practical, financial results [for Tuskegee], you are wanting . . . in ability." Underfunded and overworked by Washington, Carver nearly left Tuskegee numerous times before Washington's leadership of the school ended in 1915.[74]

The mechanization of farming as well as the national expansion of industry undermined the major economic ideas of Frederick Douglass and Booker Washington. Their key ideas were, first, that owning small plots of land equated to power and, second, that black citizens could use their spending and saving decisions to solve the problems of political and economic discrimination, problems rooted in historical and social divisions of race and class. Their romantic pastoralism was later reproposed by the likes of Marcus Garvey. Du Bois observed this serious oversight early in the twentieth century. In 1912 he wrote one of his many essays for the black press trying to expose Washington's blindness: "If now a group of people like American Negroes are advised to turn all their attention . . . to the training of laborers and servants, they are advised to commit social suicide." Du Bois warned that black Americans "will soon find that in the rapidly changing technique of industry their laborers will be displaced while they will have developed no intelligent leadership in industry or thought to guide the mass."[75] Du Bois knew that Washington could not see the larger forest of his day, but instead was wedded to one small tree—namely, vocational education and small farming.

Du Bois's major frustration with Booker Washington was that the latter seemed oblivious to the utter powerlessness that Black Belt populations faced with respect to modern industrialization. In Du Bois's view, Washington was leading blacks, of both the South and North, away from gaining a foothold in the real source of power for the U.S. nation-state: "Science and Empire." In the Black Belt, where America's black population was overwhelmingly concentrated, workers remained in operative, unskilled jobs in agriculture or lumber. Public education and colleges for southern blacks were severely underfinanced compared to white institutions. Lacking modern facilities, expert faculty, and supportive and enlightened administrators, these institutions could do little to train students for the new professional, technical, and business skills called for by big business.[76]

Du Bois's article "The Shape of Fear," in the *North American Review,* profiled the essence of Black Belt communities. Describing a rural parish in Louisiana of twenty thousand residents—fourteen thousand black—he wrote, "One can easily scent here tremendous and bitter rivalry between the rich and poor white owners, between the owners and tenants, between the white and black owners, and crushed under all of it lie the mass of black tenants." Sixty to 70 percent of the tenants could neither read nor write. Furthermore, there was "no modern wage system . . . but nearly all is barter and debt peonage." The sluggish southern economy was pushing whites into tenancy—men and women who were formerly shopkeepers, skilled workers, artisans, and small-farm owners. Any camaraderie between the new white tenants and the long-standing black ones would spell political catastrophe for the large landowners: "Secrecy, force and murder have been part of the Black Belt social economy for fifty years. The landlords lived with their hands on the trigger. Formerly, this was because of the fear of servile revolt or the hint of it. That fear is still there."[77] Du Bois explained that all of the human groups comprising the Black Belt plantations were bound to their roles by the overwhelming strength of the racial caste system. To Du Bois, the caste system was the shadow, the plantation the substance, behind the South's racial divide. While Booker Washington saw the Black Belt as a romantic Negro world, Du Bois saw the region as a sad, oppressive prison.

Segregation, Disease, and Self-Help

Despite its widespread poverty and isolation from national life, the Black Belt was not populated solely by tenants and illiterates. The communities had their own technical professionals and resources. When viewed alongside the history of Haiti and Liberia, the black subsociety of the Deep South developed a relatively large, durable professional elite and educational order. With the walls of racial segregation in the South's economic sphere so high and the disease burden especially intense in rural black communities, a black medical care network emerged. By 1906 there were 40 voluntary hospitals and health centers mainly in the South operated for and primarily by blacks. Several of these facilities trained nurses, as did five black medical schools. By 1912 the number of black community facilities rose to 65, and again in the 1920s to about 125. Also in these early decades of the twentieth century, black physicians, nurses, dentists, and hospital administrators all had established their own professional associations. The National Medical Association, the organization established by black doctors in 1895, had over five hundred members by 1912 and was growing vigorously.[78]

The black doctors and community nurses became a key element in the black South's professional strata that also included teachers, lawyers, and ministers. Collectively, these educated professionals fostered social cohesion in black communities experiencing political exclusion, economic subservience, and widespread health problems. This emergence of a black medical care network throughout the U.S. South has been described frequently by historians as educated blacks heroically trying to enter alongside whites into honorable professions like medicine and nursing. However, when looked at from the broader vantage point of the South's technological and economic development, the growing market of black consumers spurred the black medical network, as it had spurred other entrepreneurial sectors, such as the black-owned businesses and insurance companies.[79]

These consumers were the black women and men with enough income to purchase physician care and maternity services. Semiskilled workers, small-farm owners, college-educated white-collar workers, these black patients were the by-products of the sporadic industrialization in black communities throughout the South. Due to Jim Crow practices, blacks seeking physician care usually could not receive it at local public health centers or private white-owned medical facilities. Thus, lay blacks with relatively high education and wages were ready consumers for black health care services. As for the black physicians, by obtaining and selling commodified knowledge, namely professional medicine, a privileged segment of black professionals emerged outside the control of the United States' leading medical and public health centers. Black doctors, like black business and farm owners, could secure a sufficient income independent of the larger southern (and northern) economic hierarchy—a hierarchy that restricted blacks mostly to tenant farming or low-wage service or labor.

The health problems of blacks in the Deep South also became the indirect focus of federal public health initiatives as well as programs sponsored by large, northern-based philanthropies. These first efforts to control malaria among southern Black Belt populations stemmed from the centuries-old interest among U.S. and European medical and sanitation authorities in tropical disease issues.[80] Some of the consultant professionals in the Deep South had been involved in campaigns in Cuba and the Canal Zone. Federal health authorities were not so much concerned that malaria posed a critical health problem to the people of the South or that it was a serious handicap to the region's economy. The South's large business owners and planters easily replenished black and white unskilled workers. Instead, national health authorities were concerned with learning more about the disease microbes and etiology and ways to treat infected individuals.

Despite the fact that blacks developed their own medical profession, serious disease problems persisted throughout the Black Belt. In the next chapters we observe the U.S. government's frenetic campaigns to eliminate infectious killers such as yellow fever and malaria in the Canal Zone, and similarly energetic, centralized efforts to rid Haiti of yaws. By contrast, the disease problems confronting the black South were approached by federal public health authorities gradually, disease by disease. Tuberculosis was the most persistent leading cause of mortality among southern blacks. Also, malaria reached and remained at critical levels in the Black Belt in the post-Reconstruction decades. The missionary zeal of the Rockefeller Foundation led it to initiate the South's first region-wide effort to control a major disease. In 1909 it organized the Rockefeller Sanitary Commission for the Eradication of Hookworm Disease. Within a few years the commission estimated that more than two million people in the South were infected. Despite some opposition from southern civic and health leaders, by 1914 the commission had treated close to half a million persons.[81]

Hookworms entered the body through the skin, lodged for years, and left the victim anemic, emaciated, and mentally debilitated. The ailment was rampant among southern whites, especially the children. As for the South's blacks, some medical surveys found prevalence much lower compared to whites living under similar conditions. For example, one survey in Jones County, Mississippi, estimated that black children had an infection rate of about 23 percent, compared to 76 percent for whites.[82] However, in 1921 another major survey of black children living in severe social conditions discovered high rates of parasitic infections. In a 1921 survey of preschool children in the largely black and rural Charleston County, South Carolina, 45 percent of the black children were found suffering from intestinal parasites such as hookworm and ascaris.[83] Despite this mixed evidence, many in the public health community maintained that hookworm was not a major problem among blacks. Consequently, the hookworm campaign gained little importance to health activists concerned with the black South, since these activists were not included in the hookworm effort.[84]

Compared to the scarcity or lack of medical professionals in most areas of Haiti and Liberia, the Black Belt South benefited from a substantial supply of physicians and community nurses. Several preconditions were advantageous to the rise of the U.S. South's black medical network. First, primary education is basic to the development of an ample supply of students for all forms of higher education, including medicine and nursing schools. It was widely available throughout the black South, compared to the situations in Haiti and Liberia. Literacy levels of newly freed blacks in the Deep South immediately following the Civil War were very low, about equal to those of Haiti's children in 1905. In 1870, illiteracy for blacks in southern states ranged between 81 and 92 percent. The figures for individual states were as follows: North Carolina, 84.8 percent illiterate; South Carolina, 81.1 percent; Tennessee, 82.4 percent; Florida, 84.1 percent; Louisiana, 85.9 percent; Mississippi, 87.0 percent; Alabama, 88.1 percent; Texas, 88.7 percent; Virginia, 88.9 percent; and Georgia, 92.1 percent.[85]

However, during the 1870s and 1880s literacy throughout the black South began to climb. Reconstruction state governments as well as northern missionaries and philanthropists established hundreds of common schools for blacks. By 1900 some 31 percent of blacks in the South aged five to twenty years were attending school (compared to 54 percent of southern whites). Ten years later the percentage had increased to 45 percent (compared to 61 percent for whites).[86] These rates were substantially lower than those for southern whites. Nonetheless, these literacy levels in the black South were substantially higher compared to those in Haiti and Liberia. In Haiti, as of 1905, less than one-tenth of the approximately 350,000 school-age children were attending formal schools. Literacy rates for Liberian children were probably even lower than Haiti's. The basic literacy rates for blacks and the non-English-speaking "colored population" in the Panama Canal Zone were the highest of all, about 85 percent. This is not surprising. The population in the Canal Zone during the construction period had been recruited for its physical and mental potential to perform (or learn to perform if need be) work skills. This specially selected population made up most of the Canal Zone's residents.[87]

With the support of the black community and northern philanthropy as well as the federal government, a network of black colleges located primarily throughout the black South had emerged by the beginning of the twentieth century. By 1900 there were thirty-four institutions providing collegiate training. Most had been established to assist newly emancipated slaves by missionary aid societies and the federal government's Freedmen's Bureau. Also, black and white churches founded some of these colleges, and, finally, federal land grants for agricultural colleges resulted in a few such colleges for southern blacks.[88]

The most influential school movement in the Deep South was Booker T. Washington's industrial education or "Tuskegee" model. Washington's idea that blacks were best served by special education stressing basic subjects and vocational job skills became widely popular among U.S. philanthropists. Between the 1890s and World War I, thousands of black youths were trained at his Tuskegee Institute in farming, building skills, nursing, and teaching and church occupations. Moreover, the large philanthropic organizations active in the South adopted the Tuskegee approach for the hundreds of black public schools these organizations were building. Most notable among these organizations was the Julius Rosenwald Foundation.[89]

Promoting the education of the South's blacks for small-scale farming or semiskilled work had universal appeal to the nation's philanthropic and government leaders. These individuals tended to be politically moderate or liberal on the matter of improving the life of black Americans. Wealthy capitalists and government leaders strongly endorsed the idea that basic education was the vital means for improving black American community life. In their view, primary and trade school education engendered middle-class values of service and thrift while lessening tendencies toward labor unrest, strident black nationalism, or rampant crime. Such basic education also produced a constant supply of unskilled and semiskilled black workers available for lower-level industrial work as well as for service fields in the segregated public institutions reserved for black communities.

Community nurses and an abundant supply of lay midwives also were used widely throughout the South. As black medical schools and hospitals expanded, so did the number of trained black nurses. Prior to the Civil War, the slaveholding class of the South had encouraged black population growth. Black midwives used indigenous knowledge to deliver infants for low fees throughout black communities. These midwives practiced their skills in communities outside the reach of the medical profession and the regulation of local public health officials. The midwife thrived in the social stability and isolation of the South's rural black communities. Strong rural community and Christian values bonded these women with their populations.[90] During the Jim Crow era, when blacks and whites were largely spatially segregated throughout the South, the "granny midwife" tradition from the antebellum era continued unabated as the dominant means for childbirth throughout the Black Belt South. Also by the early twentieth century, a network of black public health nurses working in ones and twos throughout the rural black counties flourished in the South.[91]

The widespread use of midwives prevailed in Haiti and Liberia. However, compared to the Black Belt South, a growing segment of educated and wage-earning populations did not emerge in early twentieth-century Haiti and Liberia. These populations are crucial to the development of medical trainees and professionals as well as a large clientele of paying patients. In addition, the midwives throughout the Black Belt frequently developed complementary relationships with trained black nurses and public officials. Local black health professionals tolerated and occasionally attempted to upgrade the services of midwives.

The black South's indigenous medical profession provided a substantial advantage to its population compared to the Haitian, Liberian, and Canal Zone health settings. A medical profession, medical school, and health care institutions had emerged in pre–1900 Haiti. Prior to the U.S. occupation that started in 1915, Haiti historically had a small but viable community of well-trained physicians, a few state-run hospitals, and professional medical societies. These doctors had learned medicine in France or under French-style schools in Haiti. However, the U.S. occupation would all but eliminate the autonomy of Haiti's medical professionals. As for Liberia, as we will see in later chapters, through the 1960s this nation lacked the preconditions to educate even a few score of Liberian physicians. In the Canal Zone, health care resources were provided directly by canal authorities, precluding the emergence of an indigenous Latino or black medical profession with its own institutions.

As the United States embarked on the twentieth century, black American leaders and leaders of Haiti and Liberia sought many paths to elevate the race in their respective lands. Throughout the United States virtually all black leaders maintained that thrift, hard work, literacy, and voting would inevitably open the gates of opportunity for black Americans. The most prominent race leaders—the Booker Washingtons, Thomas Fortunes, and Ida Wellses—shared in particular one belief: economic security or political well-being were inevitable for America's blacks, if only blacks located the right formula for capturing this wealth, votes, and, ultimately, social respect. Other leaders were more inclined to leave America for Africa. Henry McNeal Turner and Marcus Garvey, most notably, believed that black Americans only needed to transfer start-up businesses, Christianity, and the work ethic to African settings; and powerful African nations would ensue.

However, most all of these leaders did not fully recognize the wheels of technology transforming urban America and poised to overrun the South's paternal agrarian economy. Social leaders of black Americans could not grasp the implications of the internal revolutions in industrial management, engineering, medicine, and the laboratory sciences—revolutions that were increasing efficiency in heavy industry and farm production as well as in hospital care and transportation. In this context of technological change, the scope of opportunity for post-Reconstruction black Americans depended only superficially on political rights, styles of popular education, social idealism, and small Pan-African commercial ventures. Instead, the fate of the black masses in the United States, indeed in the Atlantic world, lay in their social and economic relationship to large-scale technologies, transportation networks, and plantations.

As for the core population of the Black Belt—the destitute sharecroppers— they relied on Christian messianism and church life, folk health care, and social survival skills to shield them from death at the hands of hazardous work and living conditions. These basic community institutions and concrete knowledge, more than Washington's Tuskegee schools or Du Bois's NAACP (National Association for the Advancement of Colored People) branches, most immediately stood between life and death for the typical worker and family in the southern Black Belt. As the Black Belt drifted through the dismal era of Jim Crow politics and plantation economics, U.S. industrial and foreign affairs leaders moved forward excitedly into other parts of the Atlantic world. These leaders were anxious to test abroad the tools and promises of scientific capability.

CHAPTER 2

Industry on the Isthmus

THE PANAMA CANAL ZONE

*W*ith its victory in the Civil War, the industrial North and the Union military took control of the South's political institutions. At the same time, industry's railroads joined the East and western territories. While the plantation economy of the South had been disrupted by the Civil War, it gained new life as the federal political and military constraints over the region were lifted. Following Reconstruction most of the nation's blacks still lived in the South and worked under the low plantation technology of the Black Belt. By the opening of the twentieth century, the same drive for technical efficiency forging heavy industry and large plantations pushed the reunited nation into the Age of Energy. The United States emerged as an industrial leader on the world stage. In coal, steel, and railway production, the United States pulled ahead of former leaders Britain and Germany. Steam-powered railroads, massive production of coal and steel, and large supplies of manufactured commodities were rapidly increasing America's urban populations and built landscape. This U.S.-led "second" industrial revolution also expanded electric power, lights, telegraph communication, and factories run on electricity. Finally, the United States had developed speedier and larger shipping lines rivaled only by those of Britain and Germany. To the American public, these advances in communication and transportation seemed destined to put all parts of the world within instant reach.

In this outpouring of industrial competitiveness during the 1890s, U.S. interest in building a canal across the Isthmus of Panama intensified. To patriotic politicians, military officials, business leaders, and newspaper editors, this great canal would be the crown jewel for U.S. domestic and international prosperity. A canal in the Caribbean basin would open coast-to-coast commercial shipping for the United States, linking the Atlantic and Pacific ports. The canal would

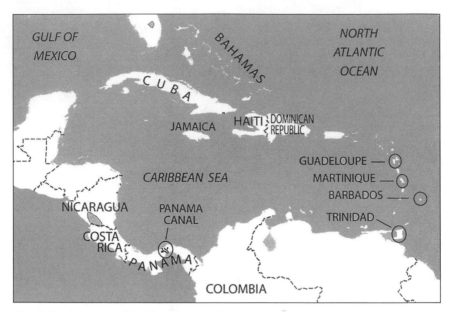

MAP 2.1.   Panama and the Caribbean regions

also mean a gargantuan leap for U.S. military power by giving the nation a "two-ocean" navy.[1]

Technological innovation and investment had swept forward huge factories, railroads, and early, large-scale mechanized farming throughout the United States' northern, midwestern, and western states. Now, as the twentieth century opened, the Panama Canal loomed as a key challenge to U.S. investors, engineers, and tropical disease specialists. Just as the technology of the plantation shaped the African Americans of the U.S. South, the Panama Canal project would play the dominant role in reshaping the Africa-origin populations of Central America and the Caribbean. This industrial colony on the Isthmus of Panama would be worked by a huge, racially segregated, mostly black West Indian workforce. The U.S. officials who moved to Panama to take on managing the canal construction were as committed as ever to testing and expanding their prowess in engineering, tropical medicine, and race relations. In its wake, the completed canal left the surrounding region a growing multiracial, multicultural society—a society divided sharply into a white-black racial hierarchy.

Early Commerce and Labor

Mercenaries and trade speculators from Europe and the United States had envisioned the economic potential of a sea route traversing the Isthmus of Panama long before 1904, when U.S. military and corporate engineers arrived to build the canal. The Spanish conquistadors coming into the New World planned such a route as early as the 1530s. Countless unsuccessful starts were made at a canal project by subsequent European leaders and commercial speculators. U.S. politicians began dreaming of an interoceanic canal across Central America early in this nation's history. It was first pursued by Secretary of State Henry Clay in 1825. But the concession to U.S. merchants for this purpose never got off the ground because adequate private capital could not be raised for the project. Ten years later President Andrew Jackson once again explored the feasibility of the canal project. Subsequently, innumerable proposals and schemes emerged throughout the United States and Europe.[2]

African peoples and cultural enclaves appeared throughout settlements along the Caribbean shores of Central America. From the days blacks assisted Balboa in discovering the Pacific Ocean, throughout the span of the slave trade, indigenous black Panamanians laid roots in this largely Hispanic colonial domain. So-called "Libertos" were slaves allowed to live and work in places like the La Triana district of Portobelo and in La Guinea. By the mid–1800s French- and American-owned railroads, shipping lines, and plantations had clawed onto the Isthmus of Panama. Once again, ebbs and flows of black laborers and indigenous people were drawn to these newly developing production sites on and surrounding the Isthmus. The flow of West Indian workers migrating into the Panama region especially thickened after Emancipation occurred on the British West Indies during the 1830s. British planters used West Indian laborers in the region of Bocas del Toro (northwest Panama). In the meantime, U.S. investors and firms developing trade, railroads, and steamship lines throughout Central America, including Nicaragua and Panama, also used heavy numbers of migrant laborers from the West Indies.[3]

The same time that California was experiencing the gold rush, a U.S. firm was building a railroad to run across the Isthmus of Panama. The Panama Railroad at first employed mostly Irish and Chinese workers. But these men perished at alarming rates. West Indian laborers were then called upon.[4] Antedating the first U.S. transcontinental railroad by fourteen years, the Panama Railroad lines traversing the Isthmus were completed in 1855. Said one of its chief civil engineers, "The railroad paralleled the trail of the old Spanish mule trains, and for many years did a very lucrative business." Sensational stories in newspapers throughout the United States had promised gold and wealth for those cultured individuals who ventured into exotic Central America. But actually the riders of the Panama Railroad were quite different. "Instead of carrying golden treasure and adventure seekers," the engineer recounted, it shuttled "hardy pioneers and many thousand tons of manufactured goods from ocean to ocean."[5]

Black workers, either as slaves, so-called free blacks, or emigrés from the West Indies, had also become rooted in Central America because of the spread of commercial banana growing throughout the region. In the mid-nineteenth century, U.S. firms began overtaking traditional banana cultivation throughout Central America. As with their railroads, U.S. companies organized plantations that depended heavily on long-standing and new immigrant black workers and native peoples. These workers were various indigenous people—American Indians, mestizos, and "Caribs"—as well as the descendants of runaway enslaved Africans. Banana-growing sites developed near the railroads constructed along the coastal lowlands.[6]

Shipping companies involved with the banana trade were another key aspect of early U.S. commercial growth in Central America and the Caribbean. Beginning in 1866 schooners from New England brought shipments of yellow bananas from Central America. Soon U.S. steamship firms emerged, most notably the Boston Fruit Company, that trafficked even larger supplies of bananas. Finally, several of these smaller firms combined to form the United Fruit Company, incorporated in 1899.[7]

Along the rail lines on Costa Rica's Atlantic coast, English-speaking communities of black Jamaicans and original Americans developed.[8] Charles Kepner, an authority on the banana industry, described the influx of black labor into Honduras and Panama: "At the dawn of the banana industry, the low-lying Caribbean jungles were so unhealthy that with the exception of certain courageous Honduran and Panamanian planters, few Central American nationals were willing to live and work in them. Hence the fruit companies introduced large numbers of West Indian Negroes who were immune to tropical fevers and more anxious for new opportunities for work. With the Negroes worked the Caribs, shore dwellers in whose veins are mingled Indian and African blood."[9] United Fruit used mostly black labor for the backbreaking work of laying rail lines from port to port throughout rugged stretches of Central America. In Guatemala, for example, by 1912 some 2,825 of the 3,000 United Fruit employees were black.[10]

In the meantime French investors, engineers, and labor crews took the first concrete steps to excavate a channel across the Isthmus of Panama and build the canal. France's work on the canal started in 1881, when its first work crews arrived, and ended in 1893. The canal workforce for the French reached 19,000 employees by 1884. Mortality and disability caused by yellow fever and malaria among these immigrants were constant throughout France's canal effort. As financial problems rose, as well as deaths of French managers, engineers, workers, and their families, so did the dissatisfaction among investors and the French public for the canal venture. By the late 1890s, the French had terminated their canal work and sold off their property interest in the region.[11] An estimated 20,000 black workers from the West Indies were laboring on the French canal project when it closed.[12]

Politics and Civilization

While France was largely abandoning Central America, fear of competition in the Caribbean from Spain and other European nations triggered the United States' involvement in the Spanish-American War. Not only was Spain swiftly and easily defeated, but the United States under an aggressive President McKinley emerged controlling for itself a new territorial empire. Cuba, Guam, Puerto Rico, the Philippines, and Hawaii, all fell into America's hands as a direct or indirect result of the war. Central to the United States' victory and early occupation of these nations was the effectiveness of U.S. authorities in using public health measures to protect their soldiers from tropical diseases. Owing to its strong military, industrial capacities and military health services, the United States was now a threatening young upstart to the armies and navies of the older imperial powers of Britain, France, Germany, and Belgium.[13]

Anglo-Saxonism was a strong rationale in U.S. government, commercial, and public health circles for expanding throughout Panama, the Caribbean, and Asia. From McKinley through Wilson, U.S. presidents and congressional leaders proclaimed that societies of black, brown, and yellow people were unhealthy political entities, unsuited for self-government and laissez-faire capitalist economies. They urged their constituents that only America, as an Anglo-Saxon republic, could successfully build and oversee the canal across the Isthmus of Panama. In their view, military intervention, followed by enlightened, industrially organized administration, was the only means for such societies of dark people to obtain democratic nationhood. While these leaders were social Darwinists, they embellished this theory with the idea that the heavy hand of modern industry could tame even the most naturally unfit tropical races.[14]

Theodore Roosevelt and other foreign affairs leaders saw in social Darwinism a scientific explanation for the success of U.S. expansionism. To Roosevelt and his political and commercial supporters, physical struggle between civilized and "backward" societies was a natural phenomenon rooted in biological hierarchy. The darker races were mired in political backwardness. Moreover, these tropical societies were destined to hold a subordinate place in the burgeoning free-enterprise, international economies of the United States and European nations. As historians Freidel and Brinkley explain, Roosevelt believed that "'Civilized' nations . . . were predominantly white and Anglo-Saxon or Teutonic; 'uncivilized' nations were generally nonwhite. . . . Civilized nations were, by Roosevelt's definitions, producers of industrial goods; uncivilized nations were suppliers of raw materials and markets."[15] Western nations and white races had an obligation to remain ever prepared for military conflict. "In the long run," Roosevelt stated, "civilized man finds he can keep the peace only by subduing his barbarian neighbor, for the barbarian will yield only to force."[16]

With growing U.S. investments in Central America, the social and political rationale to build and control the Panama Canal gained momentum. The original Monroe Doctrine was expanded by Roosevelt's famous Corollary. This new policy permitted U.S. intervention in American countries that were considered too unstable, in the judgment of the United States, in order to prevent political involvement by European nations.[17] U.S. military forces began to dig in throughout the new territories in the Americas, standing ready to enter other countries such as Haiti and the Dominican Republic. The Panama Canal was to be the hub from which U.S. naval and military forces could speedily reach vital locations both on the Atlantic and Pacific Oceans.[18]

In his 1904 annual message, spelling out America's new expansionist foreign policy, Roosevelt emphasized that "chronic wrongdoing [or] an impotence which results in a general loosening of the ties of civilized society, may . . . ultimately require intervention by some civilized nation." In the Americas, "the adherence of the United States to the Monroe Doctrine" required this nation to exercise "an international police power."[19] To Roosevelt, the tendency of nonwhite people toward cultural backwardness was evident in U.S. blacks as well. Blacks in the Jim Crow South were facing widespread political disfranchisement and destitution. Roosevelt believed southern whites were justly recoiling from the socially backward blacks because of "the entire inability of immense masses of Southern blacks to meet any test which requires both intelligence and moral vigor."[20]

The biological-political imperative—that is, the merging of the campaign to protect modern man from tropical diseases with the drive by the United States and European nations to expand abroad economically—had gained strong momentum during and after the Spanish-American War. This political ideology coated the more deeply seated cultural mission to expand science-oriented industry and medicine. Together they formed the ideological and cultural drives behind U.S. technical projects developed abroad prior to World War II in areas of public health and civil and military administration. This ideal of technical superiority was transplanted not only to the Canal Zone, but later to Haiti and Liberia as well. The ideas of Orville Platt personified the bio-political impulse shaping U.S. public health activities in the tropics. The Connecticut senator authored the famous Platt Amendment that was the basis for the U.S. occupation of Cuba.

Platt believed that a successful democratic polity could never be initially all black. In 1901 Platt's essay "America's Race Problem" appeared in the *Annals of the American Academy of Political and Social Science*. "To insure the success of free government," he stated, "certain conditions seem indispensable." Above all, "[t]here must be a homogenous people possessed of a high degree of virtue and intelligence." But "[t]he colored population in Cuba presents a most interesting sociological problem," Platt wrote, because the "admixture of blood in his veins exceeds, perhaps, that of the mulatto in any other part of the world." According to Platt, the different classes and racial groups "of Cuban population have little in common, except a desire for liberty, as yet scarcely understood."[21] To leaders like Roosevelt, Platt, and American officials and professionals in tropical countries such as Cuba, Haiti, and Jamaica, the black-skinned populations of these islands were humanity's least civilized, possessing tendencies toward moral anarchy.

Wolfred Nelson, a prominent New York physician had traveled to Cuba years before the Spanish-American War. He spoke about his visit and a U.S. occupation of Cuba to the American Association for the Advancement of Science. Nelson described Cuba if left ungoverned by Anglo-European nations: "The Cuba Libre of the blacks would be a veritable hell upon earth, a blot upon Christian civilization. . . . Knowing the island as I do, I fear that an independent Cuba will be an impossibility." To him, the only solution, without doubt, was the swift imposition of U.S. control. "As an American colony she [Cuba] will blossom and bring forth her increase," he stated. "Then, and then only, will the black plague of . . . Cuba cease to be a nightmare." Once placed under U.S. administration, "Cuba will be the brightest spot in the colonial possessions of the United States."[22]

But the push by U.S. forces into Cuba and other parts of the tropical world proved much less smooth than U.S. expansionists, such as Platt, Roosevelt, and Nelson, had anticipated. Following its defeat of Spain in Cuba during 1898, the United States set about improving courts, banking, public works, education, and health care facilities throughout the island. However, a yellow fever scourge that killed officers, troops, and civilians in huge numbers greeted the U.S. forces garrisoning throughout Cuba. By 1899 the Cuban medical researcher Dr. Carlos Juan Finlay had identified the mosquito species that carried yellow fever. In 1901 the laboratory work of Walter Reed and his associates began gaining acceptance as absolute proof that yellow fever was transmitted by the *Aedes aëgypti* mosquito from an infected person to a healthy one. Reed's work led the U.S. military occupying Cuba to shift its disease-control measures. William C. Gorgas, the chief sanitary officer for the United States in Havana, implemented sanitation and quarantine efforts so that the brunt of the yellow fever outbreak was quickly eliminated.[23]

In 1903 General Leonard Wood, the U.S. governor-general of Cuba, exalted the results of the U.S. occupation of the region. He called the outcomes a virtual miracle. "The work called for and accomplished was the building up of a REPUBLIC," he wrote. The United States had managed to "transfer to the Cuban people [a] republic free from debt, healthy, orderly, well equipped, and with a good balance in the treasury." The United States completed its work "in a little over three years, in a Latin military colony, in one of the most unhealthy countries of the world." According to Wood, the United States had transformed Cuba into "a republic modeled closely upon the lines of our own great Anglo-Saxon republic."[24]

The Search for Immune Workers

Ensuring the fitness of its managers and workers was of paramount importance to U.S. Panama Canal authorities. U.S. military heads called the region where the canal was to be built "the unhealthiest place on earth." The Isthmus of Panama was a convergence point for thick jungles, high temperatures, long rainy seasons, and low-lying bodies of water. They pointed to the dreadful encounter the French had had with disease in the area. From 1881 to 1888 the French hospitals in Ancon and Colon had recorded 5,500 deaths due to disease, or a rate of about 63 deaths per 1,000 employees. In 1885 alone some 687 malaria deaths occurred in the population of 20,276 workers. From 1884 to

1893 the city of Panama, which averaged about 20,000 to 22,000 residents, experienced over 3,500 deaths due to malarial fever.[25]

In addition to the influence of social Darwinism on political ideas, it reinforced a view popular in medical and sociological circles at the turn of the century, a view that blacks in Africa and America possessed inborn racial immunity from tropical disease. Moreover, blacks were believed to possess unique inherited or acquired physical stamina. Their racial durability enabled blacks to withstand heat and deadly effects of diseases such as malaria and yellow fever.[26] This medical and social perception was based on general impressions. Key sources for the racial durability idea were the nineteenth-century travelogues and medical accounts of European explorers and missionaries in the African interior. While Western travelers newly arriving in Africa often died from fevers, indigenous populations seemed impervious to severe fever.[27] A second root for the idea that blacks had special immunity against tropical disease stemmed from accounts of slave physicians and social thinkers of the antebellum South.[28]

By the time the United States began to plan construction of the Panama Canal, there was plenty of information that contradicted this idea that people of black African background possessed greater immunity to tropical disease. Better than the anecdotal impressions of missionaries and planter physicians were the mortality and sickness records on black workers in the Panama region immediately prior to the start of canal construction. These should have demonstrated clearly that black workers could die just as rapidly as whites from tropical diseases. And conversely, black laborers whose work sites in Central American jungles were insulated from mosquito vectors experienced high survival rates.

The presumably immune black laborers of the Caribbean had suffered extremely high death rates from diseases while toiling in the harsh jungle regions of Central America. During the 1870s, some four thousand West Indian workers died laying a British-owned railroad in Central America. An estimated twenty thousand more had perished working on the French's Panama Canal project. Most of the canal area deaths were from malaria and yellow fever. Those black workers who survived their work ordeals, according to Conniff, usually returned home "in broken health, unable to enjoy their savings from abroad."[29]

By contrast, the health of black railroad workers in the Panama region was much better. Before U.S. engineers started constructing the main structural components of the canal—the walls, gates, and locks of the canal—they had to entirely rebuild the worn, lightweight rail lines of the Panama Railroad. Also, the railroad's tracks had to be placed on much higher levels of ground. The Panama Railroad provided the sole means for hauling away the millions of tons of spoil from the land cuts the canal required. This railroad also transported tens of thousands of workers and their construction equipment across the canal building sites and handled the distribution of commercial freight to the Canal Zone region.[30] These railroad workers included black workers—from both the West Indies and local communities—as well as white laborers. Malaria and yellow fever deaths among the railroad laborers were infrequent because, years before the canal was constructed, railroad authorities learned that by using thoroughly screened cars as construction camps for their workers, infectious mosquitoes were kept at minuscule numbers.[31]

Regardless of these prior health episodes, U.S. canal authorities continued to believe steadfastly that West Indian blacks possessed greater immunity to diseases endemic to the Isthmus of Panama. For President Roosevelt and the teams of industrial managers and engineers laying out the Panama Canal project, two practical challenges overrode all others. First, they needed to assemble the black workforce for the strenuous jobs at the canal construction site. Second, sanitary living and work conditions for canal managers and, if need be, for their workers had to be built to prevent mass disability or death from tropical diseases. U.S. officials and canal planners decided to organize the Canal Zone. This Zone encompassed a ten-mile strip of land that ran along the canal dig from one end of the Isthmus of Panama to the other. By the time the canal was completed, a new industrial black Diaspora in Panama would emerge inside and surrounding the Zone.

Charles Pepper, a leading missionary from the United States, visited the Isthmus of Panama in 1904. He wanted to assess the Canal Zone and the early stages of the building project. In his travelogue, he described the Canal Zone as "a colony within the Republic of Panama, yet not of it. This colony, which includes laborers, civilian officials, occasional detachments of marines, and a police force . . . will be a conglomerate mass—Jamaican and other West Indian negroes [*sic*], Chinese coolies, Mexican and Central American peons, possibly a few American blacks, Italian railway workers, and similar elements." Pepper assured his American audience that, like the canal construction managers, he knew black laborers had special stamina to do the hardest work constructing the canal. Black workers had done "the most of what was accomplished by the French company [and] built the railroads along the unhealthy coast of Costa Rica" In the end, the "Panama Canal will be the monumental contribution of the despised black race to civilization."[32]

Indeed, once the United States took over the excavation of the channel for the canal and the operation of the Canal Zone, it set in motion a dramatic enlargement of this region's long-standing African Diaspora labor. The labor demands both at the canal building site and at the Zone settlement attracted tens of thousands of Caribbean and Central American blacks as well as indigenes. Like the French before them, U.S. Canal Zone employers recruited these mostly black West Indian workers for the hardest and most hazardous jobs on the canal.[33]

Since the economics of the Caribbean nations could not support their growing domestic populations, immigrant labor from this region was in abundant supply for canal employers. Moreover, West Indian blacks had a reputation among North Americans as hard working and courageous and desperate enough to labor and live under the toughest of environments. There also were specific benefits to West Indian nation-states to "push" their most productive workers into Central America. The constant emigration of large numbers of West Indian

TABLE 2.1   *Nationality and Number of Panama Canal Employees, 1904–1913*

Nation	1904	1905	1906	1907	1908	1909	1910	1911	1912	1913	Total
Spain			1,174	5,293	1,831						8,298
Cuba			500								500
Italy			909	1,032							1,941
Greece				1,101							1,101
France			19								19
Armenia			14								14
Total Europeans			2,616	7,426	1,831						11,873
Fortune Island			1,317						785		2,102
Barbados	404	3,019	6,510	3,242	2,592	3,605				528	19,900
Guadeloupe				2,039					14		2,053
Martinique		2,733	585	2,224							5,542
Jamaica		47									47
Trinidad			1,079					205	143		1,427
Total West Indians	404	5,799	9,491	7,505	2,592	3,605		205	942	528	31,071
Costa Rica		244									244
Colombia		1,077	416								1,493
Panama		334	10	13							357
Grand Total	404	7,454	12,533	14,944	4,423	3,605		205	942	528	45,038

Source: R. E. Wood, "The Working Force of the Panama Canal," Transactions of the International Engineering Congress, 1915, San Francisco, Calif., September 20–25, 1915.

workers rewarded their indigenous island communities with "diminished unemployment [and] reduced demand for food and shelter." Emigration to work sites of European firms also in Central America "brought cash remittances to families of workers employed abroad, and gave workers training in construction or farming techniques."[34]

U.S. canal authorities negotiated treaties with West Indian governments and agents to bring in black laborers for work on the canal. The labor population that prepared the Canal Zone and constructed the canal was composed primarily of West Indian blacks, European contract workers, and engineers and managers from the United States. By 1906 the population of the Canal Zone was over 10,000, about 7,400 of whom were canal employees, and 5,800 of those workers were West Indian (see table 2.1 and figs. 1–4).

Early Construction

Out of this swirl of excavation and construction, a new multiracial community forged itself. Black work crews labored alongside those of indigenous peoples and immigrant Asian and European laborers. Dispersed throughout the work sites were the white American managers. Harry Franck, a popular travel author from the United States, described the fury of the construction site: "Everywhere are gangs of men. . . . Shovel gangs, track gangs, surfacing gangs, dynamite gangs, gangs doing everything imaginable with shovel and pick and crow-bar, gangs down on the floor of the canal, gangs far up the steep walls of cut rock, gangs stretching away in either direction till those far off look like up-right bands of the leaf-cutting ants of Panamanian jungles; gangs nearly all, what-ever their nationality."[35] U.S. visitors such as Franck were usually most displeased encountering the black workers, especially since many of the workers spoke non-English languages. "Here are Basques in the boinas [berets], preferring their na-tive 'Euscarra' to Spanish; French 'niggers' and English 'niggers' whom it is to the interest of peace and order to keep as far apart as possible, . . . laborers of every color and degree." White American laborers were "more conspicuous by their absence" because of "the caste system that forbids white Americans from engaging in common labor side by side with negroes [*sic*] in [this] enterprise of which the leaders are not only military men but largely southerners."[36] Mostly black laborers had done the heaviest work on the railway lines that fed into and out of the canal site. Now with construction on the canal itself started, blacks toiled as "pick and shovel" workers inside the huge excavation pits.

Canal construction had barely begun when labor discord broke out. Black workers complained about their measly pay and poor work conditions in spite of frequently doing the exact same jobs as white employees. Many Panamani-ans living locally detested the U.S. presence. The canal destabilized their local, small market communities and cut them off from their historic links throughout Colombia. Finally, white "gold" employees complained about the poor quality of living and frequent injustices, from various Zone officials, resulting in short-changes in pay, vacation, and services.[37]

By the spring of 1908, an embarrassed President Roosevelt had to con-vene a special commission "to investigate conditions, especially as regards la-bor and accommodations, on the Isthmus of Panama." U.S. officials discovered that the European workers were an unstable source for employment beyond a few years. Since they were unable to bring their families, low morale among this group was a constant problem. Moreover, the language differences between them and their English-speaking American supervisors made it difficult for them to understand work orders, so frustration and conflicts with on-site foremen fre-quently set in. Finally, some European workers chafed at working and living in the midst of blacks.[38]

As construction on the canal intensified, to reduce this friction, personnel authorities increasingly relied on black labor for low-level "silver" jobs. At the same time, Zone administrators privileged white American and European work-ers, segregating them, their families, and their housing from black workers throughout the Canal Zone "island community." Housing for married gold em-ployees was designed "to secure the maximum comfort possible in a tropical climate [with] modern plumbing, including bathrooms; water, fuel, and light are also free." Furthermore, Zone authorities or private employers arranged services and facilities for white gold employees and their families, from schools and so-cial and recreation clubhouses to women's clubs, churches, and theaters.[39]

By organizing the black laborers into large units of low-skilled workers, productivity at the canal site was maintained while formal and informal physical association between blacks and white European coworkers was minimized. Camps for the workers' living quarters were also segregated. The special commission found it best to remove black policemen from the camps, to prevent black "Zoners" from having any formal authority over whites in daily work or community life: "The elimination of the colored policemen from the camp has proved a very wise measure, they being supplanted by white policemen. Most of the European laborers brought here have not been accustomed to being ordered around by colored men, and the natural feeling of rebellion would break out among them when one proved that it is practically impossible to put a colored man in charge of white labor, owing to the almost invariable abuse of authority on the part of the colored man when given opportunity."[40] The wages for gold employees were usually double those of silver employees. Canal employers hired and kept black labor in the lower-paid silver positions. This practice reinforced segregation of Negroes (or "coloreds") in all other aspects of public life. When Canal Zone authorities segregated housing and schools, they were simply bolstering the racial separatism originally established by the "gold-silver" employment practice.[41]

To attract and maintain engineers and other highly skilled civilian and military personnel, the Canal Zone authorities established schools, recreational resources, and women's activities for their families. For example, by 1906 canal employers needed more than the 20,000 laborers they had hired. Citing the inefficiency of the initial West Indian black workers, they began to bring in European contract labor via Cuba and Spain. Their idea was to cut back "fresh arrivals" of "irresponsible persons in various West Indian islands from coming here in search of employment."[42] Between 1906 and 1908 about 12,000 contract workers arrived from Europe—75 percent from Spain and the remaining workers from Italy and Greece. According to canal officials, "the European laborer was paid twice as much as the West Indian laborer for exactly the class of work, but his efficiency at first was rated as approximately 3 to 1 as compared to the Negro."[43] (see table 2.1).

During the first years of the canal construction black laborers were restricted to the lowest-rung technical occupations throughout the Zone. The black workers in the Canal Zone differed from those in the agrarian Black Belt South. In fact, the employment segregation practiced by the Zone's officials and managers made the canal's employment hierarchy more analogous to the urban Northeastern and Midwestern regions of the United States. In northern U.S. cities and industrial towns, blacks were shunted into unskilled jobs since white skilled workers were increasingly preferred over black ones. The employment discrimination in the North prevented blacks from entering occupations in manufacturing, trade, and transportation sectors. Moreover, black labor in the North faced stringent exclusion from apprenticeships and unions.[44]

In 1905 Dr. William DeBerry, a black Congregational minister, surveyed the black working population of his city, Springfield, Massachusetts. He described conditions common throughout the North. Among Springfield's 808 employable black adults, he asked, "Why is it that 86 percent . . . is confined to the lower strata of industry?" These black workers have the skills, "but . . . race prejudice has closed the door of industrial opportunity against these men as a class." DeBerry decried white employers for practicing "merciless industrial ostracism which shuts out the capable and worthy Negro because God chose to create him black."[45] Similar employment segregation, we have observed, was widespread in the Canal Zone. However, by the interwar period, Zone officials and managers had gone further than northern industrial segregationists. They had replicated the practice of Jim Crow segregation in social institutions characteristic of the Black Belt South in the United States.[46]

The experiences of the laborers on the Panama Canal are a study of remarkable endurance, adaptability, and skill. And the work of the black laborers was even more remarkable because they faced daily poverty and social discrimination. With railroads bringing in greater numbers of workers, the camp houses for this population became much larger structures that were more difficult to maintain safely. Professional construction and military personnel from the United States expected and in general received clean, fashionable living quarters fit for their families and their recreational lifestyles. But to keep overall housing costs low, canal authorities initially placed the canal labor crews in leaky or improperly sealed, screened, or ventilated barracks. Their early tasks involved laying out construction sites, which meant installing roads, water, and sewage systems. Over 200 million cubic yards of earth had to be excavated, earth dams raised, and special plants built for making the concrete and other materials for constructing the canal locks. At the peak of the U.S. construction work in 1908, thousands of workers, digging with shovels, operating cranes and steams shovels, and loading trains, excavated 37 million cubic yards of rock and earth.[47] By 1912 the Canal Zone population had swelled to over sixty thousand (see table 2.2).

A canal worker, later a union organizer, recounted his early experience to W.E.B. Du Bois. Unable to obtain school education, he began working at the canal as an older boy and became a pick and shovel worker.[48] Black canal workers "lived in hovels, with poor food and wretched sanitation, their schools were poor and teachers underpaid." The United States in the Canal Zone "was discriminating against colored workers (called silver workers) and paying white workers (gold workers) wages often 100 percent higher, with good homes, schools and opportunity for promotion."[49]

Black women and infants living in the Canal Zone region also suffered deprivation. Official surveys of health conditions in the Zone area revealed an "enormous infantile death rate among the negroes [*sic*], due to ignorance and indifference on the part of the mothers."[50] Panama City in 1914 had a population of 53,948, which suffered 1,863 deaths. Infectious diseases caused most of these deaths (1,772), and about half of the total deaths were among children under

TABLE 2.2   *Canal Zone Population, by Color* and Sex: 1912 and 1920*

| | Color and Census Year | | | | | |
| | 1912 | | | 1920 | | |
	White	Black*	Total	White	Black	Total
Both sexes	19,413	43,397	62,810	12,370	10,488	22,858
Male	14,959	30,204	45,163	8,555	6,013	14,586
Female	4,454	13,193	17,647	3,815	4,457	8,272
% of total						
Both sexes	30.90	69.10	100	54.10	45.90	100
Male	33.10	66.90	100	58.70	41.30	100
Female	25.20	74.80	100	46.10	53.90	100

Source: U.S. Bureau of the Census, *Fourteenth Census—1920,* vol. 3, *Composition and Character-istics, Population by State, Panama Canal Zone* (Washington, D.C. 1922), table 10, 1247.
* Chinese, Filipinos, and Japanese are included in the Black category. In 1920 they comprised less than 1 percent of the Zone's nonwhites, while Negroes (or blacks) comprised over 99 percent of this category.

five years of age. A similar situation prevailed in Colon, a city of 23,265. Some 590 deaths occurred in Colon, 43 percent among children less than five years old. While health authorities blamed poor care by mothers for these deaths, they admitted unsafe living conditions were also involved. They found "serious over-crowding in the tenement districts," where it was "a frequent occurrence to find six or more persons sleeping in one unventilated room, 10 by 10 feet." The bar-racks type of housing for "colored" families living in the Zone persisted through the 1930s.[51]

As labor demands at the canal site intensified, gradually black workers were allowed into the more skilled jobs. Major R. E. Wood, stationed at the ca-nal from 1905 to 1915, was an expert on the project's labor supply. At first, his expectation about the skills of the black workers was pessimistic and misin-formed. Speaking about the canal workforce at the International Engineering Congress in 1915 held in San Francisco, Wood remarked: "Practically all of the West Indians had worked in fields and had never done any construction work. Many of them had never seen a railroad; few of them were acquainted with ex-plosives. Outside of the building trades, none of them were of any value as arti-sans. All were slow and stupid, and altogether they were most unpromising material."[52] According to Wood, the workers from Europe, by contrast, "were as a rule much more intelligent, [even though] few of them had ever worked on a construction job before." "They stuck more steadily to the work than the West Indians and were, in the beginning at least, much more to be depended upon."[53]

But during the course of the canal's construction, Wood found that black workers improved dramatically. "The West Indian, while slow, has learned many of the trades and many of them have developed into first class construction men." By the end of the project, the black skilled workers "developed steadiness, and toward the end of the construction period . . . remained on the job as steadily as the Spaniard or even the American." Wood gave most credit for the improved quality of the West Indian canal workers to U.S. foremen "who developed and educated this labor." Nonetheless, he admitted, "The bulk of the building work on the Canal has been done by West Indian carpenters, masons, and painters, under the direction of American foremen."[54]

The Sanitarians

The project to build the Panama Canal had generated a massive convergence of railroads, construction equipment, laborers, and their families into the Canal Zone. But it was the success of American public health experts in reducing malaria and yellow fever that most fed the missionary enthusiasm for the canal endeavor. The entire outpouring of U.S. investments and engineering and construction personnel onto the Isthmus would have been worthless were it not for the accomplishments of certain key U.S. public health planners. These health professionals and engineers were "sanitarians" in the sense that they stressed improving the most detailed aspect of the Canal Zone environment in order to reduce the presence of mosquito vectors and unsanitary conditions. The knowledge that reducing the breeding areas for mosquitoes and improving sanitary conditions would eliminate malaria and yellow fever epidemics emerged in the late nineteenth century. It derived from the early work of leading experimentalists, frequently military medical officers, and self-styled malariologists. Prominent among this group were Patrick Manson, Ronald Ross, Walter Reed, and William Gorgas.[55]

The arsenal the public health sanitarians used against mosquito-borne diseases centered on improving environmental sanitation. These steps included draining exposed bodies of water (swamps, puddles, and the like); screening windows and doors of houses and work sites; destroying breeding areas of mosquitoes; paving streets; and cleaning and enclosing the water supply in pipes to minimize storage of water in barrels and cisterns. This group of early public health experts also stressed maritime quarantines. Ships were monitored to prevent them from releasing infected staff or travelers onto the Isthmus. Also, the ships and their cargoes were methodically screened for rodents or insects that could reintroduce the pathogens. Finally, the health program for the Canal Zone provided isolation and treatment for those already in the community who were discovered with infection.

The Canal Zone initially was a fertile environment for epidemics. The multitudes of transient workers were at greatest risk because of their harsh work conditions. They labored immersed in tropical forests with a hot, rainy climate and lived in unhealthy housing conditions. Panama City, with about twenty-five thousand residents in 1904, and Colon, populated by about four thousand, were the two seaports that became key transport sites for the workers and materials needed at the canal project. Both cities lacked sanitation and quarantine infrastructures and, therefore, were recurrently struck by yellow fever outbreaks.[56] The huge political and financial investment in the canal project was threatened by the productivity loss that could ensue were Canal Zone managers and workers struck down by such outbreaks. Hence, launching a successful public health campaign for the Canal Zone became the central concern for the United States' leading public health experts and military officials.

The success of American public health authorities with yellow fever and malaria control throughout the United States' construction of the canal was due largely to methods developed by Majors Walter Reed and William Gorgas to stem the great yellow fever epidemics of Cuba.[57] During the Spanish-American War when American troops occupied Cuba, a broad program to eliminate yellow fever mosquitoes had been successfully launched by Gorgas. He served as the chief sanitary officer of this campaign. Similar measures were found effective for the control of *Anopheles* mosquitoes, a primary vector for malaria.

When U.S. officials arrived in Panama to initiate the canal construction, the disease peril in the region was horrifying to them, but epidemic diseases were already proving deadly to the blacks and Asians who had settled in Panama well before these Americans arrived. About fifteen thousand people lived in the continuous stretch of villages forty-seven miles long running from the towns of Panama to Colon. In 1904 *Public Health Reports,* the federal government's leading public health organ, described the dismal health environment in these villages:

> The population is almost entirely negro [*sic*] and Chinese. All the villages are filthy, without regulations or restrictions, without sewers, and having the usual water supply of the country, viz., rainwater during the wet season, and water from streams during the dry season. No attention is paid to the wholesomeness of the source of the water supply. Mosquitoes are prevalent in all these villages, breeding in rainwater barrels, in the swamps, along the streams and in ponds. Malaria, elephantiasis, and beriberi are always to be found, and yellow fever and smallpox will occur when favorable clinical material presents itself, unless proper precautions are taken.[58]

Gorgas began planning in the Canal Zone the same mosquito-control measures that had proven successful in Cuba. According to Gorgas, in Panama "the anti-malarial work in order of its importance, consisted of drainage, brush and grass cutting, oiling, use of soluble larvicide, prophylactic quinine, screening and the killing of adult mosquitoes in the quarters of laborers."[59]

Three other public health measures greatly improved the health of those living in the Canal Zone and the surrounding Panama port cities. First, piping for filtered, chlorinated water was laid in the Canal Zone as well as in Colon and Panama City. This eliminated the stagnant water in tanks, pots, and reservoirs where mosquitoes bred ferociously. Communities could now be cleaned with sterilized water. Also, by providing drinking water and food clean of harmful bacteria, the numerous "dysenteries," such as typhoid fever, amoebic and bacillary dysentery, and diarrheas, were largely eliminated (see figs. 5 and 6).[60]

Second, Gorgas was able to have the housing for the black workers improved, which included closing the unhealthy camps used for black workers. In 1914 Gorgas was brought to South Africa by mining officials to advise them about ways of lowering the high death rates of their black miners. Gorgas described the situation first encountered in Panama. During the opening years of canal construction, the black workers slept crowded in bunk units, often in damp cloths and without proper blankets: "Our men worked continuously in the rain, and came home soaked through at night. The majority of them were so poor that they did not have a change of clothing. They therefore slept in their wet clothes."[61]

Housing and meal conditions were so cramped and unventilated that even black workers who had dry clothes and blankets were experiencing high death rates from pneumonia. Gorgas managed to have the practice of quartering black workers in cramped bunks ended. As he explained to South African mine officials, "[t]he fall in death-rate from pneumonia was coincident with allowing laborers to leave the barracks provided by the [Canal Zone] commission and build for themselves cabins on the hills, or to live in the towns of the Canal Zone, and the bringing in of their families." In 1906 the death rate among black canal workers was 18.8 per 1,000, but dropped to 10.6 and then 2.6 in 1907 and 1908, respectively. To keep these pneumonia rates low, the canal commission began requiring workers to provide proper bedding and underclothing, but provided them to workers who could not afford them.[62]

Finally, Canal Zone health officials oversaw the most detailed hygienic conduct and recreation of the various work groups. Regular examinations, including urine and feces tests, were done on all food handlers at all food establishments connected with the canal. Quinine tonic and pills were made available at all restaurants.[63] Canal Zone officials even monitored recreational hiking and camping trips made to nearby areas by teachers and scoutmasters to minimize possible infections among these travelers outside the "sanitated" Canal Zone.[64]

As these public health measures were implemented throughout the Zone, Gorgas saw to it that those sick with malaria received hospital care, while quinine (then called "quinin") was disseminated throughout the entire population. An elaborate, densely packed network of hospitals and dispensaries for isolating and treating those sick with malaria was implemented. Two large hospitals left behind by the French were renovated—one at the northern end of the Isthmus, in Colon (Colon Hospital), and the other at the southern end, in Panama City (Ancon Hospital). Between these two base hospitals, Zone authorities organized many medical and sanitary districts. Some eighteen small hospitals of twenty to one hundred beds each were then erected inside these districts. District hospitals were connected to Colon and Ancon Hospitals by hospital trains that, in turn, were met by hospital ambulances. Further still, in small villages within the districts, about forty smaller, subdistrict hospitals or so-called rest camps, each of five to fifteen beds, were set up. These cared for patients until they were transferred to the district hospitals. The district networks also included outreach staff who, under the direction of district physicians, were responsible for getting laborers to take daily quinine doses (see map 2.2).[65]

In addition to malaria control, the health authorities in the Canal Zone were intensely concerned with preventing the spread of any other diseases that would immediately sicken or hospitalize a worker or diseases that created panic because the symptoms broke out quickly, leaving horrifying appearances on an individual. Smallpox was one such disease. Canal authorities made vaccinations compulsory for all Zone residents. They instituted a vaccination program in the Canal Zone, Colon, and Panama City with the cooperation of the Panamanian authorities. Also, on some occasions towns in more remote areas were sites of mass smallpox vaccination programs. Revaccination was conducted every five years. In addition, school children and infants were vaccinated, and immunizations were given when appropriate to persons coming from abroad into the canal region.[66] The success of Gorgas's campaign became a much-heralded aspect of U.S. governance of the Canal Zone. In particular, Gorgas's measures against malaria mosquitoes in the Zone became recognized as one of the West's most effective malaria-control campaigns in tropical regions.[67]

Building a Segregated Society—Education

Back on the U.S. mainland, racial segregation had become a common fixture in the New South. Known popularly as Jim Crow segregation, this caste order stemmed from centuries of animosity among white slave owners and small planters of the South toward African slaves. Slave labor undermined the economic livelihood of the South's white wage-earning and small planter class. The white South's losses in the Civil War and Reconstruction further intensified antagonism toward blacks throughout the region.[68] Racial segregation was swiftly duplicated in Canal Zone society. As mentioned earlier, the occupational structure of the Canal Zone workforce was strictly hierarchized by race. This segregation within the Zone's industrial framework was the foundation for, and later reinforced by, offshoot segregation in schools, housing, and resources for women throughout the Canal Zone region.

From the first day U.S. workers started canal construction, the provision of schools became a persistent issue for government and residents on the Isthmus. Panama established public schools in its Constitution of 1904 and raised special revenues for this purpose. Prior to the advent of the Canal Zone and the arrival of U.S. employees, municipalities and communities of West Indian workers in Panama had established small schools throughout the Isthmus. The municipalities of Gorgona, San Pablo, Bas Obispo, Matachin, and Cristobal all had opened schools. Moreover, West Indians had started schools for primary instruc-

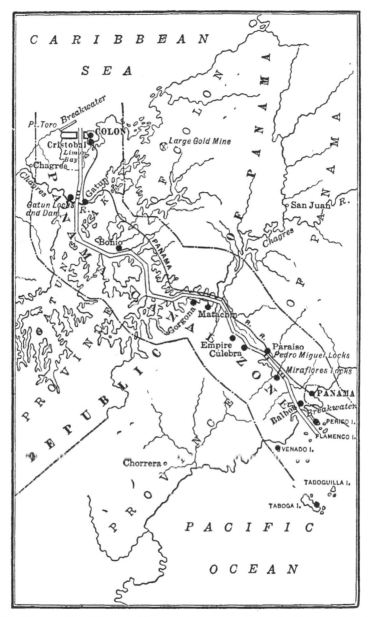

Map 2.2.   Panama Canal Zone and Sanitary Department hospitals, 1915. Reprinted from William C. Gorgas, *Sanitation in Panama* (New York: D. Appleton, 1915).

tion in private homes and other places.[69] What U.S. authorities would bring to the Canal Zone region was a public school regime that was systematically segregated along racial lines. Education resources and goals were designed to reproduce the subordinate employment roles of blacks in the canal region, with skilled and professional roles reserved for the region's whites.

In 1905 the Canal Zone had about two thousand children of school age, more than half of whom were illiterate. The following year Zone authorities established their own school system and incorporated the small municipal schools. By the year's end the Zone administration operated eighteen schools that enrolled 611 students taught by twenty-one teachers. The next year the number of schools increased to thirty with 1,237 students and thirty-four teachers. The schools had been initiated primarily to stem the turnover of the professional and skilled U.S. employees coming to work on the canal. At first, many would leave within one year. However, to encourage workers to stay in the Zone permanently, each year canal authorities provided more funds and personnel to the Zone's school system.[70]

Several Canal Zone policies relating to race, nationality, and occupational rank welded together the racial division of the Zone's schools. In legal standing, Zone residents were divided along the lines of two broad nationality categories—U.S. citizen and non-U.S. citizen. Virtually all of the U.S. citizens coming into the Zone were white Americans. By contrast, the non-U.S. citizens were mostly West Indians of African descent. Thus, the policy of dividing school sites on the basis of nationality in effect racially segregated the educational institutions of the Canal Zone.

As the number of children in Zone schools grew, so did the extent of racial segregation throughout the Zone's education program. Schools for the children of white U.S. citizens were intended to prepare these students for successful lives in their native United States. The other children of the Zone, mostly of West Indian background and referred to as "colored," attended separate schools taught by colored teachers usually trained at West Indian colleges. The rationale for the dualistic arrangement for the Canal Zone school system was stated by the director of civil administration in 1907. He pointed out that the purpose of the Zone's schools was "to provide for the children of American parents instruction similar to that which they would receive in the [U.S.] public schools." The Zone administration offered these schools "in order that they [i.e., American students] may not be handicapped by reason of their temporary residence on the Isthmus." As for "native children and children of the Commission's West Indian laborers," the director stated they would receive "such schooling as they might obtain in the Republic of Panama or in their native lands."[71]

Each year the canal construction progressed, the population of workers' families and children multiplied in the Canal Zone. Both the number and quality of resources of schools for whites increased. But in the so-called colored schools, the resources and grades were comparatively limited. From the construction years through World War II, the public schools for white Canal Zone residents included twelve grade levels, but for black residents schools only went to the eighth grade. Education authorities considered the need to adopt a Booker T. Washington model for the colored schools in the Canal Zone. No new Tuskegee school involving advisers from Booker Washington's institution was implemented. Still, "practical" training such as gardening became the emphasis for the colored schools.[72]

School policy required that teachers in white schools be white females, and, by 1914, unmarried white females. In the colored schools, teachers were males of black West Indian background. Salaries for teachers at white schools were more than three times higher than those of black teachers.[73] As for the allocation of teachers, the colored schools were greatly disadvantaged. The number and ratio of teachers per pupil bear out this discrepancy. In 1908 Zone schools employed twenty-three white teachers for 721 students in its schools for whites: a ratio of one teacher per 31 students. By contrast only twenty teachers were employed for the colored schools that enrolled 2,146 students, a ratio of one teacher for every 107 students. By 1914 Zone schools had forty-three white teachers for 1,270 students, a ratio of one teacher per 30 students. However, for the Zone's colored population, twenty-three colored teachers were instructing 1,492 students, or a ratio of one teacher for each 65 students.[74]

During the early decades of the Canal Zone, U.S. authorities continued to limit the schooling of the area's nonwhites to vocational education. By 1930 Zone school officials adopted the "Service Technique" approach inaugurated by the United States in Haiti during the early 1920s. The curriculum for nonwhite residents of the Zone would be not high school academics, but skills such as weaving, tailoring, and woodworking.[75] Even when secondary education was finally established for blacks in the late 1940s, the curricular focus was on "occupational" training.[76]

Canal Zone Women

The racial division of labor for the Panama Canal and its transportation terminals predetermined the social and occupational lives of women in the Canal Zone region. From its inception, the female population of the Zone expanded steadily. As early as 1907, when the U.S. white labor force was estimated at 5,000 to 6,000, about 1,200 were women and children. By 1912 there were about 15,000 white males and 4,400 white females (3,100 over the age of fifteen). As for the Canal Zone's black population, in 1912 there were approximately 30,200 males and 13,200 females of which 9,250 were over fifteen (see table 2.2).

The U.S. women in the Canal Zone came to raise families or work among the professional personnel of the Canal Zone. Many of the U.S. women had come specifically to provide family life for a spouse who was a professional employee for the canal. Many had substantial formal education and were middle-class oriented. Regardless of marital status, these women were allowed to live in the modern housing for gold employees. The homemakers among the more privileged

U.S. women focused on establishing domestic activities such as churches, social clubs, and gardens. Other U.S. women were employed as nurses, hotel managers, and teachers.

As for West Indian women in the Canal Zone, once canal construction was
underway, local and immigrant women from the West Indies worked both in the
formal and nonformal employment sectors. Some were domestic workers for
U.S. gold employees and their families or did various forms of service work for
local kitchens and hotels. Other women immigrants from the West Indies ran
fruit shops, worked on nearby plantations, or were laundresses. Many sent their
earnings or cloths they purchased locally to their families back home in the West
Indies (see figs. 3 and 4).[77]

Thus, a racial division of labor was embedded within the technical (occupational) hierarchies constructing and, later, operating the canal. This racial divide in the technical occupations in the Canal Zone also reinforced the way
gender privileges and subordination were institutionalized throughout the region.
Recent anthropological and cultural studies of colonialism present much evidence linking the effectiveness of settler domination not simply to military and
economic institutions. Colonies had to become miniature societies that re-created
the social and cultural identity categories of the colonizer's homeland. During
the nineteenth and early twentieth centuries, Europeans from the cities strove to
reconstitute and normalize within their tropical settler populations the middle-
class identity of their homelands. Colonial leaders were trying to prevent the
loss of cultural loyalty from the lower-class European settlers moving to the larger
environment. Colonial authorities rebuilt inside the colonies the social institutions that would inculcate, for European women, the culture of domesticity and
wifery and, for the European men, values of industriousness and legitimate
fatherhood.[78]

In the Canal Zone, similar social dynamics prevailed. These social practices were propagated in the name of maintaining the racial order characteristic
of the then legally segregated, Jim Crow U.S. South. Even though virtually all
of the nonwhite canal workforce were West Indians or indigenous Latinos and
did not originate from the U.S. South, Zone administrators consistently overlooked this fact in this aggressive process of cultural reinvention.

A stratified, triangular relationship emerged among the women in the Zone
region. Women living in or near the Canal Zone were composed of three primary groups: indigenous Panamanians, emigré West Indians, and U.S. whites.
Traditional roles of indigenous rural women centered on food production and
selling at local markets, gathering fuel, and child care. But traditional farming
and markets throughout Panama were altered by the influx of manufactured commodities for consumption by the immigrant workers and professionals of the
Canal Zone. Local farming communities did not have time to naturally diversify since canal authorities imported food produce and other household necessities. Thus Panamanian women living in the traditional, informal economy
without access to cash remained in relatively more harsh living conditions than

their U.S.-born counterparts who were either homemakers or employed in the formal economy of the Canal Zone.[79]

American women in the Canal Zone frequently wrote about their daily lives in the official Zone newspaper, the *Star and Herald*.[80] One such writer, Rose Van Hardeveld, describes how U.S. women worked hard to replicate a version of the upper-class domestic worlds of their communities back home and recounts her revulsion at the dark-skinned locals and canal workers. Her utmost concern was with spreading the ideal of the middle-class household. Such domesticity, she believed, assured families of high morality even in the face of the constant threat of local physical dangers and cultural exoticism: "Shortly after my arrival my husband took me to see a village dance. A big barny room crammed with negroes and natives of every shade from pale cream to ebony black and a small sprinkling of white men; all the dancing vulgar and suggestive made me long to at once withdraw. 'Here is where influence makes history,' I thought, 'and as far as in me lies my influence shall be for home and decency.'"[81]

Most U.S. women in the Zone had to do a much greater amount of housework. Some could not tolerate black women servants. Still, eventually such servants became more widely used and their tasks differentiated. According to Van Hardeveld, "[w]asherwomen were asked to clean the kitchen floor after they finished the washing, which seemed to them an outrage. Girls who ironed were given to understand that a drop of water would not kill them also that they must finish in at least two or three days instead of taking a week or ten days to do a washing."[82]

The unpredictability of the rural Panama environment reinforced the need for the U.S. immigrant women to build even higher class distance between themselves and local women of color.[83] Once the necessities for a comfortable home were secured, more expensive cloths and household accouterments were sought and made available at Canal Zone supply stores. "After a while, whether due to the demands of imperative women or to a change of heart or head [by the authorities at] headquarters, I cannot say which, we had, first necessities and then luxuries along the line of ladies furnishings." "The buying of such things and fancy dishes, became an obsession with some, who bought and bought, and stored away and carried to the states, in the course of a few months, things which would ordinarily take a life time to accumulate."[84]

Social leaders of middle-class U.S. women saw West Indian and Panamanian women with much the same upper-class elitism as did male officials of the Canal Zone. West Indian and local Indian women were considered necessary nuisances. "The number and difference of type in the colored woman sometimes reminded me of the rats [back home] in Brunswick," wrote Van Hardeveld, "and I am afraid few of us found much that was praise worthy among these different West Indians. They were poor servants, indifferent mothers, and disloyal companions to their men, though, of course, there were exceptions." As for the indigenous women, she called them the "little brown mothers of the jungle," who "for the most part [are] ignorant, not married, ill-treated by their men, their lot seem[s] hard indeed."[85] The absence of sisterly ideology between the wives of U.S. professionals and skilled female employees, on the one hand, and the larger female population of the Zone region, on the other, reflected the wide racial gulf built into the canal employment hierarchy.[86]

Disease and Depopulation

Throughout the building and early operation of the Panama Canal, U.S. officials gloated that, whereas the French's effort on a Panama canal had been brought to a standstill by epidemics, the United States had worked a public health miracle. However, the health program of the Panama Canal project did not so much permanently eliminate tropical diseases as seal key diseases off from the canal construction site and spur others at locales outside the Zone.

In 1909 Herbert Clark, a young pathologist, arrived on the Isthmus to begin work at the board of health in Ancon, just north of Panama City. He was shocked at what he encountered. He did not find the disease-free canal community projected in the press. "So much publicity had been given this region for many years that it seemed likely that yellow fever and plague would crowd out most of the other causes of death," he wrote. "Imagine my surprise to find that malaria, dysentery, tuberculosis and pneumonia were the high ranking causes of death."[87] The fact that major infectious diseases were frequently reemerging in and around the Canal Zone went largely unpublicized by the canal sponsors on the U.S. homeland. Moreover, in countries immediately outside Panama, epidemic disease was spiraling up due to the flows of people and ships to and from the Canal Zone.

This diffusion of disease throughout the larger Caribbean region was a side effect linked to the public health measures and maritime quarantines enforced firmly by officials inside the Canal Zone. Health authorities had focused their public health measures specifically on mosquito- and rodent-borne diseases inside the Zone, the port cities serving the canal project (Cristobal-Colon, Balboa–Panama City, and Bocas del Toro), and the rail lines to and from the canal site.[88] The Zone's health officers had made the mosquito controls and quarantines inside the Zone and its transit spokes airtight, resulting in the near eradication of yellow fever, malaria, and bubonic fever. However, a substantial health care structure had not developed for the populations living in the land and ports adjacent to the Zone. Thus, epidemic diseases constantly accumulated in populations living in outlying Panamanian communities and nearby nations.

U.S. officials were lethargic about implementing permanent health care institutions because their larger intention was to "depopulate" the Canal Zone once the construction phase was completed. President Taft reaffirmed this policy in his 1911 annual message to Congress. On the Canal Zone, he remarked, "Now, it has a population of 50,000 to 60,000, but as soon as the work of construction is completed, the towns which make up this population will be deserted, and only comparatively few natives will continue their residence there." The Zone would continue being administered by U.S. presidential orders issued through the War Department. According to Taft, for the Canal Zone, "the government should be that of a military reservation, managed in connection with this great highway of trade."[89]

Following the end of construction in 1913, canal authorities attempted the depopulation policy. They expected to eliminate all but a small number of the Zone's black construction workers and their families as well as local Panamanians who had settled inside the Zone. In 1912 the Canal Zone had about 61,000 residents, 31,000 who lived in government housing. But late in 1913 and 1914 officials had begun sending workers who were not American citizens back to their native countries and closing many of the schools as villages and encampments emptied. By 1917 the relocation policy had reduced the number of residents to 31,000, all housed in government quarters. Thousands of black workers were taken up by fruit companies to work at their sites in Central and South America. Moreover, Chinese attempting to come into the Republic of Panama via the Canal Zone in general were excluded.[90]

At first, the Panama Canal Commission announced its success in removing the now unneeded foreign-born labor population from the Canal Zone. The depopulation policy had removed "practically all of the native squatters from the land within [Canal Zone] boundaries." Officials attempted to resettle some of the released West Indian workers into farming on the shores and islands of Gatun Lake and other nearby areas. The idea was that these farmers would produce food for the area's stable population. But once outside the malaria-free districts adjacent to the canal, this population began to harbor the disease once again. Consequently, officials stopped issuing settlement permits. Also, as foreign workers left the Canal Zone and outlying port cities on outbound steamships, an almost equal number of new immigrants arrived on incoming steamships. In June of 1914 canal officials noticed a particularly large number of new arrivals. "The immigration from Central America has been greater than usual during the past few months," they reported, "due to returning laborers, who have been working on the banana plantations and on railroad construction in Guatemala and Honduras."[91] By the early 1920s canal authorities had no choice but to allow farmers, mostly black, to settle on the vacant land throughout the Canal Zone.[92]

As when black immigrant workers came in droves to the Canal Zone during the 1900s and 1910s, during the interwar decades social relations became strained between the different native and immigrant black, Asian, and indigenous Latino populations. Moreover, this ethnic and racial conflict still varied from place to place and from year to year. In the mid–1920s one United Fruit Company overseer witnessed serious antagonism and personal violence between the different black and Asian ethnic groups on the company's farm in eastern Guatemala. Workers and residents included black railway construction workers from the United States, English-oriented blacks from British Honduras and the West Indies, Francophone blacks from Martinique and Guadeloupe, and Chinese merchants. He found these groups unable to "abide in harmony. The Negro from the United States has no use for the British subjects. The Jamaican has no regard for the black from Belize or Barbados, and still less for the French-speaking Negroes and the blacks from the United States."[93]

Both indigenous Central American blacks and those newer black immigrants faced racial discrimination from governments throughout the region. Government officials turned their nationalism into racism. For example, of the 32,000 residents in the Limon province of Costa Rica, 18,000 were blacks. This growing black population raised concern among the government and commercial leaders. They viewed their black residents as social "infiltrators" who threatened their "white republics." Central American governments even opposed U.S. firms attempting to gain concessions in their countries from fear that the Americans would bring in local or immigrant black employees. For example, in 1930 Costa Rica opposed the United Fruit Company, fearing that, like the Caribbean coast in prior decades, the Pacific coast would become "Africanized."[94]

At the same time that the heroic building of the canal faded from public consciousness as a distinct event of international renown, the Zone authorities were attempting its depopulation policy—a policy that completely ignored the established communities of black Panamanians and West Indian immigrant workers rooted around and inside the Zone. These officials viewed the Zone residents not born in the United States as floating, hired hands coming and going on the Isthmus as employers wished. The 1920 U.S. census, for example, stated that the "population of the Panama Canal Zone may be said to have been created or called into existence by the construction of the Panama Canal [and] the growth and decline of the population have been mainly determined by the labor requirements for that work."[95] But since the early years of the construction project, sociological evidence existed about the permanence of the black communities of the Canal Zone.

By the late 1910s families formed about one-fourth of the Zone's residents. In 1917 there were a total of 7,180 Americans in Zone housing quarters: 3,708 men, 1,706 women, and 1,766 children. Of the remaining occupants, 342 were Europeans (274 men, 27 women, and 41 children) and 10,372 were West Indian (5,927 men, 1,807 women, and 2,638 children).[96]

Not only did the policy to depopulate the Canal Zone fail, but Panama's population in general began growing relentlessly. The West Indian immigrants were increasing in numbers and indigenous Panamanians were beginning to feel uncomfortable about their permanent presence. As Conniff writes, "Panamanians, realizing that 40,000 to 50,000 West Indians had decided to stay in their country, developed a love-hate attitude toward them. They needed the West Indians' business but resented their presence."[97]

Unwilling or unable to tolerate Panama, thousands of blacks who had formerly worked on the canal sought to resettle in other areas in Central America. Still others journeyed further north to the United States. Writing around 1925, one United Fruit Company worker described the tensions encountered by black workers let go from canal work: "At the completion of the canal, a large number of the mass of laborers was turned loose in Central America. The thrusting of such a group of labor upon the various countries of the mainland led to serious trouble, in racial lines. Numerous fights and small riots occurred in various districts. One of the most serious riots took place on the plantation of Zehuana in the Quirigua district of the Guatemala division. Trouble between the natives and Jamaicans arose which was not quelled without bloodshed."[98]

The stringent controls and quarantines implemented in the Canal Zone resulted in infected and obviously unhealthy persons settling in surrounding areas that had little or no public health resources. Ironically, as the canal construction sped forward and public jubilation in the United States grew, most of the surrounding regions experienced growing problems with not only mosquito-borne diseases but other ones. In rural lands just short distances outside the Zone, malaria persisted. In Panama City and Colon during the early 1920s death rates were double those of the Canal Zone.[99] By 1927 Zone medical officials were open about malaria's persistence. In a report in an army medical journal, the Zone's chief medical officer, Colonel Weston Chamberlain, gave an update of the health program. "Loose statements have sometimes been made to the effect that malaria has been eradicated from the Canal Zone," he wrote. But Chamberlain emphasized, "These statements are erroneous. To free the entire zone from malaria would require years of effort backed by millions of dollars."[100]

Moreover, in the nations surrounding the Canal Zone, yellow fever remained a peril. In its annual summary for 1919, field personnel for the International Health Board of the Rockefeller Foundation reported "[o]utbreaks of yellow fever . . . during 1919 from Peru and Brazil in South America, from Honduras, Salvador, and Nicaragua in Central America, and from Mexico." A few years later yellow fever outbreaks were reported in Colombia.[101]

Tropical diseases largely eliminated from the Canal Zone were festering anew also because of out-migration from the Zone. Infected canal workers returning home to Barbados, Jamaica, and other nearby areas were sources for increased disease incidence in these neighboring regions. These workers must have had latent infections while living in the Zone or picked up infections in their journeys from the Zone. The Caribbean and Central American nations lacked ample public health resources. At the same time these nations were experiencing growing populations, including workers with no employment available in their local economies. Consequently, mortality among their black, Asian, and white European populations rose considerably. In his seminal study *The Prevention of Malaria* (1910), the British malariologist Ronald Ross presented surveys of the diseases in the Caribbean nations. He discovered that malaria mortality and incidence were increasing throughout this region. In Barbados, for example, "malaria is not endemic," his surveyor wrote; however, "imported cases of malaria are common amongst the laborers returning from Panama."[102] Barbados also was struck by a yellow fever epidemic in 1908 and 1909.[103]

In Jamaica, the population pressures were intense, while health conditions worsened. From 1881 to 1911 the population almost tripled from 83,526 to

235,730. Ross's investigators found that serious surveys of malaria had yet to occur and, moreover, "preventive measures were practically non-existent."[104] In the summer of 1908 bubonic plague broke out in Kingston, the Jamaican port that regularly trafficked travelers and goods to Colon. Canal officials traced the Kingston outbreak to cases already verified in Venezuela and Trinidad.[105] In 1909, an expedition from the School of Tropical Medicine of Liverpool made a brief health survey of the island of Jamaica. It revealed that malaria was responsible for about one-fifth of the nation's 680 annual deaths, and one-third of the roughly nineteen thousand admissions each year to the government hospitals.[106]

Zone health authorities made virtually no efforts to reduce diseases that did not affect the immediate ability of their work crews. In particular, hookworm, tuberculosis, and venereal disease were never attacked with the scale of resources Zone authorities had used to conduct their highly effective campaigns against yellow fever and malaria. The campaign in the Republic of Panama against hookworm did not start until 1914, when help for conducting this work was received from the International Health Commission.[107]

Syphilis also posed a major problem to the Canal Zone region. Due to the conservative social morals in U.S. public life during this period, it is not surprising that Zone officials and political supporters in the United States tended to keep discussion of this disease out of the public limelight. Nonetheless, it spread widely throughout the Zone region because of this area's complex social and population dynamics as well as limitations in its medical care resources. Neither the control of population influxes nor medical screening for venereal disease were within the purview of the Gorgas public health campaign.

In 1914 Walter Baetz, a physician at Ancon Hospital, wrote a medical report for the *New York Medical Journal.* It summarized his hospital's work with syphilitic workers, covering 1911 to 1913. Baetz was alarmed by "the inroads this disease [was] making among our laborers." He discovered at least 500 cases of syphilis (or roughly 6 percent) among the total of 8,226 patients that had been admitted to the wards set aside for black males. Baetz was certain the total proportion of patients with syphilis in all phases of infection coming to the hospital was much larger. He gave several reasons for this situation. First, Ancon treated mostly malaria patients and did not necessarily test consistently for syphilis infection. Second, serological diagnosis for the disease was still in development. He had been able to secure the new, reliable Wassermann test to confirm most (466) of his hospital's cases, but colleagues at his hospital were not so fortunate or were not using any other screening for syphilis diagnosis. Baetz believed, for example, that there were likely innumerable undiagnosed syphilitics in the hospital's surgery wards.[108]

Like the U.S. soldiers who remained on duty throughout Panama, most canal workers with syphilis could and did continue to work. Those who came to the hospital were in stages of the infection that caused obvious lesions or extreme discomfort such as severe muscle and joint pain. But the majority, whether diagnosed by new tests or not, experienced no grave symptoms. Hence, they were not systematically isolated from their employment sites or communities. Many, after several weeks in the hospital, even were returned back to the workplace. "In the building of this canal," Baetz wrote, "the stress of work has usually necessitated a rapid diagnosis and an intensive treatment, so as to return the laborer to his work as soon as possible."[109]

Despite the evidence that syphilis was growing among canal workers, canal physicians such as Baetz found it difficult to get resources to address this problem. They tried to attract the concern of American and Canal Zone authorities by presenting their medical information in terms of racial phobias. Baetz cautioned that syphilis in the canal workforce was a relevant health issue for whites in the United States as well as in the Canal Zone and the rest of Central and South America. "From a social viewpoint alone," Baetz wrote, "it may at first sight seem of little consequence to the Caucasian how much the negro [sic] is infected with syphilis." But he stressed, "As numerous sociologists have pointed out, the negro will always remain a fertile supply depot for venereal infections." Baetz urged, "One has to think of the large number of mixed breeds of all shades in the Americas, to realize at once that . . . treponema in the body of a full blood negro are often transferred by successive infections through all grades and shades of human beings, to the fairest representatives of the white race."[110]

As for the black workers, Baetz tried to reemphasize their importance to the canal construction. "No one who knows the history of the building of this canal will underestimate the value of the negro race [sic] for this class of work in tropical jungles," Baetz stated. Of the 500 patient cases he studied, 366 had come to Ancon Hospital specifically for treatment of syphilis symptoms. It was no great economic loss that "366 of our negro laborers earned nothing and produced nothing for a time of 7,595 days because of syphilitic infection, during the period of 23 months and that their care necessitated additional expense." He tried to show that there would be severe economic loss were these numbers to grow. First, resistance among such workers to other, more virulent diseases was lowered, and thus new epidemics were possible. Second, diseased workers evinced "lowered efficiency before and after admission to the hospital." Third, the cost to other social institutions in the canal region would rise since "a certain number of them will eventually become public charges in asylums, hospitals, and jails."[111]

In the eyes of Zone employers, however, none of the above points about the need to protect the black workers from syphilis were relevant. The fact was that workers infected with syphilis usually were able to perform muscular work for years before symptoms such as arthritis disabled them. Zone authorities provided free hospital treatment for canal and Panama Railroad employees for all immediate illnesses and injuries, except venereal disease and alcoholism. They also maintained a policy allowing infected men who had received hospital treatment to return to work once released from the hospital.[112]

In sum, it appears that racial arguments by medical researchers only tended to reinforce segregation in the Zone community. First, these views were rationale for political unresponsiveness to the idea of increasing health services for black workers. Second, they bolstered the practice of racially segregating public accommodations, schoolchildren, and housing throughout the Zone. If Baetz and his medical colleagues were correct about the idea that black canal workers were the natural harbingers of epidemic diseases, then providing more syphilis screening and treatment services for blacks was fruitless. Also, the warning that the health of white canal employees could be affected by the spread of syphilis from the black to the interracial and the white social groups was of little alarm to canal authorities because syphilis infection did not immediately disable workers.

The fact that American soldiers were another source for an increase in venereal disease in the canal region also was of little importance to Canal Zone authorities. Throughout the completion of the canal, a growing number of American troops were stationed in Panama. In 1911 only 204 officers and soldiers were in Panama. But by the start of World War I the influx of U.S. troops into Panama had increased to over 7,000. This military personnel was mostly white, although a small number of Puerto Rican troops were sometimes assigned to this region. Venereal disease rates for U.S. troops in Panama climbed steadily during three periods: 1911–1912, 1914–1916, and 1920–1924.[113]

Like syphilis, tuberculosis was another infectious disease that did not prevent canal employees from performing work for short, intense periods, yet it was quietly wreaking havoc in the Zone and surrounding communities. This disease had a significant presence in and near the Canal Zone, beginning as early as 1908. That year authorities reported one death from pulmonary tuberculosis contracted inside the Zone.[114] In 1912 a medical researcher (McCarthy) reported that tuberculosis "plays havoc there with the mixed tropical races" and that "whole families [are] being sometimes infected simultaneously in their unhealthful huts to the complete extermination of groups of natives." He studied over 600 tuberculosis deaths of West Indian blacks in Panama.[115] Later in 1915, another researcher (Clark) analyzed the deaths, caused by primary tuberculosis, of 452 West Indian blacks. The autopsies revealed that these individuals had died from miliary tuberculosis, one of the most severe forms of the disease. In cases of miliary tuberculosis, the infection spreads beyond the lungs into other vital organs. Eighty-three of the autopsied cases had had infected hearts (myocardium, 15; mural endocarditis, 6; pericarditis, 62). Tuberculosis was also found in the spleen (in 263 cases), liver (in 238 cases), and kidney (in 193 cases).[116]

By 1918 tuberculosis and pneumonia were the leading causes of death in the Canal Zone. Also, the death rate due to tuberculosis in Panama was among the highest in the world. An active campaign was underway in the Canal Zone to locate infectious individuals. But authorities admitted, "In present circumstances in Panama little can be done for most of the cases, but the health office is at least able to instruct the victims so that they will be less of a menace to others."[117] Tuberculosis and pneumonia were also the leading causes of death in the cities of Panama and Colon. For example, in August 1923, 144 deaths occurred in the Canal Zone and what officials called the "sanitated" cities of Panama and Colon. Twenty-one of these deaths were due to tuberculosis, another 21 were due to pneumonia. There were 162 cases of malaria reported: 117 in the Canal Zone, 9 in Panama City, 7 in Colon, and "29 originated outside of our sanitated area."[118]

Tuberculosis control programs in the Canal Zone and Panama City were not started until 1933. They were planned not by the Zone authorities, but under the supervision of the International Health Division of the Rockefeller Foundation.[119] Urban growth connected with the operation of the canal nurtured even more tuberculosis in the regions surrounding the Canal Zone. By 1950 nearly one-third of Panama's 800,000 people lived in or near the cities of Panama and Colon. In these areas, tuberculosis rates were most high and, overall, the disease was the eighth leading cause of death (27 per 100,000) and first among the infectious diseases causing death.[120]

Overall the depopulation effort largely failed. During the 1920s and 1930s the black population inside the Canal Zone and not living in government housing began to grow. Also immediately outside the Canal Zone laborers and their families returned, increased, and spread out. Most resettled in the cities of Panama and Colon, in new settlements along the Atlantic and Pacific coasts, or just outside of Zone boundaries. In the years immediately following the canal's opening, more than one-half of the black canal workers remained and became part of permanent communities on Panamanian soil. For example, of the estimated 45,000 black Barbadians who emigrated to the Canal Zone and Panama from 1900 to 1920, only 20,300 returned to Barbados.[121]

Yellow Fever and Malaria Resurgent

A final negative outgrowth of the Canal Zone was the failure by Zone authorities to develop health services adequate for the huge new populations of workers, civilians, and military personnel traveling into and near the Zone. While ample health services had been established inside the Zone, the larger region of Panama faced serious health care deficiencies. In the winter of 1919–1920, the famous surgeon and hospital administrator William Mayo journeyed to Central and South America to survey the region's medical education, hospitals, physicians, and surgeons. In an extensive article in the *Journal of the American Medical Association,* the prominent physician gave a sweeping, idyllic overview of the people and geography of Panama, the West Indies, Central American countries, and several South American nations. But he also reluctantly had to raise troubling questions about the organizational narrowness and inadequacies in the U.S.-operated local hospital and medical corps.[122]

The tranquillity, diversity of cultural folkways, and industriousness of the Latin American people pleasantly surprised Dr. Mayo. Reflecting the racial lens of his segregated U.S. society, Mayo saw no positive black African presence in anything he encountered. He assured readers that Central and South America were not uniformly vexed by lowly dark-skinned people. "Generally speaking, the South American natives were a superior type of Indian, and when they mixed with the white race many of the good characteristics of both races were retained." He emphasized that "Negroes are to be found in the southern half of the United States and in the West Indies; that they are of an alien race brought over from Africa, and that there has been less mixture of the Africans with the Indians of South America than with the whites of North America." As for the Canal Zone, "to the vision, courage and political sagacity of Theodore Roosevelt the world owes the Panama Canal," he wrote. He described Colon as a "clean, wholesome city," and "the beautifully kept port of that part of Panama City which lies within the Zone inspire[s] pride in the American and the wish that the government could be as well enacted at home." In these sections, "[e]verything is spotlessly clean," he commented, "and everywhere the dreaded mosquitoes and flies are absent, an imperishable monument to our revered Gorgas."[123]

Mayo strove to paint U.S.-run medical facilities in the area in the best possible light. However, while Dr. Mayo called the U.S.-run hospitals in the Canal Zone "splendid" and "up to date," he also pointed out they had serious staffing and managerial deficiencies. During the years of canal construction, U.S. hospitals in the Zone and nearby were carefully administered to keep out infectious mosquitoes as well as to provide diagnosis and isolation wards for malaria and yellow fever cases. When Mayo visited the eight-hundred-bed Ancon Hospital (later named Gorgas Hospital) in the Zone, he found physicians employed at the hospital constantly turning over. The hospital had to use contract surgeons since there were insufficient permanent staff physicians. Also, there were not enough army medical doctors on the hospital staff because the hospital's army doctors were "inadequately paid, without authority in their work, and without facilities for study and observation." The military staff physicians were usually much better educated than their supervisory lay army officers. Nonetheless, the army staff doctors were "under the iron-clad and often unsympathetic rule of [these] purely military men, who [were] as yet little touched by social progress."[124]

In addition, there were problems with nursing services at the Ancon Hospital. Mayo emphasized that trained nurses were vital to modern hospital care, but Ancon had no nursing schools. "The high-standard registered nurse is one of the greatest blessings of modern civilization," he wrote. Yet he found the hospital's nurses were underpaid considering the "high character of [their] training" and the fact that they "represented the best type of human machine for the care of the sick." The hospital also lacked support nurses. "We need other types of nurses," Mayo wrote, "less highly trained but nevertheless important social services vehicles, the Fords, so to speak, of the nursing world." As for the five-hundred-bed hospital in Panama City operated by the Republic of Panama, it was loosely managed. It had a nursing school, but it was "for nurses of various tints; very few are white." Since any licensed physician could admit patients to this hospital, the quality of its clinical care was unclear.[125] Other conditions compounded the inadequate hospitals in the canal region. Most important, Panama lacked a functional national health department, which was not established until 1926.[126]

Given such deficiencies in public health in the region immediately adjacent to the Canal Zone, it is not surprising that during the early 1920s yellow fever began to creep back into the case reports at Ancon Hospital. Furthermore, as one official of this hospital wrote, "[t]here is no doubt that many ambulatory cases escaped the records."[127] By the 1930s it was clear that there were other strains of and sanctuaries for the yellow fever virus. These had not been eliminated by the U.S. health work during the construction and early operation of the canal. Officials for the governor of the Canal Zone bemoaned the discovery of "jungle yellow fever" and "several varieties of jungle monkeys . . . found to be infected with the virus." With yellow fever poised to rear its head once again in the canal region, these officials took the issue gravely: "Whether this jungle yellow fever is of recent origin or has persisted for centuries in tropical America is largely of academic interest; it is of considerable significance, however, that when introduced into urban populations, in the presence of the *Aedes aegypti* [mosquito], the disease may assume its classical epidemic form."[128] As the biologist Rene Dubos pointed out in his classic *Man Adapting*, the heralded yellow-fever campaigns of the turn of the twentieth century had had limited, local effectiveness after all. "The discovery of jungle yellow fever revealed that the epidemiology of the disease is far more complex than the one on which the programs of 'eradication' had been based."[129]

Malaria also presented a consistent problem throughout Panama as well as inside the Canal Zone. Between 1924 and 1944 the Canal Zone health department estimated that twelve hospital admissions out of one thousand were attributed to malaria each year. By 1944, when an average of twenty-two thousand persons of color and seven thousand whites were civilian employees in the Zone, malaria was the second leading cause of hospital admissions. The malaria situation was much worse outside the Zone. According to one leading U.S. health researcher studying Panama in the mid–1950s, since 1931 malaria had been "the most serious disease of rural Panama [and] only a fraction of the cases, new or recurrent, come to medical attention and are reported."[130]

A special committee appointed by the governor of the Canal Zone in 1939 to study progress of the canal called for a practical, realistic assessment of the malaria situation. "The sanitarian's attack upon malaria has been carried on with one practically unattainable object—the complete riddance of malaria from the Zone." The singular dedication of Zone health authorities to this goal had netted results, but the battle against malaria had to be waged anew each year. The committee spelled out the "plan of attack" in use since the canal opened: "Anti-mosquito screening; removal, in so far as possible, of entire settlements of infected natives to sites outside the Zone; prompt treatment of all persons suffering from malaria; attempted medical prophylaxis; and attempts at mosquito eradication by preventing their breeding." These measures, when implemented as a "combined method" resulted in an annual, "demonstrable, diminution in the rate of infection" in the Zone. But still, the committee emphasized, the Zone was located within the vast geography of natural and man-made bodies of water throughout Panama. Also, the Zone area attracted a constant flow of hinterland populations drawn toward it for economic reasons. Thus, the complete eradication of the malaria-laden mosquitoes from the Canal Zone and nearby Panamanian communities "is almost humanly impossible."[131]

Dual Legacy

During the 1930s the epidemiological and sanitation issues regarding yellow fever and malaria were starting to burn anew. Yet many among the new generation of tropical health investigators and Zone officials clung to the old paradigm that the disease experiences of racial groups were distinct and, therefore, prevention programs should be different for these groups as well. By the early 1930s, in the words of a notable medical geographer, A. G. Price, Western tropical disease specialists had "recognized the American success at Panama as epoch making as regards health and sanitation." But they were apprehensive about the Canal Zone "as a geographical and economic . . . experiment." The black and indigenous population in the Zone region was growing. Price was concerned that this larger black and native population in the Canal Zone and Panama would rekindle malaria and other tropical health problems for whites living in these areas. According to Price and malariologists, any persons of black African background in regions such as Panama, the American South, and the Caribbean stood up to tropical diseases more sturdily than whites and, therefore, could reintroduce these diseases to the less-exposed whites in the Canal Zone and other similar settlements.[132]

The approach to black West Indians as disease carriers and not disease victims is pronounced in the work of a leading American malariologist, Marshall A. Barber. In an overview of his research work in the southern United States, Central America, West Africa, and other parts of the world, from the 1910s through the 1940s, Barber emphasized that, "with respect to malaria, Negroes are a particularly efficient reservoir." He stressed repeatedly that the Africans transported in the slave trade were a key "factor in carrying malaria from Africa to the Americas." Since blacks were the "source of much of the malaria which gained a hold in . . . the Americas, the parasites undoubtedly found an abundance of mosquito vectors ready on the ground." In Panama, Barber studied the "Spanish Americans and West Indian Negroes." He concluded, "The Negroes seemed to suffer less with malaria, although often infected with parasites."[133]

Broad historical and social conditions as well as the original Canal Zone health program should have demonstrated to Barber and other malariologists that their idea of a tropical disease response unique to black African people was myopic. First, black West Indian workers had experienced dreadfully high numbers of deaths in Panama as workers on the railroads, the plantations, and the French and U.S. construction crews building the Panama Canal. Second, the public health campaign in the Canal Zone had been enormously successful for the specific populations and geographic sites it targeted. Epidemics in the Zone region were a function of the degree to which public health measures were in place and sustained over time as well as the availability of isolation and treatment facilities for known cases. Nonetheless, the continued popularity of the racial view of tropical disease during the 1930s and 1940s demonstrates how easily the actual methods of the public health campaign used in the early Canal Zone, and the effectiveness of these methods, were scuttled or forgotten.

In 1938 President Franklin D. Roosevelt called the Panama Canal "to my mind far and away, the greatest, engineering work in the world."[134] While Franklin Roosevelt and later U.S. presidents and foreign policy leaders saw the canal as stupendous engineering technology, the canal feat was not marvelous in terms of its social impact on surrounding communities. The legal and managerial practices of this very same project locked public segregation in place throughout the Canal Zone. Public schools established throughout the Deep South protected and propagated the larger racial caste of Jim Crow segregation, a social stratification that reinforced the plantation economy. Similarly, in the Canal Zone, public schools had been designed to reproduce the supply of Zone workers stratified by race into white-collar, skilled, and unskilled labor. Throughout the 1930s and 1940s, the segregated school system in the Canal Zone reinforced the racial hierarchy put in place in hiring and work sites by management during the earliest phase of the canal construction. Employment regulations at the canal required workers to reach twenty-one years old before hiring. But most black youth were completing eighth-grade education at the age of sixteen. With no secondary schools available, these youths drifted socially, forced to take informal, nonskilled jobs.

In Pittsburgh in 1934 Charles Frederick Reid, a leading authority on U.S. education in the overseas territories, spoke before the American Association for the Advancement of Science. He denounced the educational situation in the Canal Zone and other U.S. territories. Reid urged the federal government to close the education gap in the Zone "with a definite policy for the Negro people in this area." He emphasized that the U.S. government must make a clear choice: "Is the Panama Canal Zone to remain a mere reservation for laborers who are engaged in repairing and maintaining the Zone, or is it to be made into a community of human beings?"[135]

Race relation dynamics in the Canal Zone were intermittently compounded by political crises in Panama. In periods of economic depression, racial xenophobia blended with political nationalism. During the Great Depression white labor unions in the Zone, especially the Metal Trades Council, strongly supported by the American Federation of Labor, called for the elimination of blacks from skilled jobs in the Zone. In the meantime some native Panamanians, especially those in powerful families who were members of the Panama National Assembly, made laws to curtail the entrance and even speed the expulsion of blacks from Panama.[136] In later years, Panamanian nationalists targeted U.S. domination of the canal and the country's economy but also voiced anti–West Indian or anti-"Antillean" sentiment. Panama's elite and aspiring middle class came to distinguish blacks of West Indian descent, known as "antillanos" or "chombos," from the black Panamanians whose roots preceded immigration during the canal construction, and who were known as the "negros nativos."[137]

By 1940 the Canal Zone had about fifty-two thousand residents—thirty-three thousand whites and nineteen thousand blacks. Many residents were U.S. Army and Navy personnel and their families. During World War II, the Zone still remained primarily a segregated society. The persistence of social and political segregation in the Canal Zone contrasted with the multicultural, Spanish-speaking populations growing throughout larger Panama. A sociological investigation of the region, published in 1950, emphasized that the Zone operated under a deceptive system: it was based ostensibly on "skill and citizenship" but actually used "a caste system based on race." Schools, stores, recreational facilities, hotels, hospitals, and housing all remained segregated, except post offices, "which dropped Jim Crow in 1946."[138] In general, racial prejudice was less extreme in Panama than in the U.S. Canal Zone. Schools run by the Catholic Church on the Isthmus made no racial distinctions. Newer generations of Afro-Panamanians with West Indian backgrounds were assimilating into Panama's broader Spanish-language cultural and social fabric. This made the racial segregation in the Canal Zone, indeed, the U.S. presence in the Zone altogether, even more unpopular to Panamanians.[139]

Following World War II, opposition to segregated public schools and discrimination in employment in the Canal Zone was a thorny political issue for U.S. foreign policy. Union drives by canal workers targeted racial and anti-Panamanian inequities in wages, work conditions, and public accommodations. Black U.S. military personnel sent to the Zone resented its segregated facilities and schools. Since U.S. foreign policy entering the Cold War era stressed opposition to Communist-bloc politics, government-sanctioned segregation became less defensible. As in the U.S. mainland, federal officials begrudgingly accepted equal access to public schools and government facilities in the Canal Zone and other U.S. territories.[140]

Although hammered into an environmental Zone largely rid of the twin threats of malaria and yellow fever, the Canal Zone community became a dualistic social order. On the one side were the engineers, the military, doctors, school officials, other white Americans in skilled occupations, and their families. On the other side were the laborers, semiskilled workers, small merchants, and itinerant ministers, who were mostly black, colored Latino, or Asian. In the end, canal authorities could boast about industrial progress and invincibility of scientific medicine, pointing to the fantastic canal as well as the effective campaigns against yellow fever and malaria. However, overshadowed by these successes was the fact that the canal population was highly divided by economic status and segregated schools and housing accommodations.

In short, the black population amassing in the Panama region in the early twentieth century and through the world wars prospered in certain key aspects of physical health but remained subordinated economically and educationally. On one hand, the spirit of science in U.S. foreign affairs, engineering, public health, and medical research had gained enormous intensity and confidence from the successful canal project. The canal had also earned glowing headlines for the feats by U.S. health authorities against yellow fever and malaria during the building of the canal. On the other hand, an unevenly modernized, largely subordinated Afro-Caribbean Diaspora had developed inside the Canal Zone and the surrounding region. This population was bursting with the work ethic, Christian evangelism, flare-ups of black nationalist identity movements such as Garveyism, and hopes for a prosperous future. They were looking forward to the prosperity that their contributions to Central America's railroads, banana plantations, and now the Panama Canal had yet to yield.

CHAPTER 3

Curing the Caribbean

HAITI THROUGH THE OCCUPATION

*B*y the early twentieth century, U.S. political and business leaders brimmed with zeal about the nation's progress in industry, commerce, and science. Several events abroad helped stoke this enthusiasm: the military successes in the Spanish-American War, the Asian annexations, the completion of the Panama Canal and start of bicoastal shipping, and the amazing triumphs against tropical diseases in the Panama Canal Zone. At the same time the U.S. government was growing increasingly suspicious about Europe's political designs in the Americas. The nation's trust in its military defense and industrial might, as well as its public health expertise in controlling tropical diseases, was put to a new test in Haiti. It was in this nation, the one the abolitionists called proudly the "Black Republic," that the United States attempted to protect its stakes in the Caribbean against the backdrop of the oncoming World War I.

Few among prominent black Americans believed in the inherent good of U.S. industrial progress as strongly as Frederick Douglass. In 1889 Douglass was appointed minister resident and consul general to Haiti and chargé d'affaires for Santo Domingo. He was elated with the opportunity to work for Haiti's welfare. To him, Haiti's improvement would bring racial dignity to blacks the world over. To Douglass, the pace of technical advances was burning stronger each day, advances that could not but help the cause of the Haitian nation. "Happily . . . the spirit of the age powerfully assists in establishing a sentiment of universal brotherhood," he wrote in one of his early diplomatic dispatches. "Steam, electricity and enterprise are linking together all the oceans, islands, capes and continents," he urged, "disclosing more and more the common interests and interdependence of nations." Douglass was convinced this "growing commerce and intercommunications of various nationalities" would help elimi-

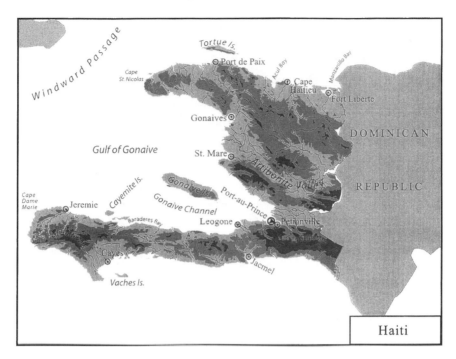

Haiti

MAP 3.1.   Haiti. Courtesy of Perry-Castaneda Library Map Collection, University of Texas at Austin Library (www.lib.utexas.edu/maps).

nate "hurtful prejudices" widely held in the United States about Haiti. The surge of new communication and transportation, Douglass predicted, would lead Haiti and the other nations of the world to "peace and fraternal relations between them."[1]

But in reality, over the next twenty-five years from the time Douglass wrote these words, little of the United States' technical prowess reached Haiti. At the start of the twentieth century, Haiti was a nation of two million people, mostly peasants, wrestling with severe poverty and political disorder. In the United States, politicians, journalists, and academics viewed Haiti as historically the least civilized nation in the West. They held out Haiti, even more than nations such as Cuba, the Philippines, and Hawaii, as proof of the inherent incapacity of darker races to govern themselves.[2]

The negative political stereotypes about Haiti, widespread throughout the United States and Europe, were woven with popular perceptions about Haiti's "primitive," non-English culture. French Creole was Haiti's major language, a linguistic anathema to many in U.S. politics and academics. Furthermore, Catholicism, syncretic Catholicism, and Voodoo were the dominant religions of Haitians. Most American politicians and other public leaders believed Haitian Voodoo was wanton superstition. As for the traditional Catholicism, nativists in the United States also viewed this with suspicion. Finally, through the early twentieth century many medical scholars and natural scientists in the United States propagated ideas about Haiti as an island of rampant disease, the source for Europe's syphilis scourge.[3]

By 1910 the many decades of political ostracism by the United States, commercial pillage from abroad, and political and class fragmentation from within had brought Haiti to a boiling crisis. Haiti's leaders and intellectuals searched feverishly for answers. Many stressed that Haiti's inability to industrialize was at the heart of its turmoil and poverty. Speaking at an international conference on race in London during 1911, General F. Déus Légitime, a former president of Haiti, was asked to address why his nation consistently performed poorly in its national life. He emphasized that leaders in Haiti fell from power rapidly because of economic stagnation. The domestic economic institutions of Haiti were insufficiently organized to increase production for worldwide commerce. Haiti's critical condition, Légitime said, "is not a racial question, but simply a problem of political economy." He called agriculture, commerce, and industry the "vital tripod" for any modern nation. However, in Haiti these resources "are exhausted and ruined under the burden of taxes and paper-money." Revenues from Haitian agriculture and business simply were woefully inadequate to eliminate huge deficits facing the government each year.[4]

From 1911 to 1915, the year that U.S. intervention started, Haiti had six presidents, all of whom were overthrown or assassinated in political unrest. U.S. government and military leaders perceived strong threats to American financial and security interests were Haiti to fall into Germany's hands.[5] By 1915 U.S. Marines were fanning out in Haiti for what would become two decades of occupation and stabilization efforts.

The U.S. occupation in Haiti was initiated at a time when the international security concerns for the United States had broadened beyond just the Western Hemisphere. Indeed, each successful military operation or economic venture by America abroad brought a need for more resources to secure old outposts and bring forward new ones. In the case of Haiti, U.S. leaders favored intervention in this nation because it would fortify U.S. military presence throughout the Americas. Haiti could serve as a site for naval forces to stave off encroachments by European powers. Also, the United States needed to protect its control of the new Panama Canal—a pathway that opened the Pacific not only to the United States but also to its allies.[6] Once it assumed power in Haiti, the United States, like it or not, buckled down to confront Haiti's political and economic malaise head on.

While prompted foremost by strategic military objectives, the U.S. occupation of Haiti was implemented as much by lay professionals as soldiers. The administrative heads of these professionals believed that modern engineering, administration, and medicine would work miracles even in the Americas' most backward of lands. To these U.S. nationals, scientific management and mass physician care would rescue this unstable land. Indeed, the framework of U.S. control during the occupation was not the supply and maneuvers of its soldiers. It was the more subtle strands of technical and medical programs that U.S. experts feverishly tried to knit together. Their hope was that by using the latest developments in tropical medicine and social management, Haitians from port cities to its rugged mountains and vast valleys could be fused healthfully together.

The Sweep of Techno-Nationalism

Along with their military partners, American health and engineering professionals in Haiti brought with them the mission of techno-nationalism—that is, the idea that U.S. science could impart miraculous progress in the "backward" society that was Haiti. Modern science and engineering—in the forms of electrification, telephones, highways, and new medicines to control diseases—could gradually bring order and prosperity to this politically unstable, illness-ridden republic. The U.S. experts landing in Haiti had another underlying interest: the technical and medical experience in Haiti would ultimately be useful to American military and medical science back home and in other parts of the world.

U.S. technical involvement that resulted in the Canal Zone in the Panama region had a pinpoint focus. In Cuba and Panama, U.S. technical and public health missions fulfilled three specific objectives: to expand physical access of the U.S. military and commercial interests to these regions; to establish and maintain disease-safe zones for U.S. employees and soldiers, so that shipping and construction projects could proceed; and to siphon low-cost labor from local populations into U.S.-owned and managed operations such as railroads, agricultural businesses, and canal projects. The creation of the Canal Zone had obviated the need for U.S. canal officials to negotiate with the Panamanian and Colombian states.

As the presidency of Woodrow Wilson dawned, U.S. foreign policy was driven even more strongly by the mission of technical progress and the impulse of the "crusader state."[7] The United States targeted the whole nation of Haiti for modern renovation, administered centrally by U.S. military experts. This policy was comprehensive in scope. Direct manipulation or curtailment of the Haitian national government and the nation's professional elites was only one side of the United States' Haitian initiative. Reducing the Haitian people's burden of disease and introducing a school system that would remake the average educated Haitian to fit the new techno-medical utopia were equally important to U.S. authorities seeking to stabilize Haiti.

Neither the leaders of the U.S. military occupation nor the subsequent "comprador" Haitian rulers could maintain civil order without constantly trying to erect two infrastructures: one concrete and the other symbolic. First, they had to provide living and work facilities as well as health resources adequate to protect U.S. military operations and support personnel from tropical disease dangers. As in Cuba and the Canal Zone, the U.S. officials in Haiti placed urgency on organizing sanitation and public health to guard its military personnel. U.S. authorities were confident that epidemic diseases and other tropical health problems in Haiti could be controlled by the U.S. military both for its personnel and eventually for Haiti overall. By 1915 the American foreign forces had the experiences of Cuba, the Canal Zone, and the Philippines under their belt. Thus, U.S. public health experts in Haiti were drawing on nearly two decades of experience obtained from these regions on disease control campaigns and civil administration of foreign populations.[8]

The second aspect of the groundwork that the United States tried to build in Haiti was cultural. U.S. intervention officials strove to disseminate the symbol or culture of industrial order, public health hygiene, and pharmaceutical medicine. Pain and sickness are the physical hardships that are the most personally felt. Other health events are also experienced as a natural part of the life cycle. These include growth, childbirth, death, injury, infection. Sickness, pain, and natural health processes—regardless of their microbiological properties—all are phenomena that are socially and culturally constructed.[9] In order to cultivate popular loyalty throughout Haiti, the United States attempted to promote the virtues of medical science over traditional, African (i.e., Voodoo) constructions of sickness and health events, and industrial education and skills over what U.S. authorities believed was pointless education based on effete French intellectualism. The public health campaign was a central tool for the U.S. interlopers because it provided a vehicle into and for controlling the microworld of the Haitian peasant.

Indeed, the overriding vision of the foreign affairs and the occupation leaders of the United States was that scientific administration and medical breakthroughs would rid Haiti of its political and health scourges. By terminating the grip of infectious diseases such as malaria, syphilis, and yaws and installing modern management in government, economics, and public works, U.S. officials believed that the entire Haitian civil society would be made orderly and civilized. It was this intention to impose top-down scientific regulation on a civilian population that made the U.S. mission in Haiti more protracted and risky than what had occurred in the Canal Zone. As a leading U.S. scholar of Latin America wrote after the U.S. occupation was underway, in Haiti "the United States is making a great laboratory experiment."[10]

Public Works, Public Health, and the Gendarmerie

Modernization has been defined as the transformation of society through technology.[11] A more fitting description for what occurred in Haiti under the U.S. occupation is "expatriate" modernization. Under the command of a military commissioner, between 1915 and 1935, the U.S. Army exercised direct rule over Haiti, including Haiti's national funds. U.S. high commissioner John H. Russell controlled much more than Haiti's armed forces. According to political historian Robert Rotberg, Commissioner Russell "drafted all Haitian legislation, negotiated contracts with foreign (especially American) companies, took charge of all educational, agricultural, and sanitary matters, constructed the roads, and collected the revenues. He was omnipotent, a dictator constrained only by harassed officials in the State Department."[12]

Another leading scholar of Haiti, Michel Laguerre, has called the U.S. occupation "a demarcation between the past and present" because the U.S. military "completely destroyed the old army and gave birth to a new military structure." The ultimate priority of the U.S. occupation was to inflict a long-term invasion and "rule through a puppet president."[13] However, the view that the American occupation in Haiti sustained itself primarily by the heavy hand of U.S. military personnel is too narrow. For example, in the late 1920s, when Haiti's population was estimated at 2.2 million residents, the U.S. brigade in Haiti consisted of merely five hundred enlistees and thirty-seven officers. Its Haitian counterpart, the Garde d'Haiti or "Gendarmerie," had between 2,500 to 2,700 men.[14] It was U.S. technical experts more than military personnel who drummed up discipline and cooperation in U.S. policies throughout Haiti's social sector. Like their military partners, American health and engineering professionals in Haiti brought with them the confidence of technology nationalism—that is, the belief that industrial and medical sciences would work miracles in the "backward" society that was Haiti. And it was these professionals who wove the more subtle, influential strands of U.S. control during American intervention throughout Haiti's civilian sector.

To bypass Haiti's long-standing professional sectors and educated social elite, U.S. administrators gained control over Haiti's disheveled quasi-military. U.S. authorities remade the Gendarmerie, Haiti's traditional military police. The idea was to mold the Gendarmerie into a centralized network of supervisors. U.S. experts would plan projects, but the Gendarmerie would mobilize local Haitians for these U.S. military, aid, and education programs in Haiti. By bringing non-elite populations into the ranks of the Gendarmerie, U.S. authority eased its ability to govern both in the political and military sense while undercutting Haiti's traditional social elite.[15]

During the initial years of the occupation, U.S. Marine major Smedley Butler handled the control of the Gendarmerie. As the major general and commandant of the Gendarmerie, Butler made one of his highest priorities the construction of public roads using the Gendarmerie. As he explained to one legislator, "I am most desirous of keeping . . . the building of the public roads of Haiti." Such construction "makes us a power in the land, for peace and order."[16] Smedley and other U.S. military in Haiti also politicked hard to control the mail and telegraph systems and other public works. Butler and the other U.S. Marine officials believed that controlling these networks "nailed down the occupation as a marine enterprise" in public opinion.[17]

Soldiers from the U.S. military and the Gendarmerie under Butler's command began to round up civilian crews for the laborious work of clearing and laying roads. In the meantime, Butler sent news of the great public works projects to politicians back in the United States. In 1916, Butler conveyed a happy scene to an Illinois congressman. He informed the politician that the marines were "doing everything in their power to assist the native population in rebuilding their roads, their irrigation works, their bridges, to clean up their towns, and generally better the condition of the people at large." By doing these improvements, Butler remarked, "we hope to absolutely do away with the desire on the part of any Haitian to revolt against his government.[18]

By the mid-1920s, the U.S. mission to bring Haiti onto the path of technical progress was at full steam. American military forces had engineered a rudimentary public works network for Haiti. By 1924 some 650 miles of roadway, although much of it haphazard, linked all the major towns of Haiti. At the end of 1929, 1,600 miles of road work had been finished. Also during this period some 15 steel, 68 concrete and 127 wooden bridges as well as several wharves were built; airfields and commercial airline routes were opened; hundreds of miles of telephone cables were strung; and numerous irrigation canals were built.[19]

For U.S. officials and builders in the Canal Zone, the public health and economic crises occurring in civilian populations outside of the Zone and its vicinity were of secondary concern. However, in Haiti the health of the entire population was central to U.S. interests. Putting to use its engineering and technical expertise to implement a nationwide public health campaign in Haiti became the central strategy for U.S. occupation officials.[20] American officials believed they needed to and literally could cure the physical ills ravaging many among the Haitian masses. Such a feat would dampen Haiti's potential to fall into anti-American politics and build support for the U.S. military. Providing medical services to average Haitians would depoliticize the population by introducing the mission and power of medical science as an alternative to the widespread practice of Voodoo. A U.S. public health campaign would also deter Haitians from seeking disease control assistance from competing foreign powers (such as France or Germany) quite capable, building on their imperial experiences, of public health campaigns in a nation like Haiti.

The medical needs in Haiti at the beginning of the occupation decades were assessed at that time as immense. Medical experts estimated that about 70 percent of the Haitian population was infected with either yaws or syphilis. In some districts malaria infections were found in 100 percent of the population. Finally, soil-related diseases such as dysentery, typhoid fever, and hookworm were widespread. The health conditions incapacitated many of Haiti's people and, consequently, much of the nation's economic resources.[21] U.S. public health work in Haiti had started with the original intervention treaty of 1915. In accordance with the agreement, the United States appointed a sanitary engineer to head the public health work. Within two years this official had seven assistants and was responsible for the U.S. occupation's health initiatives, including sanitation, hospitals, and quarantines. By 1919 the unit had become legally established as the Public Health Service (i.e., Service d'Hygiène) and had grown to fifteen assistants as well as two nurses. The country was divided into nine health districts with major cities and towns serving as headquarters for the various districts. In addition to sanitation measures, such as building private latrines, the Public Health Service was given control over the few hospitals throughout the nation.[22]

The rural health unit became the main arm of the U.S. public health initiative in Haiti. These teams were not large enough to conduct serious health promotion programs such as community education about sanitation practices. Treponematosis, malaria, and intestinal parasites were widespread throughout the interior population of Haiti. Thus, it was believed that health teams in rural sites would facilitate the dissemination of new pharmaceuticals to this population as well as promote public health education, but as a secondary priority. Between 1926 and 1929 the United States initiated 147 such clinics, located in a variety of dispensaries or rented or charitable buildings.[23] As with public works projects, U.S. authorities often employed members of the Gendarmerie in lieu of health professionals and trained health workers. Through 1929 the Gendarmerie handled the care of patients with "mental diseases" or "delinquent minors." After that year, the Public Health Service was given this responsibility.[24]

Opposition

Despite the aggressive start-up of medical and public work projects throughout Haiti, there were repeated demonstrations and violent outbreaks against U.S. military presence. Each wave of political opposition was followed by exposés by liberal U.S. officials and civil rights activists. During the occupation serious disorders against the U.S. policy broke out—especially in 1919 and again in 1929—and had to be suppressed. These included persistent Cacos revolts among the peasant sector and student strikes, the most serious (in 1929) requiring the imposition of martial law. There were also strikes among laborers, such as stevedores and emigrant workers, and even a strike by schoolchildren.[25] As one leading U.S. Marine in charge of the Gendarmerie wrote of the occupation in his autobiography: "We had put down one revolution and then built a road by forced labor. It was lawful, but in this case a bit overworked. In resentment at this, the Haitians had started another revolution [i.e., the Cacos revolt] ten times worse than the one we had just suppressed. We were cordially disliked and mistrusted by all elements."[26]

In November and December of 1919 major conflict broke out. Evidence of harsh treatment of Haitians by U.S. forces assisted by the Gendarmerie began to surface throughout the U.S. and foreign press. Finally, in February of 1920 James Weldon Johnson of the NAACP traveled to Haiti for the association. In his autobiography, *Along This Way*, Johnson recalled his visit: "My aim was to gather information and shades of opinion from as many sources as I possibly could."[27]

When he arrived, Johnson was surprised at the scenic beauty of Port-au-Prince and the many nice sections in this city. But as for the political situation, Johnson remarked, "All the Haitians I talked to complained bitterly of conditions."

Johnson was shocked at the disrespect U.S. soldiers showed for both the highly educated, French-speaking Haitians and the common peasants. Johnson cited as typical the attitude of one marine who told him, "The trouble with Haiti is that these niggers down here with a little money and education think they are as good as we are."[28]

During his Haiti trip, Johnson met key elite and activist Haitians, such as Georges Sylvain, a prominent lawyer and former ambassador for Haiti to France. They frequently questioned Johnson about strategies to alleviate the harsh politics and social stress the occupation was causing. Johnson advised them to establish a protest institution similar in stature and function to the NAACP. The result was the reformation later that year of the Union Patriotique d'Haiti. The Union Patriotique had been started in 1915 by Georges Sylvain in opposition to the U.S. occupation.[29] In the meantime, Johnson and other liberal activists raised the level of public criticism at home against the occupation policy.[30]

These public inquests refuted the images, conveyed by the occupation's political sponsors and U.S. administrators, that the intervention was remaking Haiti into a lofty, untrammeled modern nation. The reality was that forced labor under the heel of the Gendarmerie was the key to the rapid construction of roads and other public works. Moreover, atrocities against Haitians suspected of resisting U.S. occupation were widespread. Finally, U.S. observers in Haiti revealed that American forces acted with racial animosity toward Haitians, utterly disrespecting the nation's own long-standing professionals and political leadership.

The Union Patriotique was in the forefront of exposing abuses by occupation personnel and lobbying U.S. officials to terminate the occupation. This organization comprised some of Haiti's leading intellectuals as well as Haitian ex-officials, doctors, lawyers, manufacturers, planters, editors, and community activists. It drew membership from all geographic sections of the country and maintained a nonpartisan policy.[31] In May of 1921 it presented a report on the effects and ongoing problems of the U.S. occupation to the U.S. Department of State and Senate Foreign Relations Committee. It was an extensively documented exposé of U.S. authoritarianism and complicity in atrocities.[32]

The Union harshly criticized the occupation for unleashing political imprisonment, civilian repression, and forced labor throughout the nation. It cited dozens of incidents involving hangings, torture, executions, and burning of houses and villages by local police or U.S. forces. Among the many abuses prisoners faced was their exploitation as forced labor by U.S. occupation forces. Citing a petition authored by a Cap-Häitien minister, Auguste Albert, the Union substantiated widespread atrocities that included the following:

- From 1918 to 1920 an estimated 4,000 prisoners had died in prisons of Cap-Häitien. During 1919, eight corpses a day were being dumped into pits.
- At the Chabert camp run by the Americans, 5,475 prisoners died during the same three years, an average of five deaths per day.
- At Cap-Häitien, which housed 500 prisoners, an average of four prisoner

deaths occurred each day; this amounted to nearly one-quarter of the prison's population each month.[33]

The Union's memorial reported that mortality in prisons was just as high in Port-au-Prince and Gonaives. The Union's figures did not specify the medical causes for these prison deaths. But it is likely that by crowding poorly fed civilians from a wide spread of rural and city neighborhoods into wretched prisons, diseases such as influenza, pneumonia, and tuberculosis ran amuck.[34]

The Union traced the source of forced labor, mass imprisonments, and political deaths to the use by U.S. authorities of the Gendarmerie. U.S. occupation leaders had assigned local police and administrative authority to the Gendarmerie, which in turn oppressed civilian populations with impunity. To the Union, what many believed had been a technical "miracle" of road and public works construction on the part of U.S. occupation forces actually had been born from forced labor and exclusion of Haitian professionals and skilled planners. The Gendarmerie had recruited and coerced laborers onto road and public works construction projects. The Union denounced this system of corvée labor, "forced unpaid labor on public roads imposed for military purposes upon the Haitian peasant." It explained that the occupation administration "carried on all public works, without any control by the Haitian Government over the nature of the works, the manner of carrying them out, their expediency, or even the amount spent on these."[35]

While political opinion concerning the U.S. role in Haiti heated up, the military training of the Gendarmerie and technical projects by U.S. forces in Haiti were stepped up. For outbreaks of unrest, U.S. administrators in Haiti developed, in their words, a "policy of putting the Garde [Gendarmerie] in the forefront in all of these disorders," such as "the banditism of 1919–1920."[36] As for long-term policy, American officials believed that administering Haiti was simply a matter of establishing the correct logistics for blanketing the country with Gendarmerie personnel. This national army would, in turn, oversee the efficient flow of national life. By 1929 U.S. administrators in Haiti reported that the Gendarmerie had 2,622 members, plus 838 auxiliaries for "the coast guard, rural police, and palace band." The total Gendarmerie force amounted to one Gendarmerie member for every 3.4 square miles and 690 residents. The Gendarmerie's responsibilities were copious: "In addition to its specifically police and military duties, the Garde has charge of the maintenance and operation of all navigational aids in Haitian waters, the registration and licensing of motor vehicles and traffic controls, and the care and guard of all prisoners and, to a great extent, the insane of the Republic."[37]

Rebuilding the Medical Sector

Another development that generated opposition to the U.S. presence was the destruction of Haiti's traditional general-practice physicians and their medical training resources. Haiti had its own medical profession, which could have functioned as another avenue to bring broader Haitian society into the modernization process—an avenue that U.S. occupation authorities promptly dismantled.[38] However, the social clash between Western medical professions and indigene medicine and practitioners, a clash that had occurred in British and French colonies such as Egypt and Tunisia, was reenacted in Haiti.[39] The medical profession of Haiti had many highly trained physicians, and their medical school had roots dating back to the early nineteenth century. Its leading physicians had been trained primarily in medical schools in France. The National School of Medicine in Haiti dated back to 1823 and was a source of great national pride. In their history of Haitian medicine, two American authors (Parsons and Stitt) documented that by 1890 a "certain real progress was made in the medical school." Under the leadership of Dr. Léon Audain, the first records of health conditions in Haiti were made available. Around this time there appeared "in Haitian medical circles, a scientific spirit in the practice and a scientific outlook in the study of medicine." From 1899 to 1911 Haitian physicians and medical academics published their own national bimonthly journal, *La Lanterne Médicale.* Numerous research tracts were also published. One such work, "L'Organisme dans les Infections" by Audain, was crowned by the Academy of Medicine in Paris.[40]

Subsequent leaders of the National School of Medicine were able to keep it afloat despite the political turmoil that occurred between 1910 and 1915. At the time of the occupation, clinical instruction by the medical school was held at the Haitian General Hospital in Port-au-Prince. However, in 1918 several of the school's best instructors were dismissed because they had voted against the U.S.-made Constitution in the plebiscite that year. Charles S. Butler, the commander of the medical corps for the U.S. Navy, was made the head of the Public Health Service. Having previously worked for twenty-five years in the Philippines, Puerto Rico, the Virgin Islands, and the South Sea Islands, Butler was considered one of the military's leading specialists in tropical medicine and parasitology. He openly opposed continuation of the National School of Medicine on the grounds that the education it provided was a farce. The nursing school was also placed under the direction of two nurses from the U.S. Navy Nurse Corps, and then in 1921 under American Red Cross nurses.[41]

Butler's earnest belief that Haiti was in dire need of modern, clinically oriented physicians was based on political and cultural reasoning. Speaking at the annual meeting of the Association of American Medical Colleges in 1927, Butler emphasized that the modern doctor had an important role to play in countries like Haiti. "We are convinced that it is a moral obligation of the white races of the world to assist this little Caribbean republic to her feet and when once on her feet to give her the moral and material backing necessary to keep her there." He urged his medical associates to "appreciate the enormous importance of our calling in helping governments to confer the benefits of civilization upon backward races." Government through effective modern doctors "wins the friendship" of the masses. If no physicians are available when "bodily ills beset" the masses, they "will turn to the stone-age man, or to the voodoo doctor rather than to die without a struggle."[42]

To Butler and American authorities, the approach to medical training by Haitian physicians did not and could not measure up to rising clinical standards. Haitian doctors favored a curriculum centered on training students inside the hospital. The U.S. medical officials viewed this approach as wholly outdated, emphasizing instead laboratory courses as central to the medical school curriculum. Intending to replace the National School of Medicine with a U.S.-designed school, the U.S. chief of hygiene blocked the medical school's students from entering the general hospital, had the medical equipment of the faculty put outside in the school yard, and turned the building into a school for nursing. In 1927 a new medical school was opened by the occupation administration and equipped through the assistance of the Rockefeller Foundation. Many of the Haitian medical professors opposed the methods of the new school. U.S. medical educators involved in the shift to U.S.-style education emphasized "actual laboratory work" as central to the curriculum. But as for the Haitian medical professors, these "'anti-occupationist' doctors set up a great protest about sacrificing so much of the student's time in the comparatively unimportant laboratory." Specialty training of handfuls of Haitian physicians was now done exclusively abroad in the United States or France and funded by the Rockefeller Foundation.[43]

Public Education for New Society

The authoritarianism of U.S. technical assistance to Haiti was also reflected in the educational program implemented by the occupation administration. From the onset of the 1916 treaty, General Russell made educational improvement a key goal of the occupation program. Haiti had had a national school program since its founding as a republic. In addition, the Catholic Church provided schools that were generally used by Haiti's "better class" throughout the country. Combined, these schools reached but a minute percentage of Haiti's population.[44] U.S. authorities implemented an agricultural-vocational school movement known as the "Service Technique."

Funds for the initiative under the occupation were raised by and from the Haitian government. Before the Service Technique movement, Haiti had six vocational schools, five of which were located in Port-au-Prince. The Ecole Elie Dubois was a girls school, operated by the Belgian Sisters of Mary, which had a contract with the Haitian government. Its purpose was to train girls in homemaking and menial vocations and as teachers. As for boys, Haiti had one reform school (Maison Central), three industrial schools, and the Technical School of Agriculture (five miles outside Port-au-Prince).[45]

The U.S. occupation administrators directed the Haitian funds to inaugurate what they envisioned would be a sweeping rural education movement, along the lines of the black South's Tuskegee Institute. The Service Technique attempted to design Haiti's schools to stress manual training and agriculture. Students and their schools were to become self-supporting through the sales of their products. Haiti's six vocational schools and many Haitian teachers were turned over to the Service Technique. The new American teachers in the Service Technique schools were to train the Haitian teachers for rural farm schools. The Service Technique schools also intended to train young assistants and researchers to become expert agronomists. When the Service Technique program started in the 1923–1924 school year, it had just fifty-one students. This figure increased to about sixteen hundred students in the 1925–1926 school year and over eleven thousand pupils by the 1928–1929 school year.[46]

Although advertised as the sure way to overhaul and expand mass education throughout Haiti, the Service Technique program became a small cluster of schools operated largely by American teachers. The majority of Haitian children were still without any schooling altogether. Only about 5 percent of Haitians were literate, and these were mostly residents of the Port-au-Prince area. During the 1920s Haiti had about four hundred thousand school-aged children, less than one-fourth of whom were in public, parochial, or private schools. Those enrolled in public schools operated by the nation's Department of Public Instruction faced deplorable conditions. Salaries and work conditions for the U.S. teachers were far superior to those of their counterparts employed in Haiti's public and parochial schools. The disproportionately low pay and overcrowding in non–Service Technique schools were seething grievances among Haiti's teachers and activists during the occupation period. Although Haiti's puppet government "approved" the Service Technique program and attempted to expand it, many capable Haitian educators refused positions involving administration of the occupation schools. Moreover, Catholic school educators also chose not to cooperate with the Service Technique initiative.[47]

The deeper criticism of the U.S. occupation education program was that it robbed the Haitian people of their intellectual leadership as well as their literate and folk cultural heritage. A team of U.S. investigators of the occupation found this from their visit in 1926: "Haitians, so far as we talked with them, dread American influence on the educational system." Haiti's educational leaders believed the United States deliberately "Anglo-Saxonized" their institutions and "turned [Haitians] away from the French cultural tradition [to] a materialist and purely utilitarian trend." The group saw clear signs that U.S. authorities were forcing the Tuskegee model forward against the support of most Haitians. "It is essentially the same issue as the old difference in the United States as to whether in the education of the negro [*sic*] all emphasis should be laid on education of the Tuskegee type, or whether the education typified by Atlanta University was also important." By all but shutting down the lycées (secondary academies) in Haiti, college-level education was precluded for the non-wealthy Haitians, who could not afford to send their children abroad for college and education in the professions.[48]

A prominent scholar and expert on foreign affairs from Howard University, Rayford Logan, was highly critical of the U.S. educational initiatives in Haiti under the occupation. General Russell's policy had aimed to make "every Haitian the master of a trade." But in actuality, "Russell's educational program not only implied the inability of Haitians to acquire professional and classical education, but it also ran foul of the Haitian tradition based upon the French emphasis on precisely these disciplines."[49]

While African Americans denounced the U.S. school program in Haiti for insulting the academic potential of Haitians, many Haitian activists were upset for more practical reasons. The United States had promised these schools would increase Haiti's agricultural productivity. But once the U.S. forces left, while Haiti was left with a small cluster of Haitian teachers, the educational system was wanting at all levels. School facilities were too few and structurally inadequate, teachers were grossly underpaid, and administrative resources for mass public schooling did not exist. As for the vaunted economic impact expected from the Service Technique program, Haiti's educational leaders watching the policy from the sidelines were critical of this program since it bore no immediate gains in agricultural production or national development for Haiti.[50]

The groundswell of Haitian opposition to U.S. control had emerged not only from the peasantry, but from Haiti's professional sector as well. The Haitian professionals, educators, and intellectuals resented the U.S. military zeal to single-handedly direct the building and maintenance of Haiti's technical and health service systems. To them, the U.S. mission was only deepening the weakness of Haiti's internal institutions, while increasing its technical and educational dependence on the United States. Most Haitians knew the public works improvements—telephones and telegraphs, airfields, and the like—were first and foremost for the service of U.S. Marines and the Gendarmerie. As for the new roads, many Haitians saw them as just a means for the U.S. authorities to improve auto and truck access for its soldiers and the Gendarmerie.

Dantes Bellegarde, the popular Haitian intellectual and educator, criticized U.S. officials for overblowing the significance of the United States' technical projects. In a political essay that appeared in Haiti and the United States in 1927, Bellegarde detailed how U.S. control had caused wages for Haitian workers, teachers, professors, and Haitian government workers to drop, while those of U.S. personnel or their selectees were exorbitant. As for the modern technical miracles that U.S. experts were reputedly constructing throughout Haiti, Bellegarde remarked, "Some roads have been restored, a few buildings have been constructed, a sanitary organization, which was really necessary, has been created." But he added, "Much ado has been made over these achievements as if they sufficed to justify the attack against Haiti and the enslaving of a whole race." Bellegarde emphasized that these projects "have cost the Haitian people . . . enormous sums." As an example, he cited forty thousand dollars the Americans used to establish a "radio broadcasting station . . . which every Friday evening carries on the wings of the air antiquated fox-trots . . . and praise of the American occupation. This sum might have better served to construct several school buildings in the rural districts or aqueducts to irrigate the land."[51]

Even U.S. intellectuals began to wonder about the efficacy of the U.S. occupation efforts to renovate an entire society and rebuild a mass education system. Clarence K. Streit, writing about progress of U.S. programs in Haiti in 1928, stated that the physical structures resulting from the twelve years of U.S. occupation would "dazzle the eyes of the [American] visitor." They would see "roads, telephone and telegraph lines, drainage systems, governmental buildings, hospitals, rural dispensaries, schools, irrigation works . . . and the list is not ended." But Haitians could never staff these new structures because they were receiving no opportunities for this purpose. "We have built an airplane for a man accustomed to riding a donkey," Streit remarked. He criticized U.S. authorities for having been "so occupied in building and operating [this machine] we have not trained the man for whom it is intended. The machine runs beautifully now while we are at the controls, but how will it run when we step out in 1936 and the man who all the time has been on the donkey steps in to fly [an airplane] alone?"[52]

Women Farmers and Indigenous Technologies

Haiti had to pay another big price because the United States insisted on attempting technical solutions to deal with Haiti's political, economic, and social problems. The Haitian people's traditional uses of human resources and technologies were displaced. U.S. technical experts in Haiti generally were oblivious to incorporating indigenous knowledge and practices historically effective in food farming, communication, house construction, and domestic care. For example, with respect to housing and community maintenance, U.S. officials reported only the worst conditions, but not the workable, effective ones. When U.S. officials started organizing Haiti's public health and medical facilities in 1919, they reported "the entire country teemed with filth and disease."[53] Thus, during his visit to Haiti on behalf of the NAACP in 1920, James Weldon Johnson expected to find rundown households at every step. But this turned out not to be the case in either the cities or rural towns.

While traveling in northern Haiti near Cap-Häitien, Johnson visited many rural communities. He found the local people living in huts "ingeniously built of thin strips of wood plaited about the heavier uprights and plastered . . . with clay." Both the insides and yards of these houses were kept remarkably clean. Johnson observed, "Nowhere did I see the filth and squalor that is hardly ever missing in and around the log cabins of the [U.S.] South."[54]

When compared to other non-modern, largely agricultural nations, the indigenous farming economy of Haiti also was efficient for household and local market needs for the majority of the nation's peasant class. In Haiti, the small farm plot was used for diverse crops of vegetables, roots, and sugarcane consumed in the home. Also these farms produced eggs, chickens, and pork sold for cash at nearby markets. With heavy involvement of women, this produce was sold at local markets within a walk or donkey ride from individual farms. According to leading international geographers, Haiti's food economy pattern was "on a tiny scale the one which existed in Eastern Europe just after the Second World War when valuable protein-rich foods were sold and starchy carbohydrates were retained on the farm to provide the basis of the diet."[55]

Traditional ethnic peoples made up the bulk of Haiti's population. Their agriculture economy, composed mostly of small farm plots, was largely in the hands of women. Typically, women controlled the preparation for planting. They also were responsible for the distribution of the farm produce at local markets. In the 1920s, American scholar-activist Emily Balch, of the Women's International League for Peace and Freedom, vividly described Haitian women and the typical market:

> Buying and selling is largely done by the women at market. Starting early in the morning . . . the long possessions come on donkey back, or afoot with amazing burdens on their splendidly poised heads, gathering to the great social and economic function of their days. . . . To ride balancing a loose slipper from the upturned toe is a sign of descent from such and such an African tribe, that to carry a jar or other burden balanced on the flat palm of the back-turned hand with the elbow pressed against the body is the mark of a different African stock. All alike are graceful, free-moving, very much alive, and often, to the seeing eye, very beautiful. They sit in the town square, perhaps in the thousands, with their small stock of wares spread out before them, and before night they are returning with their purchases.[56]

Such markets in Port-au-Prince drew about five thousand women.[57]

Haiti's coffee plantations were old, dating back to the colonial period. Yet, the coffee was of high quality and could compete on the world market. American officials wanted strongly to centralize and reorganize these coffee plantations using modern agricultural techniques. However, Haiti's rich coffee was a unique product of several indigenous qualities. First, the nation's rugged, mountainous topography provided a rich environment for coffee growth. Second, the traditional harvesting of coffee fit well with Haiti's peasant farming. This approach to coffee harvesting, known as the "dry method," required oversight and handwork that only peasants could render. In this approach the coffee berries were dried whole in the sun. According to a Haitian agricultural expert, "this drying process is twice or three times longer than that of the parchment coffee, and it must also be carried out very carefully for many days. . . . All the qualities and flavor of the coffee depend to a great extent on this drying." While this method was slower than factory-based methods, it produced a higher quality of coffee.[58]

Another characteristic of Haiti's coffee was that much of it, like most crops in Haiti, was produced on small parcels owned by peasants. U.S. authorities complained they were unable to find the concentration of local skilled and unskilled labor necessary for establishing large coffee plantations. Yet U.S. technicians did not find these indigenous factors conducive to implementing modern coffee techniques and large landholdings. Although U.S. occupation experts did not approve Haiti's traditional coffee cultivation method, it retained its dominance during and after the occupation.[59]

A similar situation prevailed with regard to cotton production in Haiti. During the occupation, American authorities made what they viewed as improvements in getting peasants to use a better selection of cotton. The U.S. experts also did experimental crops to test ways to improve quality and yield. But peasant growers refused to alter their techniques. Working along with women and children to harvest their cotton, the small-farm growers were satisfied to ignore the new, modern approaches.[60]

Similar clashes between U.S. technical aspirations and local customary practices happened in the area of communications. The communication techniques used by Haitian peasants living in remote areas baffled U.S. military personnel. They were astounded by the ability of the local people to convey information over long distances faster than military personnel with modern equipment. One U.S. military officer swore that in the Haitian countryside there was a "bamboo wireless [that] could spread news over the island almost as fast as the telegraph wire." As for surveillance of military boat operations, the officer said, "The natives would always tell me the day before she [i.e., a messenger gunboat] was coming, several hours before she came in sight."[61]

Indeed, it was their superior local knowledge of food sources, the countryside, and land communication that made the Cacos such a formidable and undefeatable opponent to U.S. forces. One Caco leader, known as Charlemagne, dealt blow after blow against U.S. occupation and Gendarmerie units. After failing to catch Charlemagne for many months, U.S. general Frederic M. Wise called the search for this rebel "a combination of Blind Man's Buff and trying to find a needle in a haystack." He went on to explain the effectiveness of Charlemagne's guerrilla tactics: "On the surface it may sound strange that a Caco with less than a thousand men could keep in the field while we had a Marine Brigade of some fifteen hundred men in Haiti and a Gendarmerie of twenty-five hundred men. But here was the trick. There was literally no beginning or no end to Charlemagne's forces." Wise stressed that Charlemagne and the Cacos "knew the country the way they knew the palms of their own hands. They needed no map and compass. Instantly they could disband, hide their weapons, and become peaceful inhabitants."[62]

Volumes have been written about the wide use of the Voodoo religion as a means for treating mental health problems. Likewise, indigenous midwives were ubiquitous and widely supported throughout Haiti's general population. Even in the cities, lay health practices were frequently effective in limiting the spread of infectious diseases. For example, city residents managed to restrict infectious yaws almost exclusively to the rural regions of Haiti. Naval doctors in yaws-control projects during the occupation reported they found that people in the large port towns had "ever since the oldest city dweller can remember . . . a rigid quarantine against persons with florid or infectious yaws remaining within the city limits." "No laws have been necessary," the doctors stated. This customary quarantine, the naval doctors said, had been practiced in Haiti since as far back as the slavery era. During these lay quarantines, "every adult city dweller . . . acts as a self-appointed health officer and . . . is supported unanimously by public opinion."[63]

Despite indigenous health practices that were ever present to promote popular health, U.S. health authorities were determined to go at it alone, using solely their modern medical techniques and drugs. Like the sanitarians in the Canal Zone, U.S. health authorities were unafraid of the diseases rampant among Haiti's multitude. In fact, they sought these menaces out because of their unshakable faith in their physicians and medical science.

The Frontal Assault on Yaws

By the mid–1920s many political groups in the United States opposed the occupation and were energetically documenting and publicizing problems with its operations. Haitian activists as well as anti-expansionist U.S. politicians and civil rights groups voiced their dislike for the political suppression associated with the occupation troops and administrators. Yet, few could criticize the medical assistance that the U.S. personnel in Haiti attempted to deliver. In fact, the extreme hardship Haitians experienced in daily living made Haitians very receptive to at least trying any medical help that U.S. workers offered. U.S. medical workers in Haiti had been jarred especially by the widespread prevalence of yaws. The disease was most rampant throughout Haiti's rural population. Thus, yaws became the focus of the U.S. rural health work from the very beginning of the occupation.

Yaws is one of a family of treponematoses that include venereal disease, pinta, and endemic syphilis. It causes a wide variety of lesions on the skin, bones, and joints. These lesions can cause from slight to extreme pain, itching, discomfort, and visible, gross disfigurement throughout many parts of the body. Current biological knowledge reveals much about why it became so widespread in a nation like Haiti. Found especially in dry, hot climates, it is spread mostly through bodily contact involving skin traumas. It tends to infect children living in unsanitary, rural, hot environments. In hot climates, children tend to go barefoot or dress scantily and develop simple hand, leg, or foot abrasions. These children transmit the disease by contacts in such everyday situations as sleeping or playing together with other children. Nursing mothers and infants can also exchange yaws infections. Once a person is infected, yaws causes ghastly lesions on the skin and bones and severe rheumatism. As time passes, the untreated yaws infection will cease causing lesions and become dormant, yet still transmissible.[64]

Beginning in the early 1920s, U.S. medical teams began to make contact with ordinary Haitians, providing treatments that at least caused the most visible and discomforting rashes to recede. In medical, political, and tabloid discourse at home in the United States, occupation officials and their political supporters trumpeted the great medical advances that the occupation was presumably rendering in Haiti. In 1925 Dr. Charles Butler presented a glowing picture of the Haitian people rallying behind the U.S. public health programs. In the *U.S. Naval Medical Bulletin*, he stated that in the service's twenty rural dispensaries, some thirty-five thousand people were seen each month. "The educational value of this massive treatment, aside from its purely medical importance, is great," he wrote. "The country districts where Voodooism was formerly rampant are now being invaded by modern ideas regarding sickness and health." In 1926 the dispensaries alone handled about two hundred thousand treatments for yaws.[65] The medical information Butler and research colleagues obtained from studying the treatment of thousands of Haitians with yaws and syphilis strengthened Butler's international reputation on this subject throughout the West's growing circle of tropical disease researchers. [66]

Owing to the vigorous work of Dr. Butler and his colleagues in Haiti, the field of public health as a biomedical science was marching on. But by the early 1930s U.S. health authorities in Haiti were not sure they could eliminate just one of Haiti's major public health menaces. Some progress had been made in bringing initial health services to rural Haitians. However, these mass eradication campaigns against diseases like yaws, even in later years when highly effective drug treatments were employed, had serious deficiencies. First, since they did not entail the training and retraining of local health workers, mass campaigns in effect ceased when the expatriate teams left the scene. Second, these campaigns diverted resources away from permanent health services. Finally, since the environmental conditions, linked to poverty, that invigorated these mass infections remained vast, so did the high levels of less dramatic major infectious diseases.[67] The leading causes of death at Haitian General Hospital in Port-au-Prince were tuberculosis, gastrointestinal ailments such as typhoid fever and hookworm, and malaria. One physician-researcher at the hospital regretted that these less visible illnesses were being overlooked: "[I]f this high mortality rate were to result from some acute disease great anxiety would be felt and remedial steps would be taken, but with tuberculosis it is not so."[68]

In 1929 P. W. Wilson, a lieutenant commander in the U.S. Navy medical corps, spoke before the Haitian Medical Society in Port-au-Prince. He called for what he believed would be a new approach in Haiti's health campaign: a "frontal attack on yaws." Early in the occupation, expectations were that serious infectious diseases would be fully eradicated from Haiti. But now, Wilson said more somberly, "we all agree that we can not hope to completely eradicate tuberculosis, epidemic meningitis, hookworm, malaria, dysentery, or typhoid fever from Haiti within the lifetime of anyone here present or for many years longer." Unlike Charles Butler, Dr. Wilson was no longer so complimentary of the Haitian people's attitude toward community health. He blamed the failure of U.S. forces to eradicate the major infectious diseases not on the limitations in the U.S.-led health programs: "The principal reason why this can not be done is the ignorance of the people."[69]

Wilson urged that yaws remain at the center of the Public Health Service's work, but believed the campaign should shift to a new treatment. During the 1920s and 1930s yaws was thought to be best treated with arsenic-bismuth injections.[70] Between 1922 and 1928 the annual number of arsenic-bismuth injections given in Haiti by the Public Health Service increased from about 23,000 to 551,000. But Wilson felt that the arsenic-bismuth treatment should not be given to patients with yaws in its infectious stage. Instead, these cases should be treated with sulpharsphenamine (injections) or acetarsone (tablets)—treatments he believed to be more effective than arsenic-bismuth. Moreover, since acetarsone was administered by pills and appeared relatively safe for children, it was more convenient for patients who tended to want to avoid the painful infections or long trips to health stations. To Wilson, this new treatment approach would decrease infectious lesions and, hence, new incidents of the disease.[71]

Wilson did not differ from Butler and U.S. health authorities about making the highest priority the continuation of massive treatment of yaws victims with drug therapy. "[Drug] treatments should be looked upon perhaps more as a public health measure than as a matter strictly in the field of therapeutics," Wilson urged. In his reasoning, by stepping up the number of persons treated with drugs, general eradication would eventually occur since, theoretically, fewer and fewer new cases would develop.

Wilson also stressed that mobile clinics should be established in all of Haiti's 552 rural sections. Rural folk infected with yaws tended to shy away from hospitals or dispensaries in cities (such as Port-au-Prince or Jacmal) for treatment due to the social ostracism they experienced. Therefore, Wilson recommended that "[i]nstead of building more dispensaries for the present, we need mobile dispensaries, i.e., tents for the work and living quarters for the personnel." He also recommended that volunteer doctors be used if possible but that the local dressers who assisted be paid to "compensate them for the work they do but also for the hardships they must undergo."[72] However, Wilson's plans were fundamentally flawed. They overlooked the need to eliminate nonhuman vectors of the disease as well as unhealthy housing and clothing (especially footwear) conditions facing the Haitian peasants.

But Wilson's new emphasis on sulpharsphenamine and acetarsone, like the period of arsenic-bismuth therapy, did not alter the essential limitations in the Public Health Service's overall policy. The service was overemphasizing the massive dissemination of treatment for yaws but essentially ignoring the more difficult task of reducing social conditions linked to continuous, widespread new exposures to diseases like yaws. These social conditions included overwork stemming from harsh living conditions, inadequate sanitation and housing that spread insect infections, and lack of public hygiene education throughout rural Haiti. Drug therapies, in general, brought immediate relief to yaws and syphilis victims by reducing the more painful and uncomfortable symptoms such as lesions.

Reducing infectious lesions also most likely lessened some contact infections. Over the long haul, however, the arsenic-bismuth and sulpharsphenamine injections and acetarsone tablets all proved of limited value in breaking the cycle of epidemic yaws in pre–World War II Haiti.

During the early and mid–1920s, the clinical treatments for yaws aroused great optimism among the medical community and infected persons alike. But as time passed, it became increasingly known that while arsenic-bismuth treatment frequently reduced the severity of symptoms, it did not destroy the infection altogether. By 1933 health researchers in Jamaica recognized that relapse rates one year after arsenic-bismuth treatment were about 15 percent, and 20 percent after two years. Treating individual yaws patients did not necessarily reduce insect vectors that simply continued to re-infect exposed populations— whether patients in these populations were under treatment or not. In addition to the discomfort that undergoing this treatment entailed, it also sometimes caused serious side effects.[73]

To administer the public health and medical care programs, the United States continued to expand Haiti's Public Health Service (i.e., Service d'Hygiène). By 1930 thirteen hospitals and 153 rural clinics were in operation. The military medical personnel and Rockefeller Foundation experts initiated the first intensive programs to control diseases that were particularly widespread in Haiti.[74] Substantial inroads were made in checking smallpox, which had spread throughout the nation. In 1920 a compulsory vaccination law was implemented. Also an estimated 850,000 residents received smallpox vaccinations. Also, street cleanings, privies, and school inspections to verify vaccinations, especially in districts with high malaria rates, were developed and expanded.[75]

The failure by the U.S.-led Haitian Public Health Service to attack on a broad scale other communicable diseases widespread throughout Haiti took the gloss off its much heralded yaws campaign and smallpox work. The all-out effort by U.S. health personnel to treat as many Haitians for yaws as possible was in contrast to the sparse work the service did to control malaria. During 1924 and 1925, the Rockefeller Foundation sponsored a medical survey of Haiti headed by G. C. Payne. Of the 4,439 people examined throughout the nation, malaria parasites were found in 67 percent. A few years later, Wilson and Clark published findings from the examination of 11,000 Haitian workers from northern and northeast Haiti on their way to work for the United Fruit Company in Cuba. Nearly one-quarter of these emigrants had malaria.[76]

American health officials implemented maritime quarantines similar to those in the United States in Haiti ports. Also, varying degrees of routine street cleaning and garbage collection were started in Haiti's cities, towns, and villages. These sanitary improvements, along with airplane dusting, helped to reduce the incidence of malaria in sections of cities such as Port-au-Prince, Cap-Häitien, Gonaives, St. Marck, Jacmel, and Jeremie. But for the nation as a whole, especially its rural population, the U.S.-occupation health program was failing to reduce insect vectors or raise overall living standards, the two vital legs for stamping out malaria permanently. In 1929 K. C. Melhorn, of the U.S. Navy medical corps in Haiti, admitted, "Malaria in rural districts continues to be a major problem."[77]

Technologies of Dependence

Despite some success by U.S. authorities in improving public health and infrastructure, the exclusion of Haiti's traditional physicians from the National Medical School engendered a permanent friction and distaste from these two hundred or so Haitian professionals.[78] Haitian physicians missed the opportunity to fashion a system of physician-centered health care similar to that which had unfolded earlier in key Western capitalist nations. In the pre–1910 period, Haitian doctors made important original contributions to medicine regarding diseases and ailments in the Haitian climatic and disease setting. But the destruction of the Haitian-based medical profession aborted any further development of bio-clinical knowledge within the circle of Haitian clinical practitioners and their hospitals, private practices, and dispensaries.

The use of solely military corps doctors and infectious disease specialists from the United States meant, in essence, that the flow of medical knowledge garnered from direct clinical experience in treating Haitian patients looped directly back to the United States, not to Haitian doctors and medical institutions. Furthermore, the U.S. medical personnel with the occupation forces were mainly military physicians or tropical disease researchers from the Rockefeller Foundation. As largely tropical disease specialists, this personnel was neither trained, experienced, nor enthusiastic about providing primary medical care for uneducated rural populations, including women and children.

In the United States, Britain, and Germany, in contrast to the Haitian situation, interspecialty networks of physicians were concentrating in hospitals as both their specialties and the hospitals in which they practiced gained exclusive license by government to render clinical care. The U.S. and European physicians were transforming themselves and the hospitals into the hub of their nations' medical-care delivery systems. With the occupation, U.S. authorities placed the military in control of the Haitian Public Health Service, and the Public Health Service, in turn, in control of the National Medical School. The U.S. medical officials wanted to focus on the new medical specialties that relied on laboratory testing. Their more "modern" approach also emphasized clinical and pharmaceutical treatments.

One leading U.S. medical educator in occupied Haiti lauded this development: "Very remarkable is the story of the coordination [in Haiti] of every type of medical activity under one head—hospitalization, sanitation, and medical education."[79] However, this new policy deprived Haiti's government and medical profession of the responsibility and power to license medical professionals and hospitals.

Usurping the control of Haiti's medical education and hospitals intensified opposition to the U.S. occupation beyond just medical professionals. When U.S. health authorities rushed in military and tropical medicine experts, these officials also shut out potential contributions from indigenous health care practitioners, especially midwives and religious healers. True, the rural dispensary was the centerpiece for the public health programs of the U.S. occupation. Using all modes of land transportation, as well as tents as work sites when necessary, Public Health Service physicians with a small team of nurses, attendants, or dressers scoured the hills of Haiti. However, these U.S. medical teams were neither designed nor staffed to provide primary care or preventive services. Instead, these medical staffs were in search of victims of very specific diseases, especially syphilis and yaws—ailments of utmost concern to Western tropical disease research.[80] The Haitian peasantry were more in need of care for injuries to women and children caused by workdays that were too long and fatiguing on farms.

A leading Haitian physician and professor at Haiti's National Medical School, Camille Lhérisson, described the direct relationship between ill health and work pressures that faced the nation's peasants. The introduction of plowing machines was making these health conditions even worse:

> Our peasants are . . . quite liable to traumatic injuries including serious falls. Ploughing machines frequently cause wounds which, on becoming infected, develop into enormous ulcers complicated by tetanus. Children and women often have to carry over poor roads, especially in the mountain regions, huge loads of lumber, water, fruits, building materials, etc. This overwork must end inevitably by damaging health. The flattening of cranial bones, the exaggerated bending of the spine, the injuries or prolapse of internal organs, muscular tremors, headaches, cannot fail to produce most harmful effects on the skeletal frame (for instance, on the female pelvis), and must affect the individual's work capacity.[81]

The Haitian physicians were keen to identify other connections between Haiti's harsh rural life and the serious health problems of the Haitian people. Lhérisson emphasized that, "in a general way, farming and long walks often compel women to stop nursing their babies. A most irrational and often inadequate feeding is then substituted, favoring digestive disturbances, nutritive disease, . . . and cachexia. These troubles will serve, then, as the starting point for serious dystrophies of a slow but fatal course." While the Public Health Service organized by the U.S. occupation "has come to remedy a sad condition," Lhérisson stated in the publication of the American Public Health Association, "there is still much to be done."[82]

United States' control of Haiti's meager public works and utilities, when viewed in international context, also damaged Haiti's long-term technological capacity and, as a consequence, its political and economic autonomy. In occupied Haiti, public utilities were engineered and administered almost exclusively by the United States' military and experts. Consequently, the Haitian state and private companies were much worse off than their Latin American neighbors. In many South American countries, mostly private British or U.S. investors and contractors had constructed public utilities. For example, between 1895 and 1913 British investments in public utilities such as gas, electric, telegraph and telephone, and waterworks grew from 5 percent to about 12 percent of Britain's total investments in Latin America.[83]

Local political and financial segments in Latin American countries frequently disapproved of foreign ownership of local public utilities. Yet lacking the capital to sustain long-term construction and maintenance that utilities required, most Latin American nations went ahead with using foreign investors for utility projects. Nonetheless, local bureaucrats, lawyers, businesses, and government officials usually benefited from these foreign-owned enterprises. Also, since these public utilities were privately owned, the European investors had a personal financial stake in ensuring their effective operation and expansion.[84] In Haiti, utility construction and administration were exclusively U.S. military operations. Thus, Haiti received neither investments nor business benefits of any sort for its indigenous professionals and entrepreneurs.

By the mid–1930s, the United States' strident technical movement inside Haiti had failed when measured in broad socioeconomic and educational terms. At the occupation's end in 1934 most of the Haitian people were still mired in poverty and the government was in even deeper foreign debt. As at the beginning of the U.S. occupation, Haiti in the late 1930s was divided politically and economically. At the top were about 5 percent of the population who were the moneyed, lighter-skinned mulatto elite and foreigners. The remaining Haitians were impoverished peasants. Furthermore, within the peasantry had developed a permanent, geographically detached subpeasantry or "underclass" consisting of several hundred thousand peasants.

This segment of underclass laborers left Haiti to work in Cuba and the Dominican Republic. By the mid–1920s about 150,000 Haitians were working in Cuba, mostly for U.S. sugar interests. Between 1935 and 1937, Haitians living and working in the frontier zone of the Dominican Republic swelled from 50,000 to about 200,000. In October 1937, a serious conflict broke out between the Haitian and Dominican Republic governments over the Haitian worker migration into the Dominican Republic. Some 9,000 to 12,000 Haitian peasants were massacred inside the western province of the Dominican Republic. It took diplomatic action by the United States to keep the situation from bursting out of control.[85]

Perhaps the most vexatious long-term effect in Haiti as a result of U.S. technical programs was in the area of education. The French and Creole culture of the Haitians had been discarded by the U.S. assistance programs to Haiti. Therefore, the U.S. education initiative ended up reaching only a minute portion of Haitian children and young adults. Moreover, educational leadership for Haiti had not been cultivated. Sufficient numbers of educated Haitians were not brought into the technical projects as peers and potential heirs to extend the objectives of these projects. Under the occupation, U.S. officials and technicians made vocational education the highest priority for Haiti. They bypassed trying to lay the groundwork for a multi-tiered, university-centered educational system. This policy doomed any hopes of the Haitian people or government of creating a national university once the U.S. troops withdrew.

Without a national school and university system to coordinate science education at all levels, as well as university-based professions, the nation had no potential to develop an ample supply of Haitian scientists, engineers, and teachers for government, business, and mass education. Haiti was facing another period in its nationhood without the minimum infrastructure of public works, a generally healthy and literate population, or a sizable economic middle class. In the meantime, U.S. commercial adventurers, tropical researchers, and religious workers were fanning out into Liberia—the United States' long-standing, yet rarely visited ally republic on the other side of the Atlantic.

CHAPTER 4

Out of the Shadows

LIBERIA BEFORE WORLD WAR II

$\mathcal{T}$he impact of U.S. industrial technologies, public health campaigns, and large plantations had been direct and forceful in the Black Belt South, the Panama Canal Zone, and Haiti. However, these forms of U.S. technical and medical activities reached Liberia only in small increments. In fact, in the late nineteenth and early twentieth centuries, Liberia was in the extreme shadows of U.S. foreign territorial contacts. Liberia was brought out of this obscurity only gradually as the events pulling the United States into World War II unfolded.

A unique friendship with Liberia has been a recurrent theme in the U.S. social imagination since the mid-nineteenth century. The movement to establish Liberia first emerged amidst the antislavery controversy in the antebellum South. The American Colonization Society, a conservative abolitionist organization led at times by James Madison, John Marshall, and James Monroe, helped start the small nation for African American emigrants. They intended it as an outpost on the other side of the Atlantic for free black Americans and manumitted slaves from the United States.

The idea that U.S. charity was the parent of Liberia echoed in America's foreign policy discourse throughout the nineteenth and twentieth centuries. Political concern in the United States for Liberia became especially strong as the Civil War approached and then during World War II and the early years of the American civil rights movement. African Americans in particular perceived Liberia as a "little brother" or "little sister" republic destined against all odds to develop into a democratic and commercial beacon for the African continent.[1] In the nineteenth century, prominent black American religious and political activists, most notably Lott Carey, Rev. Alexander Crummell, and Bishop Henry McNeal Turner, tried relentlessly to initiate education and commerce projects in the nation. Their goal was to develop Liberia into a beachhead for Christian evangelization in Africa as well as for black American emigrants generated by their back-to-Africa resettlement campaigns.[2]

Liberia's prominence in the minds of U.S. politicians, philanthropists, and black public activists could not alone generate broad-scale scientific missions to its shore. U.S. technology arrived in Liberia much more slowly compared to

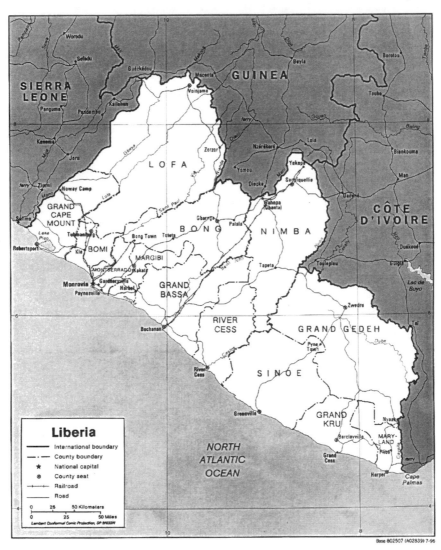

MAP 4.1. Liberia. Courtesy of Perry-Castaneda Library Map Collection, University of Texas at Austin Library (www.lib.utexas.edu/maps).

the surge of U.S. technical projects in Panama and Haiti. Prior to the early 1920s, actual aid to Liberia from America's federal government or large businesses was miniscule. No major interest group in U.S. or European politics, militaries, or heavy industries developed intentions to establish technical projects in Liberia. Consequently, before World War II, any hopes for Liberia's national development rested solely on its own shoulders. African nations that were formal colonies of European powers could count on some degree of infrastructural development from their respective imperial administrations to facilitate the exploitation of raw materials or colonial settlements. However, this was not the case for Liberia. As late as 1940, most parts of the Liberian hinterland still lacked any national infrastructure whatsoever. There was virtually no public schooling, piped water, road system, or electrification.[3]

Yet Liberia still managed to lure a small group of professionals and researchers from U.S. missionary and philanthropic bodies and commercial (rubber) manufacturers. This was the groundwork for a great step-up in U.S. technical transfer that occurred once the United States entered World War II. At that point, U.S. technical expertise and materials, sent by both its military and industrial interests, rushed into Liberia.

First Missionaries

When Liberia was founded, U.S. politicians and philanthropists involved with Africa optimistically supported the "black republic." They believed slavery could be ended in the United States by facilitating the voluntary deportment of freed slaves to Liberia. Besides providing a peaceful solution to ending slavery in America, these American leaders believed that Liberia could then serve as a successful experiment in bringing democratic government to West Africa. Liberia developed as a geographic nation from a series of isolated coastal settlements and small ports. Six large rivers run inland from its 350-mile coastline. For hundreds of years prior to its establishment, Liberia's transportation resources consisted of scattered ports. By the end of the nineteenth century there were about twenty trading stations dotting the Liberian coast, some with histories as far back as the fifteenth century. The expansion of these trading stations was mostly the result of the Americo-Liberian settlements that began appearing in 1821. The settlers were dependent on these ports to receive imports of Western goods as well as to maintain communication with Europe and America.[4]

By the mid-nineteenth century, many vessels owned by planters and traders sailed between Liverpool and New York loaded with Liberian exports. Among U.S. politicians after the Civil War, the vision grew of Liberia becoming the central trading republic between America and West Africa. They viewed Liberia as on the brink of a golden age of commerce and trade. Senator Charles Sumner of Massachusetts prophesied that "Liberia may become the metropolitan power on the whole West African coast, so that the growing commerce of that continent will be to a great degree in its hands."[5] But this prophecy never came to fruition. In fact, the path that lay ahead for Liberia's economy would be even more difficult than that of most of the African nations that fell under colonial rule.

Many historians erroneously assume that there were no noteworthy educational or technical contacts between the United States and Liberia. According to this view, the only exceptions were the romantic projects of disgruntled black Americans and evangelical zealots. Yet closer examination reveals a long string of education projects as well as medical, botanical, and agricultural endeavors sponsored inside Liberia by U.S. missions, academics, philanthropic foundations, and businesses. These contacts tended to keep Liberia from becoming totally obscured from the Western community. They also reveal the pronounced and unique difficulties Liberia presented to both Liberians and foreigners interested in modernizing this nation. Liberia's rugged natural geography—most of which is some of Africa's most rich rain forests—had been for centuries peopled by several distinct ethnolinguistic groups. These demographic and cultural factors made attempting large technical projects in this nation a challenge to even the most powerful Western interests, much less the Liberians themselves, who have had historically very little resources with which to work.

Prior to World War I, religious missions were the primary conduit of assistance from the United States to Liberia. Beginning in the early nineteenth century, U.S. missionaries and educators erected a number of small primary and Bible schools, dispensaries, and farming projects throughout Liberia.[6] Christian evangelism was at the heart of the formal ideas and psychology behind their missions. Yet these religious workers did not view themselves as hypocrites for wanting to introduce secular skills and vocational education—that is, aspects of the scientific mission—throughout Liberia's indigenous communities. These religious missionaries believed that building facilities for education, health, and worship throughout indigenous Liberian communities affirmed the practical power of Christian faith. As the fledgling republic of Liberia prospered, they believed, missionaries in Liberia would gain more new converts.

Religious and educational work was started by Episcopalians in the 1830s. Episcopalians started activities, in the Cape Palmas section of Liberia, that lasted for decades. Finally, donations by Episcopalians led to the founding in 1889 of Cuttington College in Cape Palmas. Students came from throughout West Africa to study in its agricultural and industrial departments. By teaching agricultural methods, the school administrators believed the college "had found the key to the solution of the food problem in Liberia." These educators and missionaries held to the strong belief that Christian evangelism, coupled with manual training and frugal industry, would somehow advance this nation toward national prosperity. Rev. M. P. Keda Valentine, Cuttington College's first principal, believed that farming skills learned at Cuttington would be passed on throughout the country and that "the contagion [would] go from tribe to tribe."[7]

In addition to resettlement campaigns intermittently organized by African American emigrationists such as Bishop Henry McNeal Turner, several black and white religious denominations sent medical and educational missions to Liberia. American religious evangelists believed that missionary doctors—using the power of scientific medicine in the service of the Christian mission—could light the flame of progress throughout Liberia. Also, each year historically black colleges graduated handfuls of Liberians who returned to their country and joined Liberia's small educated class.

One of the most noteworthy African American missionaries at the turn of the century was William H. Heard. An African Methodist Episcopal minister and minister resident and consul general for the United States (1895–1899), Heard conducted one of the first nationwide surveys of Liberia's people and geography. In his book based on the survey, *The Brighter Side of African Life Illustrated* (1898), Heard presented facts, photographs, and observations to depict the great promise Liberia held for black immigrants from the United States who possessed skill and drive. His book intended to refute the negative views, widespread in the United States, that he believed had soaked into the minds of black Americans and were discouraging resettlement in Liberia. Using facts about Liberia's potential mineral wealth, photographs of successful Americo-Liberians and their projects, as well as tranquil scenes with local people, Heard appealed to skilled blacks in particular to come and make new lives in Liberia: "The physician may display skill in healing a malady that carries thousands to untimely graves. . . . The merchant may reap a reward for his genius here, undreamed of elsewhere. The minister may gather a harvest such as will meet the approval of a just God. The farmer may grow wealth in fifteen years growing coffee, while he ekes out a miserable existence in the United States."[8] Heard exhorted that in Liberia "opportunities are unparalleled in any other country in the civilized world, especially for the Negro."[9]

The careers in Liberia of the missionary doctor Rev. Clinton Boone and his wife, Rachel Tharps Boone, also personified the intent of American missionaries to blend practical skills with their religious activities. U.S. evangelists and mission doctors believed that medical science and the other professions such as teaching and the ministry were the vehicles for civilization. Clinton Boone graduated from Virginia Union University in 1900 and departed immediately for the Belgian Congo. He later graduated from Leonard Medical School, a leading black institution. From 1906 through the 1920s, the Boones were missionaries in West Central Africa and Liberia. They served under the auspices of the Lott Carey Baptist Foreign Mission Society. Clinton Boone practiced medicine in Liberia and the Congo, while Rachel Boone operated a primary school in Monrovia.[10] Through the work of dedicated missionaries such as the Boones, the dream that one day Liberia would obtain a stature equal to modern nations had a pulse, even if only a very faint one (see figs. 7–9).

But in the opening decade of the twentieth century, a broader reality confronted Liberia's presidents as well as its foreign missionaries and American political friends. In terms of national development, Liberia was in abysmal condition. For one, there were almost no formal governmental operations out-

side of Monrovia. Public schools, dispensaries, roads, or railways of any significant magnitude were nonexistent. Before World War I, Liberia had no modern road that connected one county to another or Monrovia to the hinterland.
Finally, literacy was limited to the Americo-Liberians in Monrovia and the small
numbers of Liberians taught in the mission schools dotting the hinterland.[11]

Obstacles to Technical Assistance

A viable nation-state is a territory fused by common political and economic institutions. Liberia's failure to reach this level of nationhood had roots
deep in its natural ecology, ethnic character, and colonial European entangle-
ments. The lack of diversified farming and modern ports was a serious barrier
to Liberia's economic growth. Through the early twentieth century, almost all
of the crops exported from Liberia, such as palm oil and fibers, ginger, and rubber, were harvested from natural growth rather than planned cultivation. At the
same time, Liberia was receiving virtually no organized government assistance
from encroaching Western industrial nations such as France, Britain, and Germany. At the same time, traders from these nations were routinely making exorbitant profits for the raw goods they obtained from local peoples of the Liberian
hinterland. European nations were either preparing for or recovering from World
War I. They had little interest in strengthening Liberia or other parts of West
Africa beyond using these African locales as strategic outposts for European
militaries.[12]

No matter what political efforts or schemes were undertaken by the
Americo-Liberian settler-elite, the traditional ethnicity and geology of Liberia
made the development of modern transportation especially difficult. The Liberian
population was a mosaic of different ethnic societies. The traditional commerce
and political authority of each of these regional groups developed separate from
the others. This deep-rooted fragmentation, as well as the chronic fiscal weakness of Liberia's small formal government centered in Monrovia, precluded a
nationwide mobilization of either people or capital for building a modern port
or rudimentary networks of roads and railways.

In developing and industrialized nations, large ports are meeting points
for a variety of transport systems. They provide gateways for the export of goods
to foreland markets as well as entryways for goods and human capital for internal markets and communities.[13] But many of Liberia's waterfronts are naturally
shallow. They lead into large stretches of swamps or rocky ridges and then thick
rain forests. Thus, they present horrendous engineering challenges for building
ports to accommodate large ships. Before World War II, shipments by large vessels from abroad had to be reloaded to smaller boats and then brought ashore.
Moreover, as automobiles and trucking spread throughout industrial nations,
Liberia's dense topography precluded roadway development that could handle
these new motor vehicles.[14]

Throughout the late nineteenth and early twentieth centuries, the U.S. government developed a "hands-off" policy toward Liberia. Liberian representatives, U.S. black social and religious activists, and occasional business speculators made recurrent appeals for links. However, neither the U.S. government nor large firms responded with plans to attempt large-scale military or industrial projects in Liberia. American leadership was preoccupied with completing the Panama Canal and military escapades in the Caribbean, Latin America, and Asia. Once the Panama Canal was completed, trade with Central and South America sped up. It became less necessary for U.S. traders to seek agricultural produce from West African nations such as Liberia.

Speeches in U.S. foreign policy circles continued to emphasize a natural, parent-child political alliance between the United States and Liberia. But the road that lay ahead for modern Liberia to try to attract U.S. aid for development was even more difficult than that of Panama or Haiti. Indeed, during the period when the U.S. government and/or business leaders moved large technical resources into Panama (for the canal project) and Haiti (for the occupation), they sent no large technology aid to Liberia of any sort.

A crisis over the issue of U.S. government aid to Liberia occurred in 1909. At this time European colonialism was widespread throughout Africa and Asia. Great Britain, France, and Germany were maneuvering unpredictably around Liberia's borders. Internally, Liberia was still a territory of fragmented ethnic villages untouched by modern development. In 1909 it became necessary for the U.S. Congress to convene a special body, popularly known as the American Commission on Liberian Affairs, to investigate U.S. interests in Liberia.[15]

The Liberian commission pointed out that both the U.S. State Department and Liberia's political leadership felt compelled to develop U.S. political, financial, and technical aid for Liberia. But the commission concluded that this link was a long way from reality: "A review of the history of the relations of the United States with Liberia brings the commission to the conclusion that, beyond a series of notable expressions of goodwill and friendship, its positive results have been painfully meager."[16] In the meantime, the commission pointed out, Liberia remained outside the flow of modern progress. "Trades and industries languish," the commission reported. Moreover, Liberia had very little skilled labor. Instead, it had to import skilled workers from Sierra Leone. "Nearly all manufactured articles are imported," the commission stated, and, overall, "greater attention will be necessary to the development of trades and industries before a high order of national thrift can be secured."[17]

The commission recommended that the United States extend a wide range of aid to Liberia. It believed the United States needed Liberia as a strong ally. Liberia would provide a West African locale that could assist the United States against impending competition in that region from aggressive European states. It urged that the aid package include financial assistance so that Liberia could reduce its foreign debts as well as reform its internal finances. The commission also recommended assistance with the military training of Liberian forces to quell boundary disputes and the establishment of a naval coaling station and ongoing research project in Liberia. The commission emphasized that the research station would be the first step that would "enable Liberia to find herself economically [and] to enter into her own heritage."[18]

The Liberian government had originally requested that the United States help with "establishing industrial schools" in at least one of its counties "with a view of promoting a knowledge of such trades and industries as will render the Republic self-reliant." But instead, U.S. officials proposed the research station because it would enable U.S. authorities to learn about "the effect of climate on health, and the causes, treatment, and cure of tropical diseases." The United States would use the Liberian station to learn "what to do and what to favor in our own tropical possessions." But despite the commission's efforts, its recommendations were never implemented.[19]

Newspaper reporters, politicians, industrialists, tropical medicine specialists, and the general public throughout the United States had all nervously held their breaths awaiting the completion of the Panama Canal. But at the same time, the needs of Liberia were of little economic or military concern. Liberia's leaders and humanitarian advocates in the United States and abroad could find no political or industrial supporters in the United States interested in constructing a modern port on the nation's coast. President Woodrow Wilson gave brief consideration to this project as part of America's World War I military initiatives. In 1919 he tried to invoke his war powers to provide a $5 million loan to Liberia for the construction of a port in Monrovia. However, once the war ended so did U.S. interest in the project.[20]

During the interwar years, repeated setbacks occurred to Liberian officials trying to obtain U.S. government aid. Still, a deep sentiment of goodwill for Liberia persisted among African Americans and their liberal allies. Robert Moton, the head of the Tuskegee Institute, exemplified this view in 1918. Commenting to President Wilson in support of a U.S. loan to Liberia, Moton pointed out that thousands of black Americans were "deeply interested in the Negro Republic of Liberia."[21] During 1920 Moton reemphasized the special concern among African Americans for that country. Writing about black American missionary workers in Liberia, Moton stressed, "No spot in Africa is of greater interest to American Negroes than Liberia; both because the present Government was established by American Negroes who migrated to that spot under the patronage of the American Colonization Society, and because it is the one spot in Africa to which the American Negro has unrestricted admission."[22]

For now, appeals like Moton's went unheeded. However, there were two aspects of Liberia that did prove strongly attractive to U.S. interests. First, it remained a fertile field for U.S. religious missionary work. Second, its ancient forests, rich natural resources for rubber cultivation and iron ore, and variety of tropical health conditions made Liberia irresistible to U.S. scientists and industrialists.

Tropical Research and Philanthropy

Before World War I, church missions had managed to establish small groups of medical workers and small hospitals throughout Liberia. An additional thin stream of U.S. medical personnel, this time from secular sources, appeared in Liberia between the world wars, along with university researchers and foundations investigating tropical diseases, botany- and forestry-initiated geographical survey teams, and medical teams. Second, the Firestone Rubber Company, which would in the mid–1920s develop large rubber plantations in Liberia, established medical services for its employees. At first the Firestone workforce included local, seasonal workers. As Firestone plantations grew, Liberians became the source for lower skilled employees on Firestone plantations. Employees and other locals eventually had the use of the company's medical facilities. Third, loan revenues from U.S. banking interests, along with some assistance from Firestone, included small amounts to encourage the Liberian government to establish formal public health and public works resources. However, these state programs were largely failures until the United States became involved in Liberia as part of its military effort during World War II.

Public health campaigns in the Canal Zone and Haiti had been central to reinforcing U.S. industrial or military presence in these regions. However, in the case of Liberia before World War II, U.S. authorities or international industrialists had no intention of mounting large-scale sanitation and health care programs. But the impulse for more research on tropical disease and geography began to bring a small but growing number of medical scientists to Liberia's doorstep. This flow of U.S. tropical-science philanthropy reaching Liberia started in the 1920s. In 1926 and 1927 the Rockefeller Foundation, Harvard University, and the Firestone Company cosponsored an expedition to Liberia. Headed by Richard P. Strong of Harvard Medical School, the team included seven of Strong's colleagues in tropical disease and zoology specialties. The result was one of the earliest comprehensive scientific surveys of the geography, plant and animal life, and endemic diseases of interior Liberia.[23]

The surveys by the Harvard expedition uncovered evidence of possible widespread health problems among the Liberian people. For example, the expedition estimated that about one-fifth of the population had malaria. Nonetheless, the government of Liberia had little use for such information. Its agencies did not have the resources to conduct an official population census (this would not occur until the 1940s), so neither health advisors for the Liberian government nor Strong and later investigators could draw firm epidemiological information from the expedition's data. Specific disease patterns and public health needs for Liberian cities and rural communities alike would remain largely unknown for another decade.

Philanthropists also had an interest in educational projects that could expand the supply of skilled workers the Liberian economy so sorely needed. Since the days of Booker T. Washington, U.S. philanthropists were interested in expanding the Tuskegee model for education beyond the U.S. Black Belt to black societies abroad. In 1910, Olivia Egleson Phelps, a philanthropist interested in missionary programs in Africa and the success of the Tuskegee Institute in Alabama, left a substantial sum of money to establish a school like the Tuskegee Institute in Liberia. The next year the Phelps-Stokes Fund was established. It concentrated on providing studies of the social problems of blacks, scholarships for college students, and better education facilities for blacks in the United States and Africa. Gradually, the Phelps-Stokes Fund worked the Tuskegee-style school for Liberia into its plans.[24]

After years of preliminary planning and fund-raising, the Booker Washington Agricultural and Industrial Institute (BWI) opened in 1929. The school was built in Kakata on one thousand acres provided by the Liberian legislature. It was funded mostly by the Phelps-Stokes Fund and the Methodist Mission Board. Its first principal was James L. Sibley, a white educator from the U.S. South who had been an agent for rural black schools in Alabama. Sibley had traveled to Liberia in 1925 and became an advisor on Liberian affairs at the Phelps-Stokes Fund. Although in actuality it was a small vocational high school for training indigenous youth, through the years BWI received much fanfare and financial support from American donors anxious to spread industrial education throughout native Africa. Among early donors for specific projects at the school were the American and New York Colonization Societies and Harvey S. Firestone.[25]

In the United States, the Washington Institute was a popular symbol of achievement throughout philanthropic circles involved with projects in the Deep South and Africa. During the next few decades, the school produced about one hundred graduates annually. These graduates were limited to boys and young men who usually returned to their home villages as skilled workers. In the context of Liberia's overall demography—it had several hundred thousand school-age children and youth, most not in school—the BWI graduates were little contribution to the nation's overall industrial growth. However, the BWI functioned as an important emblem for the promise of technical training and science—a promise captivating to both the school's U.S. sponsors and Liberia's leaders.

Rubber: The First Industrial Link

The first substantial U.S. technical project brought to Liberia developed in the early 1920s and emerged from the private sector. In 1923 the Firestone Tire and Rubber Company negotiated a huge land concession to establish rubber plantations in Liberia. Many political and educational leaders in the United States and abroad hailed that Liberia was now entering the gates of irreversible progress. Liberia's governmental leaders also envisioned that the Firestone Company's presence in Liberia would solidify ties between the United States and the Republic of Liberia. Specialists on African affairs from the American

Advisory Committee on Education in Liberia exemplified such optimism. In their 1928 report, *Liberia—Old and New,* Sibley and Westermann predicted the Firestone plantations would upgrade surrounding regions: "Attractive villages will be built, with a sanitary water supply and sewage, neat cottages, and garden plots to assist in the production of an improved food supply. One who knows the average African village cannot but rejoice at this prospect."[26]

But African Americans were much less optimistic. When Firestone's Liberian investment was announced, most African American journalists and political activists as well as many of their counterparts in Africa denounced the concession. They believed the U.S. State Department had been strong-arming Liberian officials to accept the Firestone offer. Furthermore, they feared the Firestone plantations would just become another extension of forced labor schemes already widespread throughout colonial Africa.[27]

Typical was the view expressed in *Opportunity,* the magazine of the National Urban League. In a 1925 issue its editors put down "this gigantic transaction." They exclaimed that "Africans outside of Liberia" were denouncing the Firestone deal because, with the exception of Liberia and Ethiopia, Europeans already had the "vast rich continent" of Africa under colonialism. "Whether or not the United States is an imperialistic nation [we] point out, rather embarrassingly the example of Haiti." They also cited the United States' "imperious notes" on the oil resources of Mexico.[28] Some Liberian publicists, watching the Firestone negotiations unfold, warned their fellow Liberians that their nation should not end up losing its sovereignty as was occurring in occupied Haiti.[29]

So potentially explosive was black American political opposition to the Firestone endeavor, the U.S. State Department had to use "fire to fight fire." It enlisted into the U.S. diplomatic corps a respected AME (African Methodist Episcopal) educator and missionary, Rev. Solomon Porter Hood, and W.E.B. Du Bois, to help broker the Firestone deal. Hood was slightly acquainted with Firestone, while Du Bois was in Liberia attending the inauguration of its new president, Charles D. B. King. Du Bois's burning idealism for Liberia drew him to participate in the Firestone negotiation, even if extemporaneously. He described this moment in an essay for *Foreign Affairs.* At King's ceremony, Du Bois said, "As a gesture of good will, I was designated by cable to act as Special Minister Plenipotentiary and Envoy Extraordinary. The appointment was purely ornamental, but I did all I could to cooperate with Hood and Africa and Liberia and tell them of the tremendous interest which American colored people had in them." The culmination of the Firestone concession came in the winter of 1923–1924. The company received one million acres of land for the nominal rent of six cents per acre.[30]

Despite the misgivings about the Firestone rubber operation in Liberia, most African Americans still longed to see the venture work out for the best. At the same time the Firestone Company was setting up its rubber plantations, Liberia continued to attract moral support from many in black America, especially church mission leaders and educators whose colleges were graduating

Liberians. In 1929, J. Harvey Randolph, the head of the black Baptist Church's Lott Carey Mission Society, summarized Dr. Clinton Boone's career in Liberia as exemplary. Randolph wrote that Boone had trained himself "to be one of the best-equipped Missionaries ever sent out by any Convention to Africa." As one of the handful of trained physicians and dentists in Liberia, Randolph said, Boone had earned the highest "honor and distinction" in both Liberia and America.[31] That same year Robert Moton, head of Tuskegee Institute, introduced Boone's autobiography. He stressed Boone's career as an example of the strong historic and future links between black Americans and Liberia. Moton urged that Boone's life story "should be a part of the education of all the Negro youth of today." He called Liberia "the point of free contact between the Negro in America and the Negro in Africa, the avenue through which the American Negro can most easily make his contribution to the development of Africa."[32]

Other black American educators and civic leaders bitterly opposed Firestone's Liberian operation and the U.S. State Department's support of this venture. In 1933 a group of prominent professors and political activists met with the acting secretary of state to protest U.S. policy in Liberia. The group included Mordecai W. Johnson, president of Howard University; Walter White, the secretary of the NAACP; and Dorothy Detzer, executive secretary of the Women's International League for Peace and Freedom. By handing so much of Liberia's economic potential to "a commercial organization," the group believed the United States was in effect creating a semi-colony. They likened the Firestone concession to the U.S. occupation of Haiti: "Black America believes that this is what took place in Haiti, that with the excuse of putting down disorder and increasing the military and naval protection of the United States, we overthrew an independent government. . . . We saddled upon that country a debt so huge, a recognition of pretended obligations and contracts so vast, that the country is bound to be in economic slavery to the United States for indefinite length of time. Now, in Liberia we think we see the same process incubating."[33]

The Johnson group also told the secretary of state that the government's unbridled support for the Firestone deal offended the growing number of college-educated blacks throughout the United States. To these educated youth, the Firestone concession was one-sidedly against the interest of Liberia. The college students saw in the agreement the U.S. government's "disposition to shut the gates of opportunity in their faces and to reduce every colored country where possible to complete vassalage to white countries. Here we see no 'New Deal.'"[34]

As automobile use expanded dramatically throughout the United States and other industrial areas of the world, the United States became a huge market for Firestone rubber. During the 1929–1938 decade, U.S. rubber consumption rose yearly to more than one-half of the world's annual supply. In meeting the rubber demands of this growing domestic and international market, the Firestone plantations in Liberia grew into one of the world's largest sites for rubber production.[35]

Plantations, Local Labor, and Women

While the Firestone plantations developed, most Liberians remained in traditional, rural economies. During Liberia's earliest decades, Americo-Liberians introduced rudimentary plantations for sugar, coffee, and cotton crops. However, the Americo-Liberians could never amass enough finances and managerial expertise to duplicate the large plantations of the Black Belt South. By contrast, the Firestone Company's rubber plantations were massive in scale. The company plantations were set up for operations rapidly, and as a self-contained unit.[36]

Located about fifty miles outside of Monrovia, the infrastructure servicing the Firestone plantations had been designed and managed by expatriate personnel. Once completed, these facilities were functionally autonomous from the adjacent communities. Most of the managerial and technical employees were imported from the United States. Its transportation, living quarters for its personnel, and medical services were self-sufficient and unconnected to the local region. By 1933 Firestone had 125 miles of roads within its plantations, extensive medical services, and a radio service that operated between its Liberia sites, the United States, and other nations.[37]

In fact, the Firestone rubber producers used local populations only to perform the most harsh labor. Thousands of acres were cleared and made suitable for Firestone's rubber plants and company roads by this local labor. Abandoned stretches of highway between Monrovia, Careysburg, and the Firestone plantations were resurfaced by Firestone to connect its plantations to the capital. Local workers had to fill in deep gullies in these roadbeds, the results of recurrent heavy rains and floods. Foresters surveying Liberia at this time wrote, "At one point in this road the bed was built up from ten to twenty feet over a stretch of half a mile, all the earth being carried in head baskets by road-gang boys brought down from the interior."[38]

The relationship of the Firestone plantations to the larger regional economy of Liberia was virtually the opposite of the plantations in the U.S. South. In the U.S. South, plantations for its key crop, cotton, as well as for crops of lesser import, tobacco and rice, usually were embedded in local and statewide economic, social, and political life. These plantations became production hubs so intertwined with local economy that they formed "commodity cultures." Work life and trade in local towns, cities, and ports were fully connected to the adjacent plantations.[39] The cotton plantations in the 1920s, according to Rupert Vance, "include[d] croppers, tenants, small farmers, and planters . . . plus the banks, supply merchants, factors, and fertilizer dealers who finance the crop and the cooperatives, local buyers, general buyers, shippers, and exporters who assemble and classify cotton for sale to the mills or for export."[40] While the division of labor and plot ownership on the southern plantations were segregated by race, nonetheless the plantations as a whole had become interlaced with the southern regional economy.

By contrast, the Firestone rubber farms processed and exported their produce as essentially isolated "enclave plantations."[41] Substantial improvements in the standard of living in the regions outside of the Firestone plantation compounds or in Liberia as a whole were neither intended nor realized. In the original concession agreement, Firestone was obligated to build a port at Monrovia or Marshall, depending on the best geological location for the breakwater. However, in a few years Firestone discovered this project would require far more than the three hundred thousand dollars it had promised. The company managed to have this stipulation of the concession agreement canceled.[42] The Firestone interest had more than a free hand in the fiscal management of its relationship with the government of Liberia. From 1938 to the mid–1950s, Liberia's official depository was the Bank of Monrovia, a subsidiary of Firestone. Liberian activists disliked this arrangement from its inception.[43]

Moreover, Liberians received neither skilled employment nor vocational education from the early Firestone plantations. In general, plantations by their nature neither require nor stimulate mass education of native populations. Compared to traditional farming, according to George Beckford, the plantation has a distinct organization "in which the factors of production, primarily management and labor, are combined. The plantation substitutes supervision—supervisory and administrative skills—for skilled adaptive labour, combining the supervision with labor whose principal skill it is to follow orders."[44] The incentive for plantation owners and operators is to keep the local supply of noneducated labor for the routine, unskilled work as large as possible.[45] Indeed, the expatriate technical workers at the Firestone plantations comprised the bulk of such skilled workers found in Liberia.

From their early years of operation, Firestone plantations employed local males for unskilled labor. Indeed, each year local chiefs provided thousands of unskilled workers to the Firestone plantations with the encouragement and approval of Liberian authorities. But these plantation operations had no beneficial links directly to local skilled labor or communities. In general, the economy of hinterland Liberians revolved around small-scale farming and local markets. Hard work with bare returns was common in these traditional rural communities. Employment on the rubber plantations was attractive to many in local villages. Work at the rubber plantations provided individual cash wages, something the traditional village farming did not. Males from Liberia's rural communities were also frequently available to the plantations because among the traditional peoples women did much of the farming and food distribution. Among the Kru, for example, women customarily cleared bushes and farmed rice, while the men fished or were employed as sailors.[46]

Also, work routines familiar to local populations were used on plantations. "The production of rubber upon a plantation is a phase of industry with which the natives are more or less familiar," wrote two Western scholars studying Liberia in the 1920s. "[I]t is not like going into the mines or into a factory, which involves a violent break in the old life. Primitive agriculture is already the chief means of livelihood for the natives, the people working for the chiefs who largely control their labor."[47]

While Liberia's foreign-owned rubber plantations always found a plentiful source of local workers coerced by domineering chiefs, labor unrest developed recurrently during the 1920s and 1930s. Wages were extremely low, housing inadequate, and disease outbreaks common. President Edwin Barclay made repeated efforts to regulate Firestone's native labor practices. He tried to compel the company to raise wages, shorten hours, and establish a register of workers for Liberia's Department of Interior. Initially, the Firestone Company refused such measures. However, gradually the company improved its labor practices. In addition to keeping wages above subsistence levels, the company greatly expanded its medical facilities and services for employees and their families.[48]

Plantations and Enclave Technology

Several features of the Firestone plantations in Liberia set them apart from the plantations of the Deep South and the U.S.-owned plantations in the nations surrounding Panama. Overall, the Firestone plantation managers did not find it necessary to build ties with regional political elites as had been the case with the agricultural investors of the South's Black Belt or the United Fruit Company's plantations in Colombia. In the latter two regions, the propertied elites themselves—the South's wealthy landowners and Colombia's cattle barons—recruited employees and managed the plantation workforce. Also, supervisory personnel, artisans, and merchants were drawn from nearby local communities or small cities. In short, the entire technology of the southern United States and Central American plantations was in the hands of local, propertied political segments.[49] As such, U.S. and Central American plantations were embedded in the local socioeconomic context.

On the other hand, the Firestone rubber plantations set up in Liberia were largely enclave technologies, transplanted from the United States. They were not managed by local elites or propertied classes. Instead, a quasi-immigrant elite—the Americo-Liberians—served as the conduit to chiefs for tapping local indigenous labor. Indigenous local laborers remained employed mostly in operative jobs requiring little or no education. Skilled workers were brought in from the United States. The way personnel were hired, trained, and retained at Firestone reinforced the ethnic and class divisions that riddled pre-plantation Liberia. Merran Fraenkel, an Africanist who completed extensive fieldwork in Liberia, described the local economic stagnancy caused by the Firestone Company's hiring practices: "[Firestone] was providing wage employment on a scale hitherto unknown in the republic as well as encouraging the development of private rubber farms. But the immediate effect of all this on social relations within the Liberian population was limited by the fact that Firestone's local employees were almost all illiterate tribes-people; clerical, technical and administrative staff were, and to a large extent are, brought in from America. The correspondence between ethnic origin and occupation was therefore little affected."[50]

The commercial plantation system in Liberia and similar national settings in Africa and Asia offered some immediate benefits but many long-term burdens to economic development. Like many colonial and postcolonial African and Asian nations, Liberia was land-abundant but had a labor force of mostly substinence farmers. Liberia also had a national government infrastructure incapable of or unwilling to support modernization of utilities or economic production owned by Liberians. For Liberia and other tropical nations, commercial plantations at least utilize land for producing goods for international markets. In addition, these commercial plants provide the host nation varying degrees of tax revenues and jobs.[51]

But foreign-owned commercial plantations also offer serious drawbacks to national economic modernization. Infrastructure development of weak nations often becomes oriented primarily toward protecting the plantation owner's interest so that rents will continue to flow.[52] Also, commercial plantations like Firestone generally are neither steady nor effective conduits for advanced technology to underdeveloped nations. Instead, enclave technology retards the emergence of domestic manufacturing resources. As Daniel Headrick points out in his study of Western technology and expansion, "Western technology did not reach the tropics in a continuous stream, but in the form of discrete projects: a railway, a plantation, a telegraph network."[53]

The Western-style plantations like Firestone's transported sophisticated design and managerial techniques "lock, stock and barrel" to Liberia's rubber fields—technologies and crops that peasant farming simply cannot cultivate. Also, the Western-style plantations usually required highly skilled personnel at the managerial and supervisory levels. While underdeveloped host countries could provide abundant cheap indigenous labor, these nations could not supply these managerial or skilled personnel.[54]

Educators of black colleges were outspoken about the unfairness inherent in Western-owned and operated plantations in "black and brown" nations. To these thinkers, such as E. Franklin Frazier, Charles Johnson, and Buell Gallagher, no real transfer of skills or science was occurring between U.S. firms and the tropical settings for these firms' plantations. In his study of race and cross-cultural contacts in the modern world, Frazier singled out the commercial plantation as a key obstacle to the healthy development of African, Asian, and Latin American economies. In *Race and Cultural Contacts in the Modern World* (1957), he wrote that in the late nineteenth and early twentieth centuries, the reintroduction of Western settlers and commercial enterprises in the tropical countries had resulted in "[t]he recruitment of 'free' native or colored [plantation] labor in many parts of the world [that] has often concealed a system of forced labor."[55]

As mentioned earlier, the Americo-Liberian government helped with arrangements for Firestone to have generous supplies of local manual workers.

As Frazier pointed out the in the 1920s, "the government of Liberia which was supposed to be representative of the interests of the African natives yielded to the financial inducements offered by the Firestone Company to provide native workers for its plantations." Such assistance often entailed "an element of compulsion."[56]

In fact, throughout the 1920s, the U.S. State Department received and ignored numerous reports concerning the Liberian government's abusive labor practices. These abuses involved coercion of the ethnic peoples living in the nation's interior. By 1930 Liberia's administration was under investigation by the League of Nations. Evidence was emerging that Liberian officials had been arranging the cruel exploitation of the indigenous Liberian Kru and Grebo peoples as conscripted laborers on Spanish-owned cocoa plantations in Fernando Po. This island was located near Cameroon of then Spanish Guiana. The scandal attracted international legal proceedings.[57]

As the Firestone interests in Liberia grew, the supply of skilled workers in the local economy remained largely stagnant. The small cluster of business and commercial sites remained located primarily in Monrovia. They were divided along ethnic, nationality, religious, and linguistic lines. Government offices were staffed by Americo-Liberians. Indigenous ethnic Liberians—the Kru, Vai, and Bassa—were employed mostly as domestic workers, store hands, and ship-loading operatives. A small commercial sector in Monrovia, involving shipping and trading concerns, also developed but mostly in foreign hands. In the 1920s, Lebanese, German, and other foreign nationals set up trading businesses in the city. As for the hinterland, the traditional rural economies based on ethnicity reigned. Among these intricate village- and clan-based trade networks, the Islamic Mandingo was among the most extensive. As for the Firestone plantations, they drew most of their Liberian employees from the ethnic villages in the hinterland. These primarily lower-rung, seasonal workers returned to their homes outside of Monrovia with their earnings.[58]

Government without Substance

Try as they may, Liberia's government officials accomplished little to break down obstructions to building modern farming, industry, or social institutions (hospitals, schools, and the like) for Liberia as a whole. During the interwar period, the United States provided loan assistance to Liberia for building infrastructures; but this did not succeed. The U.S. government dispensed a loan to Liberia in 1926. The loan terms required that about 8 percent of the funds be used for sanitation and public works. However, as late as 1933 projects in these services had yet to be implemented or completed. In 1929 a yellow fever epidemic broke out. It struck both local populations and foreigners hard, killing the American foreign minister as well as the U.S. educational adviser to Liberia. As a result, U.S. authorities appointed a surgeon from the U.S. Public Health Service, Dr. Howard F. Smith, "chief medical adviser to Liberia" to help the country initiate health programs. But Smith found the Liberian officials unco-operative and was recalled.[59]

As for public works, diversion of loan funds for personal use and the in-eptitude of local supervisors stalled or doomed construction projects. U.S. offi-cials reported to the League of Nations in 1933 that one such project involved the construction of a road between Kakata and Monrovia. The road was intended to open export trade from the hinterland via Monrovia to international destina-tions. However, the project fell apart "because of the failure of the Liberian Gov-ernment to provide even the most elementary maintenance." Throughout this impassable roadway lay "road machinery and equipment, reinforcing steel, and unfinished cement work abandoned by the [Liberia] Department of Public Works."[60]

In principle, Liberia would seem to be a natural beachhead for black Americans wishing to develop commercial projects. However, political barriers obstructed these efforts, not just the Liberian government's managerial inepti-tude. Liberia's administration opposed on ideological grounds Marcus Garvey and his organization, the Universal Negro Improvement Association (UNIA). The Garvey group was seeking to make Liberia a key site for the UNIA's ship-ping line. They also planned to locate places on which to build housing for new immigrants. However, Garvey and a commission of delegates from the UNIA ran into resistance from the government in the 1920s. At first Garvey's com-mission attracted an enthusiastic reception in Monrovia. Yet Liberian officials, under pressure from the United States, returned anyone, arriving on incoming steamers, who was affiliated with the Garvey organization. Also, the Liberian government canceled a concession initially promised and deported a group of UNIA technicians. The Liberian government asserted that the aims of the UNIA personnel were to stir up political discord.[61]

In addition to the pressure from U.S. foreign officials who strongly op-posed Garvey's Africa activities, Liberia's leaders criticized Garvey for what they perceived was the clear-cut managerial impracticality of his commercial projects.[62] Liberian officials did not believe economic enterprise could be sim-ply an organization that crowds millions of one race under the umbrella of one mass business or firm. Charles King, Liberia's president, raised this criticism of Garvey's plan at a public meeting: "Take the Marcus Garvey Movement— one need not undertake impossible things. You have heard of companies being organized in Germany (Janzen, West & Company) and in England (Patterson Zokonis Company), and in America (Standard Oil Company), and the like; but you never heard of a company including the entire white race. How can any man unite all the Colored people of the world into one company? Gentlemen, it does not seem plausible."[63] In the end, the Liberian elite feared that Garvey's advances in their country would incorporate working-class elements into projects con-trolled neither by the elite themselves, nor by these officials' U.S. sponsors. Also, the French and British made it clear to Liberian officials that these two governments did not want Liberia to harbor a black nationalist organization—one these European governments perceived could destabilize colonial peace in West Africa.[64]

Discarding any notion of black nationalist commerce, Liberian officials settled for maintaining the status quo of enclave development. However, through the 1920s and 1930s the Firestone Company added little to the government revenues of Liberia. The company's original concession required payment of no income and only 1 percent on the rubber's gross value it exported. The grant also allowed Firestone up to one million acres of land. Each acre had a rental fee of six cents. But by the early 1930s world rubber prices had plummeted. At Firestone's Liberia sites, rubber tapping and production slowed, and Firestone had only about forty thousand acres of its concession under cultivation by 1940.[65]

The Pull of Medical Research

While technical development was at a standstill throughout Liberia, small U.S.-led research projects in medicine and related fields continued. Following the Harvard expedition, other American scholars involved in independent projects attempted further medical and anthropological research in Liberia and the West African region. The prominent anthropologist Melville Herskovitz conducted fieldwork. Also, in 1931, Marshall A. Barber, a researcher from the Rockefeller Foundation's International Health Division, led a major malaria study based at the Firestone plantations. That same year, a team from Yale University completed a survey of Liberia's evergreen forests.[66] In the mid–1920s and early 1930s, Firestone also made a few monetary gifts to individual academics involved in scientific and anthropological research.[67]

From these early investigations, it was becoming increasingly known throughout the West that, for Liberians living both in and outside Monrovia, malaria was the most vexing health problem. Moreover, yaws was also a widespread danger to the nation's population, especially to those living in the hinterland. The Harvard expedition attempted no national estimates for malaria prevalence in Liberia. However, its test of a small number of schoolchildren (i.e., thirty-six Kru) revealed that 87 percent were infected with malaria. The Barber study found similarly high rates throughout a range of Liberians.[68]

Barber's team was concerned primarily with tracking the output of anopheles (mosquitoes) on newly cleared rubber plantations and malaria infection rates among the plantation workers, specifically in the Du Plantations. They also surveyed some groups living nearby. They found a severe situation: "Mt. Barclay plantation, laborers and their families, 52 examined, 71.1 percent positive; a mission school, girls, 39 examined, 89.1 percent positive; Monrovia, various ages, 31 examined, 87.1 percent positive; a native village, Dobon Town, 8 adults and 16 children examined, 100 percent positive."[69]

Along with the malaria problem, the early U.S. surveys of Liberia were revealing to medical professions in America and Europe that yaws was another critical health problem for Liberia. Like malaria, yaws was endemic to the Liberia region and ravaged in particular its children and youth. Before the Harvard expedition, heavy concentrations of yaws victims in Liberia apparently had not been seen by the few Western travelers and visiting doctors.[70] The expedition found yaws widespread, especially among children. Perhaps the first American study specifically of yaws in Liberia was in 1931 by James Knott at the Firestone plantations. Based on his observations, Knott wrote in a tropical medicine journal, "Yaws is the most prevalent disease in Liberia and causes more disability than any other. I feel that yaws is mainly responsible for the stunting of growth, the premature aging, and the early death of these people."[71]

As we observed in our study of U.S. medical research in occupied Haiti (chapter 3), throughout the 1930s yaws was a burning issue in American and European tropical disease research. The puzzle of distinguishing syphilis from yaws, was especially controversial. But the published research on this subject flowed mostly from clinics and field reports in Haiti, Jamaica, and the Philippines.[72] Liberia became an African setting that broadened the research community's study of yaws. In 1933 Dr. George W. Harley, a tropical disease specialist from Columbia University, visited the Ganta Dispensary in northeast Liberia to review diagnoses at this site. When they treated a few cases of yaws successfully with the standard bismuth injections, local people "came from far and near, increasing in numbers as the fame of the 'needle' spread over the country."[73]

Harley surveyed 6,291 cases at Ganta and discovered that 5,597 cases (or 84 percent) were infected with yaws! Ninety-eight percent of the children brought to the clinic had secondary yaws. The physical facilities were so poor in Ganta that serological tests were impossible. However, Harley examined many cases with obvious lesions, so many that he developed a formal symptomatology. He collected systematically visual evidence, age of the patient, and information gleaned from interviews of older patients about whether they had had yaws symptoms in childhood. Harley's research in Ganta and his classification of the clinical types of yaws symptoms were published in leading tropical medicine journals in both the United States and England.[74]

In 1935 the Health Organization of the League of Nations sent a tropical disease specialist, Dr. Ludwik Anigstein of Poland, to Liberia. As medical adviser to the Liberian government, his objective was to assess the health conditions of the estimated 1.5 million Liberians in the hinterland. He surveyed villages along a stretch from Monrovia inland to Panto.[75] Anigstein found that malaria, yaws, smallpox, and leprosy were the most destructive health conditions in this region. Since yaws lesions frequently took away the victims' capacity to walk, the prime mode of travel in Liberia, it damaged the victims' ability to contribute to the household or village economy. "Apart from the effect of yaws on the general health," Anigstein wrote, "[a]s there are no other means of transportation and communication in the Interior than walking the fitness of the feet is a matter of great importance strictly bound up with the economic development of the country."[76]

Anigstein described the serious lack of physicians throughout Liberia: sixteen doctors, of whom three practiced in Monrovia and eleven were Europeans. As for hospitals, only one, the fifty-bed company facility operated by Firestone, was modernly equipped. The others were small facilities run by mission sisters from European counties working with African assistants.[77]

As a result of his survey, Anigstein recommended building large dispensaries staffed by qualified physicians throughout Liberia's hinterland districts. Also, in the nation's central district, a special "camp for all the lepers" should be organized. Mobile, rural health units were also needed. These traveling medical teams should be specially equipped to reach the more isolated regions of the country. Malaria and yaws should receive special focus in all of these proposed facilities. Anigstein suggested that the yaws program in Haiti be emulated in Liberia. He pointed out that in Haiti the "mass treatment" campaign run by the United States was reaching approximately four hundred thousand people annually.[78]

Finally, Anigstein urged that local traditional leaders be recruited to support the recommended health care initiatives. "Systematic propaganda among the chiefs of the towns and villages is needed in order to stimulate them to spread knowledge," Anigstein advised. He envisioned that the chiefs would work "through nurses as regards proper feeding and keeping of infants, and also to persuade the population to not resist smallpox vaccination."[79] Unfortunately, neither the League of Nations, the United States, nor the disheveled Liberian government carried out Anigstein's recommendations. Nonetheless, these surveys in the 1920s and 1930s had slowly widened consciousness throughout the tropical medicine community of the West about the diseases and sanitation problems of rural Liberia.[80] These academic projects were small but still helped prompt U.S. academic, political, and corporate circles to formally explore, as well as exploit, the natural resources and, later, the tropical medical and strategic military potential of Liberia.

The Failure in Technology Transfer

During the interwar decades, awareness of Liberia's severe health, education, and economic needs spread throughout the United States and Western nations. Yet only a few tropical health or education programs from the West or the United States trickled into Liberia. Religious missions, academics, or government agencies had sponsored these programs. Moreover, with the exception of the Firestone enterprises, U.S. commercial projects of substantial scale did not emerge in Liberia before the start of World War II.

Unlike the other three regions of this study, Liberia lacked the government resources to facilitate, either for the Liberian public or for the national elite's private reward, large engineering projects for U.S. interests. In the case of building the Panama Canal, U.S. federal government support was centralized and direct. Indeed, with the creation of the Canal Zone, the U.S. federal government had placed itself as the Zone's exclusive, hands-on administrative authority. With total control over the canal project, the United States was not compelled to share any of this authority with the governments of Panama or other nearby nations. In Haiti, the United States in effect suspended and acted in the place of the Haitian government. Having unplugged Haiti's political rulers, the United States moved full-steam ahead with its program of military-led public works projects, public health campaigns, and medical research.

The link between technology and government in the Black Belt of the U.S. South had also been fused. The South's primary technology framework—plantations—had evolved gradually, first as institutions of legalized slavery and now, in the early twentieth century, as an agricultural system of large share-cropping farms, supplemented by textile manufacturing. All levels of government protected the legal authority of plantation owners. The subservience of the black share-croppers in the plantation system, including their bare wages and living conditions, were justified by both local and regional law as well as a social or customary racial caste system.[81]

In Liberia, the nation-state apparatus was a legal mechanism reinforcing the rule of the Americo-Liberian elite. Since Liberia was a formal democracy, the United States would not install U.S. political authority over the Liberian government the same way it had been installed in Panama or Haiti. Still, the Liberian ruling elite traditionally refused to try to organize on a national scale economic, industrial, or public institutions. It had neither interest nor skills to administer national technical projects or social welfare institutions. These endeavors would require strong resources from large populations residing outside of Monrovia. Indeed, one of the elite's primary efforts was to manipulate the wealth from the local economies normally controlled by traditional rulers of Liberia's large and distinct ethnic populations.

As in many areas of colonial Africa under direct European colonial rule, in Liberia the elite were mostly government heads, clerks, lawyers, business operators, skilled workers, and civil servants. According to Chinweizu, a leading writer on colonial Africa, a nation's elite "could represent itself to the populace as products and agents of progress." In Liberia, the Americo-Liberian elite had decisive political advantage over indigenous leaders. The urban elite in Monrovia were not interested in promoting education, public health, and technical development beyond the needs of their own social strata. Instead they became "an indispensable link" between Western interests "and the human and material resources of the colonies." As for the traditional leaders, Chinweizu points out that "the traditional rulers were no longer in absolute control of African productivity [since in turn] they had lost their former control of the African economy [and] become marginal to the new political order."[82]

Given its elitist character, then, it is not surprising that the Liberian government provided no protection or support for the majority of people or lands of Liberia from Western interests.[83] Land concessions to Firestone had been most permissive. Taxation for the enterprise was so small as to amount to a tiny contribution to Liberia's public revenue—revenue which in theory could pay for government agencies, public health services, schools, roads, hospitals, and the like. The seeds for chronic civil and economic disorder, similar to the disorder that gripped Haiti before and after the occupation, had been planted forebodingly deep.

As for technology transfer, the Firestone Company bypassed working with the weak Liberian government and put into motion its own island of modern technical production and living facilities. To support the production and transport of its rubber plantations, the Firestone Company established technical services exclusively for its plantation personnel. Like the U.S. authorities in charge of the skilled and professional workers in the Panama Canal Zone, the company built public works, a large hospital, and an extensive medical program so that its personnel and the production projects could function self-contained from the larger indigenous societies.

On the Periphery

In the Panama Canal Zone and Haiti, U.S. technologies had been transplanted directly by U.S. managers and technicians, and within a short span of time. U.S. personnel installed transportation and communication projects as well as public health measures in local communities throughout these regions. By contrast, compared to Panama and Haiti, the U.S. technical link with Liberia remained thin. Until World War II, U.S. scientific expertise only occasionally reached Liberia. When the United States entered World War II, Liberia found itself within the Allies' military sphere of influence. Only at this juncture did Liberia become the target of U.S. technical assistance, including specific projects involving construction of transportation facilities as well as public health programs beyond tropical disease research.

As for disease problems in Liberia, we have seen that, as a result of the Spanish-American War and throughout the Panama Canal project, tropical medicine became an increasingly popular specialty in U.S. military and academic medicine. Panama Canal officials were bent on preventing malaria and other similar diseases from debilitating canal workers. The public health and sanitation campaigns in Panama and Haiti had been designed to be up and functioning within very short time frames. In the Canal Zone, the public health campaign began in 1904 and churned forward through the 1910s. Zone officials provided resources ample for the renovations in housing and work sites as well as for the vigilant public health surveillance that malaria control required. In Haiti, the U.S. public health movement also occurred over a relatively short period of time. It began around 1917 and gained steam throughout the 1920s.

This was not the case for Liberia. Initiatives in health or scientific areas emerged from the United States only sporadically, if at all, and were never intended to reach large populations of Liberians. With no massive investment and construction project underway, neither U.S. foreign assistance nor Liberia's meager revenues were adequate for massive public health campaigns. Tropical medical experts did investigate certain diseases rampant in Liberia—such as malaria and the dysenteries. But these investigations were more in response to America's fear for the health of its military personnel stationed in tropical regions, as well as for the American public, than for Liberian health. Malaria was already a major public health problem in the U.S. South. The U.S. public's health could be jeopardized even more by American soldiers returning home from abroad infected with contagious diseases.

Moreover, back in the United States, researchers in the public health community had become most interested in developing preventive drugs and treatments for the individual threatened by tropical diseases. They were much less concerned with the drudging work of mobilizing mass community groups and materials for altering environmental conditions.[84] Inside the United States as a whole, the entire scale of science and technology throughout its government, health institutions, and business sector was growing explosively. Led by Franklin Roosevelt's New Deal administration, Americans built titanic public works projects with "big government" agencies in employment, welfare, and industrial planning leading the way. In the meantime, Liberia would have to wait for the outbreak of World War II before U.S. military, public health, and engineering projects would come rushing over its shores.

FIGURE 1. Workers on the Panama Canal (1913) from five nations. *Left to right,* two men from Turkey, *followed by* men from Trinidad, Jamaica, Barbados, and Italy. *Reprinted from H. A. Franck,* Zone Policeman 88: A Close Range Study of the Panama Canal and Its Workers *(New York, 1913).*

FIGURE 2. Workers and Panamanians outside market building. Panama Canal Zone, 1913. *Source: E. H. Sibert Collection, United States Army Military History Institute, Photograph Archives, Carlisle, Pennsylvania.*

FIGURE 3. Construction workers erecting the Gatun Locks, Panama Canal, 1911. *Source: E. H. Sibert Collection, United States Army Military History Institute, Photograph Archives, Carlisle, Pennsylvania.*

FIGURE 4. Traditional Panamanians houses, 1907. *Source: E. H. Sibert Collection, United States Army Military History Institute, Photograph Archives, Carlisle, Pennsylvania.*

FIGURE 5. William Crawford Gorgas, chief sanitary officer of the Panama Canal Zone, ca 1914. *Source: Reproduced from W. L. Pepperman,* Who Built the Panama Canal? *(New York: E. P. Dutton, 1915).*

FIGURE 6. Worker oiling ditches to kill mosquito larva in Miraflores, 1910. *Reprinted from Joseph A. LePrince, A. J. Ornstein, and L. O. Howard,* Mosquito Control in Panama: The Eradication of Malaria and Yellow Fever in Cuba and Panama *(New York: Putnam's, 1916).*

FIGURE 7. C. C. Boone, M.D., among Kru People in Liberia, ca. 1905. *Reprinted from Clinton C. Boone,* Liberia as I Know It *(Richmond, Va., 1929).*

FIGURE 8. Mrs. Rachel Boone, teacher in Liberia, ca. 1905. *Reprinted from Clinton C. Boone,* Liberia as I Know It *(Richmond, Va., 1929).*

FIGURE 9. The Boone children, from C. C. Boone's memoir. *Reprinted from Clinton C. Boone,* Liberia as I Know It *(Richmond, Va., 1929).*

FIGURE 10. Highway construction in Liberia financed by the United States Technical Co-operation Program, 1960. *Reprinted from U.S. Department of State, Mutual Security Agency,* Report to Congress on the Mutual Security Program, Fiscal Year 1960: A Summary Presentation *(Washington, D.C.: GPO, 1961), 97.*

FIGURE 11. William S. V. Tubman, Prime Minister of Liberia, with President John F. Kennedy. *Courtesy of Audiovisual Archives, Photograph Collection, John Fitzgerald Kennedy Library, Boston, Mass.*

FIGURE 12. Medical assistance, UNESCO Haiti Pilot Project, Marbial Valley, 1949. *Reprinted from UNESCO,* The Haiti Pilot Project Phase One, 1947–1949, *pamphlet published by UNESCO in 1951.*

FIGURE 13. School, UNESCO Haiti Pilot Project, Marbial Valley, 1949. *Reprinted from UNESCO,* The Haiti Pilot Project Phase One, 1947–1949, *pamphlet published by UNESCO in 1951.*

PART II

Encountering the Science Superpower

Science is an ongoing process. It never ends.
—Carl Sagan

CHAPTER 5

Malaria and Modernization

THE DECLINE OF THE BLACK BELT

The triumph of building the Panama Canal had enamored U.S. leaders even more to the ideal of expanding industry, science, and medical research. After the canal opened, the nation's economy benefited from the interweaving of industry and transportation throughout the northern and western states with larger markets of consumers and raw materials abroad. Commercial power in the Western Hemisphere tilted significantly toward the United States. Industrial growth, ushered forward by giant companies throughout the United States, picked up dramatically, especially after World War I. Even in the still largely agricultural South, once the canal was completed, the pace of economic production churned faster.[1]

By the mid–1930s U.S. tropical researchers or government agencies had sent substantial technical projects into both of the Atlantic's black republics—Haiti on the Atlantic Ocean's western side and Liberia on its eastern side. Knowledge about Haiti's and Liberia's environments, natural resources, and health conditions was flowing back to U.S. investors and medical researchers. Also, huge supplies of rubber from Liberia began reaching the shores of the United States. But broader waves of technical transfers from the United States to these nations would have to wait for two antecedent developments. First, technical modernization and scientific research inside the United States would have to increase greatly to propagate the New Deal and the United States' successful military involvement in World War II. Second, the international role of the U.S. military and government assistance programs would have to expand as a result of the war and the global competition with the Soviet bloc.

In 1913 W.E.B Du Bois predicted, "The Southern South will be the greatest sphere of capitalist development in the United States after the opening of the Panama Canal."[2] Once the canal began operating, the South's economic dependence on plantation crops and textiles strengthened. Japan was now able to receive much larger imports of the South's raw cotton. The canal also widened access to both Atlantic and Pacific coastal markets for the South's manufactured products. Using sea-lanes opened by the canal, shipping centered in New Orleans carried southern lumber, coal, and foods. These export products reached new markets along the western coast of Central and South America. The South's major iron and steel centers—Birmingham and Chattanooga—also gained an advantage for selling their output to American locations. The canal made these industrial sites in the South closer via ships to numerous markets located along the Pacific Ocean, compared to the iron and steel centers of the Northeast and Midwest.[3]

As the South moved through the interwar decades and was pushed forward by New Deal programs and public works construction, the low technology of plantation farming gradually gave way to higher technology of machinery-based farming. Less perceptible to public and political observers of the interwar South were the growing pursuit and impact of the basic sciences and applied research. The scientific activities that were changing production techniques and products included plant biology, which accelerated plant breeding and wider use of experiment stations in agriculture. Also, the growing field of biochemistry resulted in new laboratory techniques, pharmaceuticals, and vitamins, improving public health, medicine, and nutrition strategies. Finally, laboratory and industrial chemical investigation produced fertilizers as well as new types of energy and materials production. In these three spheres—agriculture, medicine, and industry—a heightening of the mission of science was underway nationwide.

The expansion of science throughout national institutions was readying the United States for its role as a global superpower. In the late nineteenth and early twentieth centuries, industrial innovation had emerged gradually and more or less spontaneously, with trial, error, and commercial usefulness as its guide. However, by the outbreak of World War II, organized teams of research scientists were moving to the forefront of steering technologies in manufacturing and energy production as well as military invention and agribusiness.

Federal public health agencies and major foundations, excited by breakthroughs in the laboratory sciences and bacteriology, implemented large research projects on communicable diseases, especially the malaria crisis and venereal disease in the South. The nation's new public health approaches and public works projects, usually first planned and implemented by federal agencies, gradually influenced health conditions in the municipalities and counties throughout the South. Public health responsibilities, as well as improvements in public works and sanitation, gradually were taken on by southern locales. These local authorities had to accommodate the region's diversifying crops and growing industries. As the South's economic fabric changed, its population grew, adding multitudes of new manufacturing and service workers and city dwellers.

During the interwar decades, as the drive to modernize farm technology and industry gradually strengthened in the South, the demography and, more slowly, disease patterns of the Black Belt were reshaped. With the nation's consumer sector growing, the South's financial and commercial leaders became more focused on meeting new demands of this national market. In plantation economies in the Americas, agricultural mechanization was growing hand in hand with commercialization.[4] The framework of technologies that held the Black Belt economy together was gradually shifting from traditional plantations and textiles to mechanized agriculture and mixed industry. The newer technologies included light manufacturing, design and construction of factories and warehouses, printing, and paper production.

From the 1910s through World War II, the Black Belt received little of the South's new resources for community public health education or modern housing. The classic infectious diseases such as malaria, tuberculosis, and venereal disease maintained their grip on the Black Belt region. Medical scholars and ethicists are still weighing whether the notorious Tuskegee Syphilis Study was an aberration or typical of federal research involving the South's black communities. In any case, through the World War II decade, the Black Belt South had many health problems common to large population segments in the republics of Haiti and Liberia. But unlike these two foreign nations, the Black Belt community managed to improve its own educated segment, health resources, and community institutions. These health care, educational, and business professionals, in combination with therapeutic churches and music and cultural institutions, ensured not only recovery from persistent epidemics, but also broader social stability.

Some of the South's blacks managed to gain jobs at the lower rungs of the region's industries. However, increasingly, southern blacks journeyed to the North and Midwest for jobs within industry itself. Gradually, the plantation Black Belt emptied of much of its black population. This loosened population was swept into the nation's whirlpool of technologies erupting outside the Black Belt. The South's Black Belt was eroding under the weight of U.S. technological advance.

The Plantation Survives

Du Bois's prediction that following the opening of the Panama Canal the U.S. southern states would become a major target of U.S. capitalist investment would indeed come to pass, but slowly over the course of several decades. From the 1900s to the outbreak of World War II, plantation agriculture continued to dominate the South's Black Belt, as it was doing in other mainly tropical regions around the world, especially Egypt, Uganda and Tanzania in Africa, the West Indies, and India.[5] In fact the Panama Canal contributed new vigor and staying power to the South's plantation economy. The canal helped improve the flow of plantation produce from the U.S. South to national and international markets of new urban consumers. These markets demanded more raw goods and commodities such as apparel, furniture, fruits, and tobacco for middle-class tastes.

The governments of Western industrial countries also favored plantation ownership and production in the tropical nations. In the less-developed, agrarian regions both an agricultural and a manufacturing segment were the key ingredients needed for developing a substantially autonomous national economy. Thus, at some point the construction of large technical systems such as railroads, modern ports, and electrification was necessary. However, enormous undertakings to build these vital technical infrastructures did not interest large investors of the United States and Europe. To Western governments and large investors that had not developed formal imperial empires, as had nations like Britain, France, and Belgium, such industrial development in the tropical nations was too costly and required broad social intervention.

Installing large-system technologies requires huge financial investments to plan and build the project, educated and skilled workers, an advanced engineering sector, an ample public works infrastructure, and health care and housing resources for the expatriate and indigenous labor personnel for these technologies.[6] By contrast, for produce such as sugar, cotton, rubber, palm oil, and rice, the reuse of long-standing plantations or the establishment of new ones proved the most convenient and immediately profitable agriculture for investors throughout parts of Africa, Asia, and Latin America. Most labor for the plantations did not need to be literate or skilled. Moreover, investors could focus their plantations on producing and shipping one type of crop to Western markets, a task much less risky and cheaper than endeavoring to grow, process, and ship several different crops.[7]

The plantation farm method persisted in the Black Belt at the same time farming in the nation's other regions was mechanizing for several reasons. First, the growth and harvesting of cotton required hand work still not possible to mechanize and still most cheaply done by tenant farmers. As of 1900, two-thirds of all the South's farmers involved in cotton were tenants.[8] Moreover, as long as the South's commercial elite and outside investors managed to maintain control over local and state government to reinforce the supply of cheap sources of labor, the plantation prevailed. Throughout the South, locales still used crop lien laws to undergird the labor supply for the share-cropping system. Local wealthy landowners profited from commercial interests from outside the South financing the cotton sector and, thereby, giving the large plantation new resilience. Even when southern cotton prices went down, banks and, later, government agencies joined wholesale merchants as buyers of crops or underwriters offering cash incentives to cut back production.

In the words of Rupert Vance, a leading southern economist writing in 1929, in the South's cotton belt "the plantation cannot diversify; it never diversified but once and then under the stress of war [i.e., the Civil War]." In fact, the South's cotton plantations contributed to the Great Depression in the southern region. The plantation style of production had facilitated the growing, processing, shipping, and selling of cotton so efficiently, according to Vance, that "it produce[d] too much cotton," constantly depressing cotton prices."[9]

Worldwide, international market forces were also driving the price of cotton down. Unlike in the mid-nineteenth century, when southern cotton production controlled much of the international market, by the 1930s the South had to compete with numerous cotton-growing nations such as Egypt, Brazil, Uganda, and India. However, by keeping production costs as low as possible, the South's agrarian economy, especially its cotton and textile production, still managed to make profitable links to domestic and international markets.[10] In the meantime, no other technologies in the economy of the South—for example, factories, shipbuilding, railroads—could yet match the profitability of the plantation (see diagram 5.1).

During the 1920s heavy industries generated by enormous corporations came of age in the United States. The assembly-line factory first introduced by Henry Ford just before the Great War was now widespread. But as in the previous decades, large-scale technologies developed much more slowly throughout the agricultural South and Midwest. According to Rexford Tugwell, a leading economist writing in 1927, U.S. agriculture in the 1920s remained largely "a backward art, if it is compared, as a whole, with manufacturing technique."[11]

Mechanization split the two primary methods of southern agriculture. On the one hand, there was the traditional, labor-intensive farming of the large plantations or interconnected small parcels. On the other hand, there were large owner-operated farms that used automated equipment, especially tractors and combines, as well as business management techniques. In both types of large-scale farm systems, tenant farmers were key, although fewer tenants and higher

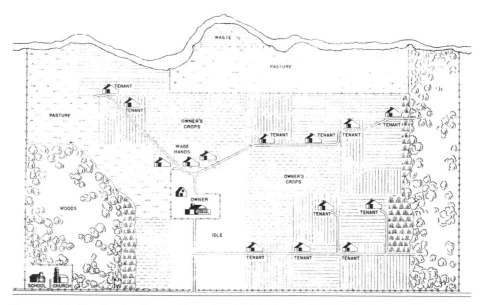

DIAGRAM 5.1 The Average Cotton Plantation (1934). Source: T.J. Woofter, Jr., *Landlord and Tenant on the Cotton Plantation* (1936).

output per acre characterized the modernly equipped farms. Ironically, the lowly black tenants frequently received much of the blame from southern whites for the inefficiency and soil erosion that occurred on southern cotton plantations. One southern governor criticized the backwardness and wastefulness of black tenant farmers. Evidencing a warped perception of the contextual relations that held together the South's plantations, he said, "The Negro skins the land and the landlord skins the Negro."[12]

In industrial manufacturing, the South lagged far behind other regions. Large-scale steel and manufacturing plants, power utilities, machine shops and foundries, and transportation equipment were at the center of the industrial growth of the urban North. By contrast, in the South industrialization emerged much more slowly. By 1925 the Northeast contained 44 percent of the nation's manufacturing establishments, the middle states 29 percent, but the Southeast just 13 percent. The major industries of the South—textiles, forestry goods, and furniture manufacturing—were all extensions of farming or lumber. As for metal-related industry, the South's small numbers of steel, copper, and machine-shop establishments amounted to less than 5 percent of the national total in these three categories. The regional power industry and the management of large-scale natural resources were reaching the South only as a result of federal projects, especially the Tennessee River Valley power complex and water-control projects along the Mississippi River. Furthermore, federal authorities largely inspired the sanitation improvements along the Gulf coast.[13]

Overall, southern blacks were kept outside the skilled and managerial segments of the South's newly expanding industries. During the 1930s blacks comprised only about 5 percent of the South's 383,000 textile workers. More than one-half of the nation's laborers in oil and gas wells were in the South. Yet blacks were only about 1 percent of these 118,000 workers. Similarly, black workers made up only a tiny fraction of the South's telephone and telegraph industry's employees. The region had 106,000 such workers, only 2.2 percent of whom were black. Blacks in the South employed outside of agriculture were heavily concentrated in construction, road maintenance, and fertilizer factories.[14]

In the interwar U.S. South, plantations produced enough profits to hold together the region's socioeconomic pyramid. On the one hand, the top layer of the southern economy—business owners, manufacturers, large plantation owners, and trading and shipping firms in the Gulf port cities—steadily prospered. On the other hand, the South's blacks, along with white share-tenants, remained locked into the lowest segments of the South's plantation society. Black Belt communities were held together socially and psychologically by networks of churches, common schools, black colleges, and social organizations. Small networks of black professionals and entrepreneurs developed within the cities and large towns in or near the Black Belt.

By sheer individual talent and perseverance, George Washington Carver continued at Tuskegee Institute through the 1920s and 1930s, developing numerous innovations for farm produce. He became a symbol throughout the South and the nation as a humble, ingenious scientist who advocated that the agrarian South and especially the black small-scale farmers had great economic potential. Carver was a deeply religious person and throughout his career regularly discussed his faith in his scientific activities. "My discoveries come like a divine revelation from God," he once told an audience of church workers. "The idea and the method of working out a new product come all together." In Carver, the South's business class and the North's philanthropists had "God's ebony scientist." To the South and nation's leaders, Carver was a fabulous "emissary of goodwill" who also held the idea that the workings of modern science and religious revelation were one and the same.[15]

But despite their self-reliance and social viability, most of the South's black population were racked by poverty, illness, and early death associated with plantation life. As Leon Litwack describes in his monumental history of the black South during the Jim Crow era, "Excluded from the white world, black Southerners drew inward and constructed their own society, with its own institutions and separate social and cultural life." But as the years passed and World War I approached, "few of them could maintain a faith in progress, trapped as they were in a web of controls that encouraged neither initiative nor hope."[16] Litwack identifies these controls as Jim Crow laws and social practices. But the more fundamental, less visible control was the plantation technology that created and was holding in place the Black Belt subsociety.

Tenants, Prisoners, and Disease

In the early decades of the twentieth century, the South's dependence on plantations and the manufacture of textiles and lumber products was nourished by the region's plentiful supply of cheap labor. Local and state officials willingly located and furnished much of this labor. Black tenant farmers, supplemented by various types of coerced and convict labor, provided the basic workforce for the plantation system. Southern states and localities allowed the widespread practice of contract peonage. These labor agreements permitted employers to purchase a laborer and then required this person to work until debts were paid. Newspapers and magazines with national audiences frequently ran exposés about the practice of peonage throughout the South. A 1903 issue of *Outlook* magazine, for example, covered a federal grand jury investigation of hundreds of indictments in Alabama. It disclosed brutal practices to discipline workers. On whippings, one manager colluding with a local constable admitted:

> We have to do it once in a while. A Negro ran away from me and hid on the next plantation, eleven miles away. I went after him with my Negro foremen. I took him out of the cabin with a revolver in my hand and drove him home. There I took it out of him with a buggy whip, while the Negro foremen held him. That sounds very shocking to you,

no doubt, but I am telling you the facts. If you were the only responsible white man on a plantation and were surrounded by more than five hundred Negroes of the most debased and ignorant character, who cannot be reached by any moral suasion, who are influenced by neither gratitude nor resentment, you would go to the field every day with a revolver in your pocket, as every one of us planters is forced to do, and you would either maintain discipline in the only way the Negro understands it, or else you would give up your plantation to your creditors or your executors, as the case may be.[17]

Peonage usually was practiced on plantations in the most remote settings, in the absence of any government agencies with field personnel. The federal workers could raise questions about illegal treatment of tenants. Thus, it was an oppressive machinery for the unschooled tenants who became ensnared in it. Indeed, peonage existed in the South through the 1930s. Often local laws allowed the planter to lock the employee in a stockade at night or to inflict any punishment viewed suitable. Other provisions charged the employee for costs incurred in the pursuit of the employee if he or she fled the work site. As for employees with families, Carter Woodson remarked, "so much the better for the employer; they must live out of the commissary and if the laborer runs away his family are detained at the camp."[18]

Southern locales also used harsh criminal and vagrancy laws more frequently on blacks than whites. These laws made the police, in effect, labor agents at the disposal of local employers and political bosses. Even Booker T. Washington, the most popular black apologist for the South's political and racial caste, found the primitive penal system somewhat troubling. In the autumn of 1912 at a rural fair in Alabama, Washington spoke of the many good aspects of life for blacks in the Black Belt. However, he also commented that the "policy of sending ignorant and poor people to jail, to the penitentiary or to the coal mine will not remove crime. We have got to go deeper and remove the cause of crime which is, in the majority of cases, ignorance and poverty."[19] Gunnar Myrdal found much evidence of "how the police and the courts were utilized to recruit forced labor." He described the process: "Convicts were hired out, sometimes in chaingangs, to planters, mine owners, road contractors and turpentine farmers. There were plantations and other enterprises that depended almost entirely on convict labor."[20]

Indeed, Myrdal described the South's prisons during the 1920s and 1930s as a blight on America recognized by the international political community. Except for the federal penitentiaries and the few new state ones, Myrdal wrote that "[t]here is no doubt that the average Southern prison is likely to make hardened criminals of all who fall into its clutches." There was irrefutable evidence of widespread flogging and physical torture, "for insubordination of any kind," by guards, themselves low-paid, poorly educated whites. Myrdal called these prisons places "where the surroundings are dirty and the food abominable, where there is a tradition of callousness and brutality, where there is not the slightest attempt to reform but only to punish and get work out of the prisoners." Since the penal systems were racially segregated, state officials purchased less food and equipment for black prisoners. Also many southern states had no juvenile reformatories for black youth offenders "as they do for white juvenile offenders, and the Negro youth must live with the hardened older criminal."[21]

The availability of cheap labor was also a primary reason textiles, lumber, and mining became the foundation of the South's slowly growing industrial sector. All these industries required large pools of unskilled, cheap workers. Southern white workers who filtered into the mills were usually "working poor," with a standard of living just above poverty. In 1924 Frank Tannenbaum wrote a field report on the social problems of the South for the Children's Bureau. To Tannenbaum, the central problem facing southern poor whites was the textile mill towns cropping up throughout the region. Former tenant or small farmers and mountain populations became mired in unhealthy, non-unionized mill jobs and inadequate housing and schooling. "You may go where you will in the cotton-mill section of the South," he wrote, and the ordinary folk will tell you the same thing: "once a mill-worker, always a mill-worker. Not only you, but your children and the children's children forever and ever." Tannenbaum observed, "By the thousands they have become drifters [who] come and go from mill village to mill village, but who never go outside of them." Tannenbaum emphasized, in his exposé, "I am discussing the burying of a complete section of the Southern community in the mill village, and the resulting loss to the life of the South."[22]

The impoverished work and living conditions facing the black and white plantation and mill workers of the Black Belt region fueled recurring waves of health problems. These populations suffered from high levels of infectious diseases, endemic to warm environments like the South, as well as nutritional diseases linked to inadequate diets. In 1912 the U.S. Public Health Service estimated that there were about one million cases of malaria annually in twelve southern states that had a total population of twenty-five million. The incident rate was believed to be as high as 40.9 percent.[23] The death rate for blacks in the South was 48 percent higher than that of whites in 1924, and 62.5 percent higher in 1925. In 1930 it was estimated that the doctor-to-population ratio in the South was one to thirty-three hundred for blacks, but one to five hundred for whites.[24]

The major avenues for health care in the Black Belt were the solo doctor practices and small community facilities or county programs staffed by black health care professionals (described in chapter 1). Also, a vast network of indigenous midwives practiced throughout local communities. In pre–1915 Haiti, as in European nations before their revolutions for bourgeois democracy in the mid- and late nineteenth century, the nation-state apparatus did not provide organized health services. The private medical profession tended to exercise political leadership. Inside the southern Black Belt region, a similar vacuum in state-sponsored services persisted. The state or local government apparatus provided or sponsored very few health services. Consequently, local black doctors, midwives, community nurses, and folk healers provided the sorely needed health care infrastructure. Since physicians in the black South were scarce, religious-oriented folk healers as well as midwives were even more important than the physicians who rarely were available.

One Children's Bureau study of deliveries in rural Mississippi revealed that in 1921 only 8 percent of the black infants but 79 percent of the white infants were delivered by doctors. As for prenatal care, only 12 percent of the black women, compared to 33 percent of the white women, had such services. In the mid–1930s in Georgia, there were more than thirty-three hundred midwives, most of them black, delivering babies for blacks and frequently "poor whites." In the two Georgia counties we mentioned earlier, considered typical of other Black Belt counties—Greene with thirty-two midwives and Macon with twenty-six—midwives delivered more than half of all newborns in 1936. In rural areas of most of the Black Belt states, 80 to 90 percent of the deliveries of black infants were performed by midwives.[25]

Labor in Search of Education

Limiting educational resources for the tenant farmer population assured the availability of this labor supply for the plantations in the Black Belt. Black tenant populations in particular received little public education, and illiteracy rates for older children and adults were high. An exhaustive analysis of the social and economic conditions in thirteen southern states for the early 1930s was conducted by the Council on Rural Education.[26] The council found public monies for education in much of the rural South "extremely limited" due to the region's general poverty. Moreover, there was a general unwillingness to raise taxes. Schools for blacks were at the bottom of the funding scale. According to the report, "[w]ith the funds available most counties have managed to maintain adequate educational facilities for white children, but this is done by decreasing standards in Negro schools practically to the vanishing point."[27]

Works Progress Administration (WPA) investigators in the cotton plantations regions found "[t]hat the Negro schools compare unfavorably with the white is common knowledge." Federal data revealed that public school expenditures for white pupils in Deep South states such as Louisiana, Mississippi, and Florida were six and seven times higher than expenditures for black pupils. Also, children of black tenant farmers attended school far fewer days than did non-tenant children. In states such as South Carolina, Mississippi, Georgia, Alabama, and Louisiana, where blacks comprised from 35 to 50 percent of the state populations, blacks received only about one-tenth of the state funds for public education.[28] (see table 5.1).

The social scientists investigating the plantation communities also emphasized that the aversion to committing public resources to close the gap between the education of black tenants and southern whites was widespread. According to one group of prominent researchers, "[m]any landlords and some members

TABLE 5.1   *Black Population and State Education Funds in Eleven Southern States, 1920 and 1930*

State	Black population by state (%)		% of state funds for education received by blacks
	All ages, 1920	*All ages, 1930*	
South Carolina	51.4	45.6	10.7
Mississippi	52.2	50.2	10.5
Georgia	41.7	36.8	13.3
Louisiana	38.9	36.9	9.9
Alabama	38.4	35.7	8.4
Florida	34.0	29.4	7.9
North Carolina	29.8	29.0	12.1
Virginia	29.9	26.8	11.1
Arkansas	27.0	25.8	16.0
Tennessee	19.3	18.3	11.9
Maryland	16.9	16.9	9.7

Source: D. A. Lane, Jr., "The Report of the National Advisory Committee on Education and the Problem of Negro Education," *Journal of Negro Education* 1 (1932): 14.

of the county school boards . . . opposed . . . the development of more efficient schools for rural colored children. They expressed the fear that the providing of even a thorough grammar-school education for colored students would make them unwilling to remain on the plantation and would end by depleting the supply of workers."[29] By severely limiting public education, health promotion requiring supportive school resources and literacy was more scarce in black tenant communities. Plantation historians have established that for the farm workers, in general, school attendance and literacy are inversely related to mortality rates. In the Black Belt South, lower literacy rates and less formal education resulted in higher death rates among blacks, from a variety of health conditions. For example, lack of education for women in the black tenant population was closely associated with stillbirths and miscarriages.[30]

It was only as a result of the efforts of northern philanthropies, and especially the Julius Rosenwald Fund, that large sections of the black community throughout the Deep South managed to obtain some formal education. By 1929, 567,000 black children of school age were attending the 4,464 schools the Rosenwald Fund had opened for rural blacks. This amounted to 26.4 percent of the South's school-aged, rural black children. Investigators estimated that about this time, one-half of the children of black tenant families were literate.[31]

The poorly funded public schools plus the better-run, more effective schools financed by the Rosenwald Fund still were not adequate for those older black youth and adults seeking more opportunities for formal education and training for modern work. The South's overall poor education system no doubt contributed to restlessness among the youth throughout Black Belt communities.

Moreover, this frustration among adults in black tenant families over living conditions in the Deep South was coupled with their growing perception that cities presented a more open job market environment. Describing the negative effects of inadequate education in Alabama, a leading Fisk Scholar wrote: "The result has not been a widespread enlightenment of the population; to the contrary, the school has to some degree served as a disruptive social agency. Those who receive what education a community has to offer become migrants from the community to the small towns and cities. The illiterate are left behind."[32]

The Black Belt: Solid but Shrinking

The hardship that plantation life posed to blacks in the Deep South led to population drops in one black county after another. These declines were gradual, virtually unobserved by the general public. The population decreases reflected not only flight from tenancy and high death rates from disease, but also technological changes. Industrial technologies were emerging steadily in the cities, in the peripheral regions surrounding the rural Black Belt, and in the form of federal public works projects throughout the South. This new industrialization was generating demand for skilled and unskilled workers of all calibers. For example, throughout Georgia's seventy Black Belt counties, blacks moved on as doors to agricultural work shut. Between 1910 and 1920 the white populations increased. In the mostly black Greene County from 1900 to 1920 the area's total population rose from 4,478 to 4,685. But the number of blacks decreased by 386 while whites increased by 595. The decline in black population occurred in those areas in which small farms owned by whites were declining due to soil erosion. Plantation owners favored hiring the displaced white farmers, who desperately needed jobs as share-tenants, in place of their black sharecroppers. Once hired, the white tenant farmers were given opportunities to own small homes or become self-reliant renters.[33]

Many historians continue to assume, misguidedly, that the majority of these displaced black sharecroppers made a "Great Migration" to the urban North during World War I and the Great Depression.[34] But the movement of black rural migrants was constant from 1910 to 1940, and the target of their initial migration was not primarily the industrial regions of the North and West. It was first the urban and industrial areas within the southern states. Between 1910 and 1930, some 50,000 blacks left Alabama's rural Black Belt counties. But many resettled in the north central industrial region of the state, encompassing Birmingham (Jefferson County).[35]

In 1917 the U.S. secretary of labor organized a study of migration from the Black Belt South. The survey team found blacks from rural regions flowing through the Birmingham region on their way to other destinations. But they also admitted that multitudes of blacks were settling in this region permanently as industrial workers. "Birmingham and Bessemer cities, owing to their railroad facilities and peculiar location in a large coal and iron industrial district, have been the most important points of distribution for Negroes going North," they reported. However, the team continued, the "Birmingham district is itself an employment center of much importance. There are approximately 25,000 employees in the coal mines of the State, about two-thirds of whom are in Jefferson County. Negroes compose much the greater part of these."[36]

In 1939 Horace Mann Bond, the prominent educator who later became president of Lincoln University, published a brilliant study of black education and industry in Alabama. He emphasized that during these decades this state "witnessed a fundamental redistribution of her population, including both whites and Negroes." Bond stressed that the changing settlement of blacks in Alabama "involved migration from the State, and within the State from rural to industrial and urban centers."[37]

Studies of specific Black Belt counties as well as oral history interviews also reveal that migrating black tenant farmers were diverse in makeup and in survival strategies. Therefore, they moved in many directions and, frequently, relatively short distances from their original rural homes. Displaced black share-croppers sometimes moved in with other families in the area. Others, if they did move their households, trudged mile by mile to the nearest large town or city and attempted to resettle there. In Depression-era Mississippi, for example, oral history research found that some black sharecroppers who stayed put in the rural areas believed they were better off than those trekking to far off cities in a risky search for jobs. One sharecropper described his situation during the Great Depression: "The Depression hurt a lot of folks, but there was nothing they could do 'bout it. It was real hard on all of the poor folks. Most of the colored folks had it real bad, but since we didn't know anything 'bout a Depression, we didn't know who to get mad at. Some folks said that it was President Hoover's fault. Lots of people lost they home and crops 'cause they couldn't pay no bills."[38] But such sharecroppers survived since they frequently kept large gardens and raised their own meats. Another Mississippi farmer reminisced: "We didn't have much, but we always ate fairly well. Evening meals consisted of meat, two vegetables, and a dessert. We never had to eat syrup, cornbread, and fat meat three times a day like a lot of other people. I guess we were lucky. . . . We had a little garden where we raised several kinds of greens, potatoes, corn, and a couple of other things. We had ducks, chicken, and a hog or two. . . . Most of the land around this area was owned by white people, and these people would let us use the land free because by farming it, we kept the land clean."[39]

The direction and endpoints for migrating black sharecroppers leaving their immediate community were strongly determined by two factors: first, the availability of railroad lines; and second, job openings either in service fields or industry in those cities and towns in which industries were growing—usually on the periphery of the Deep South's Black Belt. We see evidence of these trends in Arthur Raper's classic, Depression-era study, *Preface to Peasantry*. In this sociological survey of rural black southerners, Raper stressed that black migrants mostly sought the nearest large communities and that these movements attracted little publicity. Focusing on Georgia between 1920 and 1930, Raper wrote, "The migrants from Greene County were virtual refugees and so stopped at the first place they could, which accounts for one-third of them being in Atlanta, and the next largest numbers in Athens and Augusta—the three nearest cities with which Greene County had direct railroad connections." Another important 1930s survey of the social conditions for blacks in Georgia revealed the buildup, rather than decimation, of black households in Atlanta, Macon, and Brunswick. It found that from 30 to 40 percent of the homes of blacks had household members who were not kin and who frequently were not charged rent because of the Depression conditions.[40]

In the meantime, the geographic and economic segregation of black tenant farmers intensified. In 1920 about 39 percent of the South's white farmers were tenants, compared to 76 percent of the black farmers. Among standing renters—tenants who paid a quantity of their products to the landowner—about 88,000 were black, 27,000 white. Among the croppers—those tenant farmers who were controlled totally by landlords—334,000 were blacks, 177,000 were whites.[41] Black tenant farmers were disproportionately tied to cotton plantations—the major crop for which harvest mechanization advanced the most slowly until the 1940s. Mechanical cotton pickers—spindle pickers and strippers—had not yet been perfected to match the quality and cost benefits of hand-picked cotton.[42]

An especially large buildup of southern white farmers among the tenant population occurred during the 1930s when many southern industries closed as a result of the Depression. For example, substantial increases in white populations but drops in black populations occurred in Alabama's Black Belt counties. In counties with black populations of 70 percent or more as of 1900, the black population declined from 344,000 in 1910 to 293,000 by 1930. In the meantime, the white population grew from 89,000 to 118,000 in those decades. Between 1930 and 1935, the number of black tenants in Alabama dropped from 77,875 to 75,542; while white tenants increased from 88,545 to 100,705.[43]

The southern Black Belt population either remained locked on the technically "frozen" plantations or, if fortunate, managed to squeeze themselves into the lower rungs of the region's small industrial sector. In 1936 the Southern Regional Committee of the Social Science Research Council surveyed the South's economy. This was a group of leading university researchers whose politics were generally moderate. They emphasized that "the Negro tenant constitutes a special aspect of this [region's] agrarian situation." In the Mississippi Delta, 95 percent of the tenants were black; in the Red River Bottoms, 89 percent; in the inland Black Belt, 88 percent; and in the Cotton Piedmont, 85 percent.[44] By 1930 the population of the Black Belt, which in 1900 contained more than 4 million residents, had dropped to approximately 2.7 million, and, by 1940, to 2.6 million (see table 5.2).

Until 1930 black southern labor more frequently did not migrate to points outside of the South. Instead, they gravitated to locales nearest to their rural resi-

TABLE 5.2 *Number of Counties and Black Population in Counties 50 Percent or More Black in the South: 1900–1940*

	1900	1910	1920	1930	1940
Population	4,057,619	3,932,484	3,251,440	2,738,432	2,642,808
No. of Counties	286	264	221	191	180
State					
Alabama	22	21	18	18	18
Arkansas	15	14	11	9	9
Florida	12	10	5	4	3
Georgia	67	66	58	48	46
Louisiana	31	25	22	16	15
Maryland	2	1			
Mississippi	38	38	34	35	35
North Carolina	18	14	12	9	9
South Carolina	30	33	32	25	22
Tennessee	3	2	2	2	2
Texas	12	8	4	4	3
Virginia	36	32	23	21	18

Source: U.S. Bureau of the Census, *Population—Special Reports,* Series P-45, no. 3, March 29, 1945.

dences, such as the steel-producing regions of Alabama; Norfolk, with its ship-building industry; and cities like Jacksonville and Houston, which had small industrial sectors.[45] During the 1930s black out-migration from the South expanded rapidly. Throughout that decade 460,000 blacks and 300,000 whites left the rural South for other regions of the country.[46]

On the Industrial Periphery

Large steel-manufacturing plants had been pivotal to the industrial growth of the urban North. By contrast, in the South, its industrial sector was limited primarily to technologies tied closely to agriculture and extractive resources such as mining and lumber processing. Industrial resources were centered mostly on railroads, mills, chemical production, and construction of dams. The mills produced textile, paper, and lumber. Finally, tractor farming on large plantations began to complement the huge force of sharecroppers. We observed earlier that the overall population of the Black Belt declined between 1910 and 1940. Initially, those who left the rural Black Belt scattered mostly throughout the urban South or to the steel-producing sections of the South. As multitudes of black laborers and professionals were routinely discarded by industry, they moved to other regions of the nation.

In 1931 T. Arnold Hill, the director of industrial relations for the National

Urban League, surveyed blacks in southern industry. He found that the 2.2 million black farm workers were mostly concentrated in the South and made up about one-fifth of the nation's agricultural labor. Nearly 6.7 million southern blacks, about 75 percent of the region's black population, still lived in rural areas. However, a substantial portion of the remaining 2 million southern blacks had moved into the region's cities in search of industrial jobs. In general, except for steel and iron production, black men were turned away from urban industrial workplaces at any level. Black women made limited gains as unskilled workers in tobacco manufacturing, fish packing, and furniture making. While during the 1920s the number of black women domestics in southern cities declined somewhat, there was an increase in black males in this field. "Thus," Hill observed, "many of the Negro men who left the farm for the city have been absorbed not into industry but into the increased number of domestic servants."[47]

Southern blacks fared better in those sections of the South with iron and steel industry, namely, Birmingham and Chattanooga. Blacks found jobs, although a steep racial division of labor unfolded. The jobs that involved the dirtiest, hottest, most uncomfortable and hazardous conditions were usually reserved for black laborers. Whites comprised most of the skilled workers, while blacks were the semiskilled workers, operatives, or laborers. From 1910 to 1920 the number of black iron and steel workers in the South grew from 12,000 to 44,000, and black semiskilled workers increased from 1,400 to 11,800. During 1930 in Birmingham's iron and steel plants nearly three-quarters (5,364) of the workforce (totaling 7,222) were blacks. The black workers were heavily distributed in the operative jobs requiring heavy lifting and other physical exertion.[48]

It was the isolation of blacks on the southern farmlands that caused great frustration among black social thinkers and activists of the 1920s and 1930s. Arnold Hill pointed out that blacks had "[t]heir first association with this country . . . on the farm, to which by means of slavery, peonage, and an enforced poverty they have been held . . . closely ever since." In sum, according to Hill, "we venture this most provocative generalization, that the Negro is still a slave—not legally so, but by a more effective control than law, namely the fixity of his [racial] status . . . which renders him powerless to advance beyond the fringe of economic independence."[49]

Disease Lingers

While black southerners struggled to find ways into industrial and urban employment throughout the pre–World War II South, diseases such as tuberculosis, syphilis, and malaria continued to take their toll. U.S. authorities had carried out a drastic campaign to eliminate infectious killers such as yellow fever and malaria in the Canal Zone. Similarly, in their effort to eradicate yaws in Haiti, U.S. experts tried to implement a campaign that blanketed Haiti's entire geography. But by the mid–1930s, public health leaders in the United States had become divided by their differing methodologies as well as their disease emphases.[50] Consequently, epidemic diseases confronting the United States' black South during the 1920s and 1930s were approached by health authorities sporadically, with no unified paradigm, and in most sections of the rural South, not at all.

The nation's leading malaria-control experts began turning their attention to the South during the early 1910s. But southern local and state public health agencies were poorly financed at the very time they faced some of the most impoverished and unhealthy communities.[51] The South's business elite had no incentive to raise local finances and resources for broad public health campaigns for the black poor since there was an oversupply of black and white unskilled labor. Thus, southern health officials, compared to their northern counterparts, were less pressured by their local political and economic leaders to address epidemic disease afflicting the local working class, particularly the black working class. In 1900 in Georgia, for example, there were 1,226 malaria deaths and approximately 490,000 malaria cases reported. Yet it would not be until 1918 that the Georgia State Health Department was able to mount a substantial control program.[52]

Furthermore, most blacks lived in segregated communities inside the geographic Black Belt. Any widespread sickness among them in this region appeared to white community leadership as posing no fundamental threat to adjacent white communities and employers. Southern state and local authorities were not extensions of powerful federal and business interests with large investments in public-works development or military operations in a particular tropical region, as was the case for the Canal Zone and Haiti. Indeed, southern economic leaders were the opposite: indigenous land-based aristocrats beleaguered by both domestic and international industrial and market pressure. Large-scale industrial and financial segments inside the United States were squeezing the South's plantation owners. Also, competition from the growing number of cotton-producing nations in Latin America, Africa, and Asia constantly threatened to send the South's cotton prices down.[53]

The nation's powerful industrialists and their political circles were not substantially dependent on the South's economic production. Instead, they were preoccupied with expanding manufacturing to meet higher priorities. Industrialists wanted to tap the new urban consumer and service markets. Thus, increasing railroads, shipping, and world trade became one of their leading concerns. Also, U.S. industry became more geared toward sales to the U.S. public (i.e., government) sector. In this context, the nation's industrial and government leaders had no interest in directly designing and implementing a region-wide public health campaign for the states throughout the South.

But the South, with its hundreds of thousands of disease victims and hundreds of unhealthy counties, was attractive to northern-based medical institutions for research to expand public health science. Federal agencies as well as the Rockefeller Foundation's International Health Commission sent research teams throughout the South to work in cooperation with local authorities and researchers. Their primary aim was to expand the knowledge base in the new fields of bacteriology, parasitology, biochemistry, and nutrition. Government, foundation, and medical school leaders believed these researchers were seeking not just scientific information for the respective disciplines, but the keys for understanding and harmonizing all of society according to the principles of science.[54] In 1911 Frederick Gates, the chairman of the board of the Rockefeller Institute of Medical Research and a Baptist minister, enunciated this utopian mission: "As medical research goes on, it will find out and promulgate, as an unforeseen byproduct of its work, new moral laws and new social laws—new definitions of what is right and wrong in our relations with each other. Medical research will educate the human conscience in new directions and point out new duties. It will make us sensitive to new moral distinctions."[55]

Diseases that widely afflicted the South's masses—especially malaria, hookworm, and tuberculosis—were also prevalent throughout the larger tropical world. U.S. government and business heads knew that their institutions could not operate successfully in tropical zones without effective means for their personnel to withstand these ailments abroad. Using the newest lab techniques, federal researchers conducted study after study on populations in southern counties, dealing with infectious diseases and malnutrition. These surveys began pointing out the severity of these health problems throughout the South. Yet, while these studies revealed the need for more disease-control measures, especially to help the sharecropper communities of the Black Belt, permanent programs from local or federal social welfare agencies and public health offices rarely ensued.

Through the mid–1930s, for example, the nation's highest malaria rates were found in and surrounding the Black Belt. This area included the Mississippi River Valley, stretching from Cairo, Illinois, to Natchez, Mississippi, and through the South Carolina–Georgia-Florida "wedge." By 1914 there were approximately 600,000 malaria cases in eight southern states—about 4 percent of the 15 million residents of these states.[56] G. F. Russell, a staff member at the Rockefeller Foundation and a malaria investigator in southern Alabama and Georgia during the 1920s, recalled (in 1968), "I surveyed many communities, colored and white, in which malaria was the dominant factor in producing ill health and economic stagnation."[57]

From the 1920s through the New Deal years, national concern grew in political and public health circles about the health problems of the South's workers and farmers. Typhoid fever, hookworm, malaria, tuberculosis, and pellagra, in particular, were widespread throughout the South. New Deal politicians and health and welfare leaders knew that economic productivity in the South's farming and manufacturing sectors was tied to the lowering of disease rates throughout the region's lower classes. In order for the South to maintain its plantation agriculture as well as step up industrialization, the health and productivity of the regional working class had to be upgraded.

But from the angle of local plantation owners and operators as well as mill owners, these diseases were not necessarily damaging the local labor supply available for their plantations and mills. While federal and academic public health researchers had interests in developing more etiological information about these diseases, they were not geared to try to change the political and economic behaviors of large-plantation operators and local governments tied to maintaining the South's tenant farm population. The pursuit of pure science, not political science, became the ulterior professional value of federal and foundation health researchers in the pre–World War II South.

Malaria Control: Eclipse of the Sanitarians

During the 1920s and 1930s, federal public health authorities, assisted by southern health officials whom they enlisted, began approaching epidemiology and disease control in ways that departed from the effective turn-of-the-century programs of the likes of Reed, Gorgas, and LePrince. In both Cuba and the Canal Zone, these sanitarians had demonstrated that straightforward environmental measures were unquestionably effective in immediately lowering the number of infectious-disease victims.

However, the newer generation of public health experts was less attracted to organizing campaigns to improve the environmental conditions of homes and workplaces—measures that would break the contact between vectors and hosts. They also wanted to avoid the role of exposing publicly those social and economic segments most responsible for perpetuating unhealthy work and living conditions. Instead, public health authorities and physicians became ever more attracted to the mission of biomedical specialism—that is, a quest to understand and treat the biochemical aspects of infections and pathogens. Increasingly, the public health experts and physicians involved with populations riddled by diseases like malaria and tuberculosis emphasized microscopy and x-rays. The discourse within their professional circles centered more on laboratory data and clinical issues than on community conditions and bedside exchange with their patients. They refined diagnostic testing and followed the action of drug therapies in infected persons. By specializing in a specific disease and its pathogens, public health physicians and researchers hoped to discover a chemical means to kill the insect vectors or parasites and ultimately create a vaccine.[58]

A similar onrush into biomedicine was occurring among the medical professions involved with occupational diseases. Rosner and Markowitz's history of medicine and worker health shows that during the Great Depression "consensus broke down" in medical and public health approaches to silicosis and similar industry-linked diseases. "Different interest groups now debated how sickness should be defined and who should diagnose it." Pro-labor interest groups especially challenged the medical community to consider job conditions as a key cause of respiratory diseases like silicosis. But increasingly, physicians and public health professionals "demanded 'objective' proof in the form of laboratory tests and x-ray evidence to establish the existence of disease. It was the diagnostician, they felt, depending on these technological tools who was to define disease."[59]

The new public health science of federal experts focused increasingly on biological analysis, as opposed to the techniques of classic sanitarians. The old-style public health interventions were never applied on a broad scale inside the communities of the black poor throughout the Deep South. Consequently, the Black Belt population, like the people of Haiti and Liberia, suffered malaria rates at levels much higher than those of whites and blacks in the Canal Zone. From 1912 to 1915 a million cases of malaria were occurring each year among the 25 million people of twelve southern states. The highest incidence rate for malaria in these states, 40.9 percent, occurred in the mostly black Mississippi Delta.[60] Malaria death rates for blacks in ten southern states from 1921through 1923 were over three times greater than those of the white populations in these states (25 deaths per 100,000 persons and 7 per 100,000, respectively).[61]

Technical innovations contributed to the flourishing of the new public health science. Laboratories and medical instruments became more frequently used to investigate the microbiology of diseases and hosts by public health units as well as hospitals with patients sick from infectious diseases. These devices— in particular, x-rays, blood tests, and urinalysis—made diagnosis of disease in individual patients more precise. They also enabled researchers to amass diagnostic data from sets of patients. Public health physicians and scientists combined these two flows of information—one from their visual inspection of their patients, the other from laboratory results—in turn, generating "new fundamental problems that stimulate the course of [even more] scientific research."[62] In 1927 a leading U.S. physician and tropical disease specialist wrote about this unquenchable investigative urge. He stressed that laboratories equipped like those of modern hospitals were needed in tropical locales with high levels of infectious diseases common to the tropical world. Such laboratory facilities for studying local diseases "provide . . . for thorough clinical observation," he wrote, and they "stimulate the desire to experiment in many theoretical directions."[63]

Increasingly drawn to developing the public health profession to pursue the microbiology of infectious diseases, federal public health authorities became most interested in conducting field studies on insect vectors and means to kill microscopic parasites. They frequently acknowledged that social conditions strongly influenced what on the surface appeared to be racially determined malaria patterns. They admitted, for example, that a high risk of exposure to malaria was inherent on large cotton plantations that used the sharecropper method. Also, they recognized that race-relations practices within local, county, and state regions reinforced the disproportionate employment of blacks as impoverished sharecroppers. However, these same public health authorities demonstrated little professional interest in taking on a public role for mobilizing government and community to eliminate such social conditions and to improve public health resources for the South's black poor. They either forgot or were not interested in building public health campaigns based on environmental measures that would lower mosquito- or air-borne diseases—measures that could conveniently be modeled on the Canal Zone experience.

In fact, information about the effective steps used by the sanitarians of the Canal Zone was now widely available to both the public health profession and the general public. In 1916 Joseph A. LePrince, now chief sanitary inspector for the Canal Zone commission, and a colleague published an overview of the public health campaigns in the Canal Zone. This work, *Mosquito Control in Panama,* gave a clear description of the public health strategies unquestionably successful in eliminating malaria and yellow fever deaths during the construction of the canal. LePrince and his colleagues stated that the "methods of malaria control applied and developed on the Isthmus were put to a severe test and gave successful results." They had to apply their mosquito-control operations over fifty square miles of tropical conditions and highly diverse populations: "The topographical, meteorological conditions, and constant changes due to the construction work, together with the character and constant moving of the population and their dwellings, and social conditions, were peculiarly unfavorable to the control work undertaken. The natives and employees, infected or well, were at liberty to live where and how they chose. We had no control over their movements or methods of living."[64] Yet by drainage, screening, oiling of mosquito breeding sites, and quinine, Gorgas, LePrince, and the other canal health workers had demonstrated that "even under such adverse conditions, malaria . . . may be kept down to a minimum rate without prohibitive cost." By the mid–1910s the disease-control program of the Canal Zone was hailed internationally. Dr. Malcolm Watson, in charge of malaria control in Britain's Malay states, visited the Canal Zone to "study the details of their methods." He returned to Britain and reported to the Royal Colonial Institute that he had observed "the greatest sanitary achievement the world has seen."[65]

Paradoxically, when U.S. public health experts prominent during the interwar period advocated mosquito-control measures for the southeastern states, this was framed as a plea to the community elite, not as a political issue or a civil engineering challenge. Their concern was with convincing plantation owners and local governments that productivity would improve if malaria was reduced. In 1925, three researchers from the U.S. Public Health Service conducted a malaria study involving black communities of the delta of Arkansas and Mississippi. Published in the *Southern Medical Journal,* the study focused on twelve cotton-growing counties, including Bolivar County, reputed to be the highest cotton-producing counties in the nation. The researchers found malaria rates for black farm workers extremely high. Instead of recommending immediate measures, such as eradicating local breeding sources for mosquitoes, to reduce infections, they focused on explaining the negative effects high malaria rates had on crop productivity: "Practically the whole of the cotton crop in this region is grown on large plantations worked by negro [*sic*] labor. Any factor which interferes with the necessary supply of this labor is of great economic importance.

The prevalence of malaria, therefore, with its peculiar seasonal incidence, may be a limiting factor in the production of the crop, interfering with the cultivation, and especially with the picking of the crop."[66] But to many southern political and commercial leaders, diseases like malaria did not debilitate the black tenant to the point that cotton production slowed. So these types of admonishments were simply ignored.

Indeed, the response of the federal and local government as well as business to the malaria problem in the Black Belt was markedly lethargic compared to the protracted government-led public health campaigns in the Canal Zone and Haiti. In the Canal Zone the U.S. public health community's campaign against yellow fever and malaria had been comprehensive in magnitude. Beginning in 1904, U.S. authorities in the Panama Canal Zone had focused on quarantines and improvement of the work and housing environments for their canal managers and laborers to reduce yellow fever, malaria, and similar killer diseases. An intense, comprehensive malaria-control campaign had been implemented. In Haiti, U.S. military and administrative authorities also had placed high emphasis on a unified, protracted program to control key tropical diseases, especially yaws and smallpox.

By contrast, in the Black Belt, federal public health authorities were only advisers to local political and commercial leaders. They were much more concerned with exploring the bacteriology of infectious diseases than with helping to systematically change local social, work, and health environments of the South's black and white tenant farming communities. The federal lethargy was compounded throughout the South by local public health authorities that were slow to accept the germ theory of disease and mount state and local campaigns against diseases like yellow fever and malaria.[67]

Federal public health authorities implemented research and demonstration projects relating to tropical diseases in the South. However, they had neither the legal authority nor the interest as professionals to try to alter the health and environmental conditions of the large tenant farms, the basis for the vicious cycle of high malaria rates and poverty. Instead, they concentrated on biological research, especially on the anopheles (mosquitoes), chemotherapy to reduce infectivity in clinical cases, and theoretical aspects of malaria epidemiology. In the public health campaigns in Cuba and the Canal Zone, the focus had been on stamping out the mosquito vectors by reengineering the living and work environments, improving public health laws, and establishing networks of hospitals and dispensaries. But the issue of widely applying these effective steps went largely ignored by the northern-based public health advisers operating in the South. They were content with their peripheral civic or public roles.

A key case in point is the activities of the assistant surgeon general of the U.S. Public Health Service, Kenneth Maxcy. One of the nation's leading public health authorities, Maxcy attempted foremost to improve what he perceived were theoretical blind spots in the discipline of epidemiology by studying the South's malaria problem. Like LePrince a few years earlier, Maxcy reiterated the practical benefits for eliminating malaria. He even recognized environmental racial segregation and localities that were obviously inhabited by the poorest, most racially subjugated blacks. Yet Maxcy refrained from advocating concrete engineering, dispensary resources, and legal controls that were known since the Canal Zone experience. These steps involved improving the sanitary and medical conditions for populations in areas with a high density of infectious mosquitoes, not simply identifying and treating the population experiencing high levels of malaria.

Nor did Maxcy seek federal or local legal measures that would challenge the landowners' laxity in eliminating malarious conditions. Instead, Maxcy tried to convince the workers at the South's public health bureaus of the medical necessity to approach malarious conditions as if they were enveloped within imaginary zones or hypothetical localities. At the same time, he shied away from raising the issue of politically and legally sanctioned racial segregation and pointing out the effects of this racial caste system—higher levels of preventable diseases among the black poor.

In 1924 Maxcy conducted a major study of malaria in the South. Maxcy urged that epidemiological elements or "principles" were behind the malaria crisis in the South, not concrete social and economic conditions, behaviors, and interest groups. The title of this study reveals this obfuscation: *Epidemiological Principles Affecting the Distribution of Malaria in the Southern United States.* While he identified the harsh conditions facing black tenants, Maxcy focused more on developing his interpretation of the malaria crisis as a theoretical matter drawn in geometric terms. Maxcy urged that malaria not be viewed as an all-encompassing health problem for the South. According to his theoretical scheme, malaria did not break out as virulent epidemics the same way non-insect-borne infectious diseases seemed to do. He advised that "malaria is not an ubiquitous disease in the southern United States, as is, for instance, such a disease as measles."[68] Such a view discouraged the need for local health authorities to undertake broad mosquito-control campaigns throughout the entire Black Belt region. Instead, Maxcy went on to classify malaria "foci" as either endemic or epidemic. He gave Autauga County, Alabama, as an area representative of an endemic focus (not an epidemic!):

> Along the higher land between these sloughs are located cotton plantations cultivated by negro [*sic*] tenants, poorly housed, poorly clothed, poorly fed, living in unscreened houses, self-medicated with inadequate "chill tonics," and made miserable by the hordes of mosquitoes which swarm in from the nearby swamps at sundown each evening. During the height of the season, 20 to 30 A. quadrimaculatus can be found on the walls of a bedroom at almost any time, and in almost every house is to be found some individual suffering from "chills." Every year brings its tolls of deaths resulting directly or indirectly from the ravages of neglected malaria.[69]

Maxcy explained that the populations in Autauga County were decidedly different from those in non-endemic areas. The latter, non-malarious sections were composed of "enlightened, well-to-do white families living in well-screened houses in prosperous sections on cleared, cultivated and thoroughly drained farm land and in towns with the best sort of medical attention." By contrast, in endemic regions, "[t]he negro accepts 'chills' as a necessary evil and pay[s] it scant attention, [while] the plantation owners passively acquiesce in this shameful human and economic waste."[70]

Endemic areas made up of blameworthy blacks and morally ignorant overseers eventually spread malaria, according to Maxcy's theory, because populations that pass in and out of such zones become sources for new infections elsewhere. As an example, Maxcy cited the cotton plantations in Dunklin County, Missouri. "These plantations are worked not only by negroes [*sic*] living on them, but this labor is supplemented in times of need by an extra supply of negroes who live in the city, travel out to the plantations to work, and often spend the night on them in unoccupied tenant houses." The temporary workers then return to their homes on the edge of nearby cities, "where there is sufficient production of Anopheles to further the spread."[71]

In his description of epidemic foci for malaria, Maxcy did not totally discard human behavior and living conditions. He simply subsumed them as part of many natural causes for epidemics. Epidemic zones consisted of localities experiencing "some new conditions or circumstances" that quickened the breeding of anopheles. He signaled out weather and topography as the key factors that increased anopheles. Excessive rain and high temperatures are examples of factors leading to new mosquito breeding. He emphasized landscape changes such as dammed-up streams or blocked drainage ditches, large ponds, or poorly cleared railroad and highway construction—all could create new breeding grounds for malaria-carrying mosquitoes.

However, rather than outline and advocate specific eradication steps (launched both inside and outside endemic zones) that had been so successful in the Canal Zone, Maxcy stressed that public health authorities should dwell on designing their anti-malaria programs using his classification scheme. "The endemic focus must be the ultimate objective of any well-planned antimalaria campaign," he concluded. "[E]liminating an epidemic focus is sensational; but so long as the 'seed bed' of malaria remains, the disease will take its toll from year to year."[72] The campaign in the Canal Zone region, by contrast, sought to eliminate malaria transmissions through a comprehensive approach. Under the force of public laws, mosquito sources were wiped out for all the Zone region's residents and transient populations by permanently upgrading housing, workplaces, water systems, and dispensary resources.

Indeed, national public health experts in the U.S. South were leaving behind the pragmatic approaches of the early sanitation specialists such as Gorgas and LePrince in the Canal Zone. Instead, their mission increasingly centered on expanding the place for new microbiology in their profession. Seeking more knowledge about malaria pathogens as well as prophylactic measures that used pharmaceuticals, the new public health leadership sought wider use of their laboratory-oriented specialties, especially parasitology, hematology, and, pharmacology. New drugs, someday they hoped, could eventually keep individuals bitten by malarious mosquitoes from contracting the disease and, someday, kill the parasites. To the new public health experts, the primary causes and solutions for communicable disease epidemics of malaria, tuberculosis, and syphilis lay in the microbiological world, not the socioeconomic one.[73]

This priority of the "new" bacteriology advocates in the U.S. public health community flew in the face of the malaria-prevention paradigm arrived at by malaria experts of new international health organizations. Since World War I the emerging international health organizations stressed community prevention as the indispensable tool for effective malaria control programs. The League of Nations (and later the World Health Organization) stressed the value of "bonification," that is, the improvement of housing, farming practices, and general living conditions. The organization constantly monitored and publicized sanitation and housing improvements that had successfully reduced malaria and similar diseases, especially in the temperate-zone nations of Europe. Consequently, most of the league's programs centered on designing means to permanently eliminate infectious insects and rodents from civilian and military environments.[74]

In 1937 the malaria commission of the League of Nations issued a general report on the role of quinine and synthetic drugs in the fight against malaria. The report was based on evidence and experiences from league malariologists working in programs throughout the world. The commission concluded that programs that emphasized quinine and the search for new drugs to eliminate malaria were inherently limited in reducing overall malaria in communities: "Experience has so far shown that the eradication of malaria from a locality by the curative and prophylactic treatment, with the drugs at present available is practically impossible. To begin with, it is impossible to reach, in sufficient time, all the inhabitants of an area, or even of a small village. Moreover, while curative and prophylactic treatment may greatly diminish the morbidity, yet it cannot suppress the parasites in all the carriers."[75]

Yet many U.S. malaria specialists continued pursuing the biochemical aspects of malaria as the highest public health priority. One such researcher, M. E. Winchester, brought out this view in a study that appeared in the United States' leading tropical medicine journal shortly after the League of Nations' study. Winchester acknowledged that malaria could be eradicated by permanently destroying the anopheles' breeding areas. However, in "vast areas of the globe this ideal can not be obtained due to unfavorable topography or adverse economic conditions. The public health officer is therefore often faced with the alternative of using . . . the temporary (chemotherapeutic) methods of malaria control, or of doing nothing at all." He recommended expanding the use of Atabrine prophylaxis, a drug that reduced the accumulation of the malaria trophozoites in the body, that is, alleviated clinical symptoms.[76]

The Modern Hospital and Public Health

The unwillingness of federal and local public health authorities to provide anti-malaria care and campaigns for the South's black poor coincided with the impact of growing biomedical specialization and subsidized health care benefits. Throughout the New Deal and World War II periods, the black hospital and public health sector became less and less useful. Specialized biomedicine set within the modern mass hospital became the dominant framework for the implementation of medical education, professional specialty training, and patient care. Also, new federal programs were enabling growing numbers of working Americans to have access to hospital care. Consequently, hospital-based practice and childbirth as well as specialist-physician care expanded throughout the South. This was at the expense of the traditional black doctors, who usually were family practitioners, as well as lay midwives, who were increasingly outlawed by strict government and medical licensing bureaus.

Physician services flourishing in large hospitals reached first and foremost new blocs of insured populations—veterans, white-collar employees, the privately insured. Patients from these populations could afford hospitalization and outpatient clinical care. By contrast, the public health needs of the poorest black and white communities of the South's Black Belt and the surrounding regions offered medical institutions and physicians virtually no economic incentive. Furthermore, the traditional midwives popular throughout the South were being displaced by physicians and hospital-based maternity care.[77]

Throughout the twentieth century, Haiti and Liberia experienced a brain drain of physicians leaving these countries. In the black South, the black doctors' institutions were replaced by the larger, more modern, and well-financed health facilities near the Black Belt. Midwives remained as central means for health care in Haiti and Liberia throughout the twentieth century. But the onrush of hospital deliveries and improvements in obstetrics throughout the South spelled their demise among southern blacks. Indeed, increased training standards, the need for links with the large academic and public hospitals, and tightening of licensing restrictions spelled the rapid decline of not only the black lay midwives, but most of the black medical schools and hospitals as well.

As scientific biomedicine spread throughout the South, a mixture of lofty professionalism and social distance became evident among many southern physicians in their dealings with the rural poor. Most black doctors held the view that health problems of the rural black poor could be solved with more public health resources. But even they were increasingly critical of black folkways. Hildrus Poindexter, a prominent black physician-bacteriologist, exemplified a negative view of the social behavior of poor blacks in 1937. Surveying the health problems of rural blacks, Poindexter blamed poor diet and unsanitary living environments. But he also emphasized that the health problem of the "rural Negro" was primarily "ignorance of personal hygiene, and of the hazard of local customs and [pre]occupations."[78] The physician warned that the "rural Negro" should be more of a concern to American blacks generally than the "rural whites are to their race." He stressed: "The rural Negro migrates farther from home, having little or no property to tie him down. He migrates more blindly, and is too often a wanderer without direction or defined purpose. This type of individual is more ignorant of laws of hygiene and sanitation and is more likely to infect others or expose himself to infection."[79] The doctor urged that "unless the trend is stopped," the rural Negro will remain "the reservoir for certain diseases of ignorance."[80]

As the Great Depression period unfolded, some of the South's doctors implored counties and state health authorities as well as civic leaders to take on comprehensively the malaria menace.[81] But most federal public health authorities proclaimed themselves as helpless to change the South's poverty and racial caste. Whether malaria, tuberculosis, or venereal disease problems were discovered by health bureaus or local doctors, both local and federal public health workers approached these problems so as not to publicly disturb or politicize the need to eliminate the region's plantation labor system and sociracial caste. This was in stark contrast to the direct challenges against the South's local and state political segregationists and employment bosses that other government agencies, reform organizations, and even sharecroppers themselves raised during the interwar years. For example, experts from the Department of Labor aggressively pursued large southern employers found discriminating against black workers. Also, the National Association for the Advancement of Colored People and their congressional allies were struggling vigorously for a federal anti-lynching law as well as an end to Jim Crow laws and practices in the South's judicial and public education institutions.[82]

By contrast, many public health authorities stepped up scientific research while drifting into social inaction regarding the health effects of southern segregation. M. A. Barber, a notable public health investigator of malaria in the South, reflected an attitude now typical in the ranks of public health leadership. In 1929 he wrote in the *Public Health Reports,* "The health officer can not be expected to take on his shoulders the economic betterment of a community, although he may sometimes put health considerations into the scales when a large drainage project or other agricultural improvement is debated." To this public health authority, health officers were to *follow* not *lead* the local economic "current which . . . already sets strongly toward improvement in rural conditions of living."[83]

During the 1930s, as in earlier decades, many local and state governments throughout the South could not finance malaria-control work in poor rural counties and towns. The U.S. Public Health Service and philanthropy, especially the Rockefeller Foundation's International Health Board and the Julius Rosenwald Fund, continued to provide much of the initial technical and financial assistance for malaria investigations and pilot prevention programs in these areas.[84]

Likewise, for typical southern politicians and business leaders, disease-control resources for local black populations were of little importance. They believed, like the southern medical profession, that infectious diseases among blacks were generating outbreaks among the white populations. But these landowners and their political partners did not fear that there might be a lack of adequate self-protection for the larger white tenant communities, since these poor white farmers could easily be replenished if felled by sickness. It was in this context that black communities of the Black Belt were repeatedly and closely studied for their disease patterns or drug research but were rarely provided with ongoing preventive resources. For example, when European malariologists developed Atabrine in 1930, U.S. researchers studied it as a "mass therapeutic and prophylactic" using southern black communities.[85] And it was in this context that the Public Health Service implemented the infamous Tuskegee Syphilis Study.

Regional Development and the New Malariology

During the 1930s New Deal projects and programs began to emerge throughout the South. Among these New Deal initiatives were malaria prevention programs in endemic malaria localities in the South, sometimes inside the Black Belt. At long last, as had been the case in the Canal Zone during the 1900s and 1910s, federal health experts in cooperation with state agencies began public health programs directed centrally by federal experts. These programs were designed to alleviate malarious environmental conditions causing the spread of malaria. The federal health care initiatives stemming from the New Deal and World War II led to the gradual breakup of the traditional black health sector of the Deep South. At the same time, these initiatives improved disease control in the South and, secondarily, in its black population.

The effectiveness of federal disease-control resources was demonstrated in Tennessee as well as in the regional planning project, the Tennessee Valley Authority (TVA). In 1933 an estimated one-tenth of the schoolchildren in Shelby County (in the Memphis area of Tennessee) were infected with malaria. However, in just one year, a program involving clearing of ditches and sewers under the state's Federal Emergency Relief Administration (FERA) led to a one-third drop in malaria deaths and one-half drop in malaria cases.[86]

By 1935 FERA, along with the Civil Works Administration and the Works Progress Administration, had initiated massive drainage programs throughout the South. These agencies provided laborers for malaria control programs that resulted in the digging of about thirty-two thousand miles of drainage ditches throughout sixteen southern states. Over 620,000 acres of water-covered land was now drained, ridding many counties of these formerly mosquito-ridden tracts.[87] Also, the Social Security Act of 1935 included provisions for malaria surveys and control teams sited in twelve states. These statewide programs included teams of sanitary engineers, malariologists, entomologists, and technicians.[88]

Other federal malaria control projects in the South were implemented as a component of large-scale public works projects and involved the systematic application of environmental engineering and ongoing operational maintenance.

The most noteworthy public works projects and malaria-control programs were those of the TVA.

As the United States moved into the prosperous 1920s, politicians and corporate heads placed greater and greater faith on the capacity of new science and technology to increase productivity. In their view, technological innovations drove down the costs of production, increased output, and lowered consumer prices. The major social ills attributed to a capitalist economy, especially unemployment and poverty, would disappear, and eventually prosperity would reach all sectors of American society. In 1928 Herbert Hoover captured the euphoria that U.S. political and business leaders associated with rationalization of the nation's businesses and energy infrastructure. An engineer by profession, Hoover believed that the expansion of business, civilization, and prosperity was inevitable: "We in America today are nearer to the final triumph over poverty than ever before in the history of any land. The poorhouse is vanishing from among us. We have not yet reached the goal, but, given the chance to go forward with the policies of the last eight years, we shall soon with the help of God be in sight of the day when poverty will be banished from this nation."[89]

Electrification was a key dimension of this vision for corporate leaders and reformist politicians alike. They believed electric power held miraculous potential. To them, electric energy under modern management could so drastically reduce costs of production in industry, agriculture, and transportation that economic inequality in nations such as the United States and Britain would all but cease.[90]

During the 1920s heated political controversy arose concerning the best means to stop the destructive flooding that had historically vexed the Tennessee Valley. This region covers seven southern states from Virginia to Mississippi. President Franklin Roosevelt and his New Deal co-planners conceived the TVA as a public utility corporation that would construct and supervise a series of multipurpose dams and power plants. Instead of producing horrendous floods, the rivers flowing through the Tennessee Valley would be harnessed by the TVA to produce abundant hydroelectric power. Building of the TVA dams began in 1933. When the TVA construction projects were completed, five dams had been renovated, twenty new ones had been constructed, and an elaborate system of inland waterways had been engineered. Chronic flooding was eliminated while thousands of formerly isolated households now had cheap electricity. In fact, the TVA became the greatest producer of electricity in the United States. The TVA also supplied the power facilities for the atom bomb plant at Oak Ridge, Tennessee.[91]

The employment of blacks at the TVA and Oak Ridge followed very much the pattern of the Panama Canal Zone. Black workers were brought in to do the hardest unskilled occupations involving dam construction or maintenance work, but not the higher-level technical occupations involving skilled trades, professions, and management. Housing for black workers throughout the various construction sites was segregated, substandard, or nonexistent. Moreover, many rural black communities were eliminated by TVA land developments. The Santee-Cooper farmers of South Carolina, for example, some 840, mostly black, families, were forced to move. Federal officials found new housing for only 40 of these families.[92]

Employment rights groups and government agencies, especially the NAACP and, later, the Fair Employment Practices Committee (FEPC), struggled from the TVA's inception to improve the status of employment and housing for blacks, but generally to no avail. During the 1930s black political activists who focused on federal policy called the discrimination rampant at the TVA against blacks indicative of the black Americans' "raw deal" as opposed to New Deal. John P. Davis wrote in *Crisis* magazine that while the TVA employed a commendable number of blacks, the program had many objectionable conditions: "The payroll of Negro workers remains disproportionately lower than that of whites. While the government has maintained a trade school to train workers on the project, no Negro trainees have been admitted. Nor have any meaningful plans matured for the future of the several thousand Negro workers who . . . will be left without employment, following completion of the work on the dams being built by the TVA."[93] At its peak, blacks were 11.8 percent of the TVA workforce in 1941. But this figure dropped to 8.9 percent in 1946 and even lower, 6.2 percent, by 1963. The great majority of these employees were janitors, construction laborers, general laborers, and jackhammer operators.[94]

Despite the racial division of its labor, the TVA's disease control programs became efficient and effective, as had occurred in the Canal Zone region during the canal construction project. Measures to protect workers from mosquito-borne diseases at TVA sites benefited black workers as much as their counterpart white managers and workers. From its inception in 1933, TVA administrators incorporated a malaria-control unit as integral to its health and safety program. TVA management stressed a "closed environment" approach to mosquito-borne disease, similar to what had been done in the Canal Zone. The TVA malaria-control programs during the 1930s stressed two prongs: technical planning by teams consisting of medical malariologists, engineers, and biologists and the application of control measures under the supervision of resident sanitary engineers. Research on the biology and control of anopheles was implemented as a component of the TVA's operations work and with the specific aim of improving on-site control procedures.[95]

As TVA work crews laid out twenty-six artificial impoundments along the Tennessee River and its tributaries, they implemented wide-scale anti-larval measures. Reservoirs were prepared, water levels managed, larvicides and herbicides applied, and drift materials removed in the waterways closely supervised by TVA technicians and workers. Overall, more than ten thousand miles of shoreline and six hundred thousand acres of lake surfaces in the South came under this TVA supervision. Some 211,000 workers, many under relief programs, built anti-malaria drainage in about 250 counties. According to George Bradley, a historian and malaria specialist with the Centers for Disease Control (CDC), "The soundness of this approach soon became evident, and the TVA program became a model on which much future work elsewhere was based."[96] Inside the South, this was a new thrust in campaigns to eliminate malaria. However, actually, it was generally the same as the old-hat sanitation campaigns that had been implemented during the construction of the Panama Canal.

Outside of the federal and state projects such as the TVA, most areas of the South still lacked public health programs to systematically control diseases like malaria. Except for wealthier counties, formal disease-control campaigns were not reaching a substantial portion of the households and neighborhoods of at-risk populations throughout the South, especially in black population centers. The structural conditions underlying the region's malaria crisis were still not addressed. These conditions included the poor drainage on farm and residential lands, poorly screened housing, a largely unprotected and transient workforce, and insufficient community-level eradication resources.[97]

As the Great Depression tightened its grip in the1930s, malaria rates began to increase throughout many sections of the South. Overall, average malaria mortality rates for the individual states were declining. However, malaria deaths were occurring throughout a wider range of southern counties, indicating that new waves of high death rates for individual states and the entire region were highly probable. In 1935 there were still some four thousand deaths attributed to malaria. By 1939 the loss to the South's economy due to malaria alone was estimated at $500 million annually (see map 5.1).[98]

In 1939 E. C. Faust, a New Orleans medical professor and a leading specialist on the epidemiology of the South's malaria, wrote about this problem. "An unduly high percentage of the population in our [southern] area lives in malarious territory." He presented data showing that endemic malaria was spreading laterally, while new zones were cropping up as well. Faust was unaware of the relative paucity of local and national resources to control the disease in the South, compared to the illustrious Canal Zone project. Faust was concerned that malaria festered despite improvements in chemical agents effective in killing some of the insect vectors. Faust remarked that "[a]ll these danger signals have developed in spite of unusual efforts [by] state boards of health, assisted by Federal aid, to fight malaria both by larvicidal and by chemotherapeutic measures." Disparagingly, he cautioned the public health and malaria research community: "We do not have the dramatic picture of malaria as it exists in certain tropical countries, but we have malaria nonetheless. We cannot ignore the death toll which malaria takes every year in our midst."[99]

The Decline of Malaria

Domestic changes stemming from World War II delivered the death blow to the malaria crisis in the South, including the old Black Belt region where blacks were most heavily concentrated. First, the environmental improvements initiated by federal programs such as the TVA in the 1930s began to have long-lasting effects. Second, growth in economic opportunities during the war years

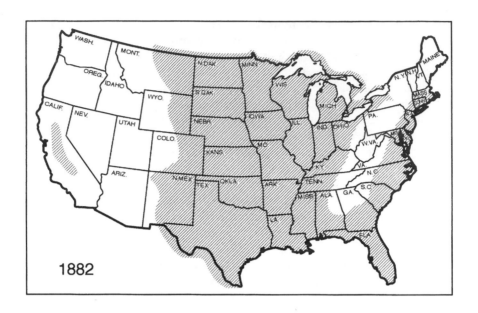

1882

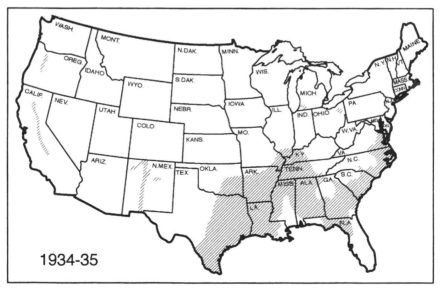

1934-35

MAP 5.1. Malaria Prevalence in the United States, 1882 and 1934–35. Reprinted from U.S. Army, Medical Department, *Preventive Medicine in World War II,* vol. 6. *Communicable Diseases: Malaria* (Washington, D.C.: Office of the Surgeon General, 1963).

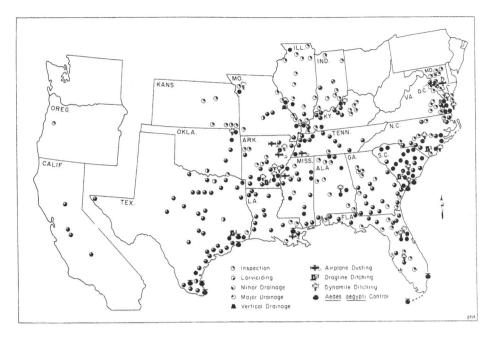

MAP 5.2.   Geographic distribution and projects by Malaria Control in War Areas, July 1, 1943. Reprinted from U.S. Army, Medical Department, *Preventive Medicine in World War II,* vol. 6, *Communicable Diseases: Malaria* (Washington, D.C.: Office of the Surgeon General, 1963).

enabled southerners to improve homes, medical care, and their agricultural and suburban environments. They purchased commercial, domestic insecticides that were applied with hand sprayers in homes and buildings. Third, population movements of southerners out of rural regions into industrial, urban areas removed this segment from malaria exposure. This migration was particularly intense during 1940 to 1944. A leading malaria expert at the Public Health Service's Communicable Disease Center in Atlanta observed shortly after World War II that "Negroes [increasingly] traveled to the northern states to escape the effects of a waning cotton economy and with the hope of finding more productive and congenial surroundings."[100]

The U.S. military and Public Health Service's environmental control measures around military bases and production sites throughout the South also contributed to the rapid falloff of malaria. With the outbreak of World War II, these agencies mobilized substantial resources to control malaria. United States and Allied troops had deadly encounters with the disease during World War I in the Near East and East Africa. So early in the World War II conflict, U.S. medical officers predicted that malaria would be the "number one medical problem of the war."[101] Key to the control effort was the development of pesticides that could be used to sanitize military facilities from mosquitoes wherever in tropical zones Allied soldiers were transported. In 1941 some of the first samples of the powerful insecticide DDT arrived in the United States. Once its efficacy for killing malaria-carrying mosquitoes was proven, the U.S. manufactured about 9.5 million pounds of DDT in 1944 and 47 million pounds a year later.[102]

The Office of Malaria Control in War Areas, headquartered in Atlanta, coordinated malaria control throughout the South. It directed drainage, filling, and the use of larvicide in nearly twenty-two hundred military-related sites throughout nineteen states. The agency's programs were far more effective than the stopgap emergency measures initiated by the federal government during the mid–1930s. From the design stage, malaria prevention facilities were incorporated into the overall construction projects at these sites, and a permanent work force was put in place to monitor the operational effectiveness of malaria control. These "protected" military sites drastically reduced malaria mosquitoes in ways reminiscent of the early Canal Zone experience (see map 5.2).[103]

The cumulative effects of these social and public health changes led to an end to the South's historic malaria crisis. Indeed, William McNeill observed drastic falls in malaria cases internationally, following World War II, where similar measures were introduced. According to McNeill, "the sudden lifting of the malarial burden . . . in the years immediately after World War II was one of the most dramatic and abrupt health changes ever experienced by humankind."[104]

Given this broad onslaught against malaria in the South by the federal government, gradually the malaria problem in the Black Belt began to subside, albeit much more slowly compared to the U.S. South as a whole. The lower Mississippi Valley, where much of the Black Belt was situated, remained the environmental region in the South with the highest malaria endemicity. Moreover, most of the Black Belt communities remained in chronic poverty and largely were not reached by public health measures of any kind. Therefore, malaria rates among southern blacks were three to five times greater than rates among whites, although the overall number of new cases had fallen precipitously by 1950.[105]

Changes in the direction of the public health profession also contributed to malaria remaining a significant problem in the Black Belt. Public health research and academic training continued to depart from the environmental and sanitation priorities that had been stressed at the turn of the century. At the same time that military measures to control malaria were expanding both domestically and in foreign military theaters, the field of molecular biology was attracting ever larger resources in the United States for malaria research. The continued growth of this new field reflected the intensification of the bioscientific mission within the U.S. medical and public health professions. Thus, an irony was unfolding in the U.S. medical and public health professions during the World War II decade. Recognized environmental approaches to reduce the malaria scourge at home were becoming even more unpopular among public health experts, but at the same time the U.S. military implemented highly effective malaria prevention measures for its troops as they fanned out across the globe.[106]

The numerous research conferences held by the nation's public health agencies and professional scientific associations stressed that the future emphasis in malaria control should center on analyzing molecular biological aspects of malaria vectors and parasites.[107] In 1941 a symposium of the nation's leading medical and entomological specialists took place. The meeting had been convened by the principal scientific and medical associations in the United States, including the American Association for the Advancement of Science and the American Society of Tropical Medicine. Louis L. Williams, the senior surgeon for the Public Health Service, addressed the symposium, speaking on the anti-malaria program in North America. Williams pointed out that "every republic of North America is engaged in some research on malaria." He cited new planning and supervision of irrigation projects as well as "airplane disinsectization" as new emerging priorities. But it was the area of chemotherapy that held great promise: "we need to develop a true causal prophylactic so that people may visit or even live in an area where anopheline control is, as yet, impossible."[108]

The search for pharmaceutical prophylactics continued to lead Williams and his generation of malariologists further away from sanitarianism. Williams urged his colleagues to find a treatment that would somehow weaken or eliminate the infectivity of the malaria parasite (the sporozoite) when it is first introduced by the mosquito into the bitten individual. "We need to discover a drug which not only cures the clinical attack but sterilizes the patient and so prevents chronic malaria [by] reducing the sporozoite rate in the mosquito," Williams stated. "For this purpose our researches should open up the almost unknown field of the exoerythrocytic phase of the plasmodium in the human host."[109] Unfortunately, the drug-treatment approach only addressed a small fragment of the scores of factors causing the spread of malaria infection and epidemics in human populations. These factors include the environmental makeup of the community, not just the types of mosquitoes or microbial parasites visited on an individual victim.[110]

Above all, as Canal Zone and then, later, League of Nations health officials had learned and demonstrated, malaria eradication hinged on a political entity's commitment to implement a comprehensive attack on environmental conditions that sustained the insects and sick populations infected with the disease. In the midst of the years that the Panama Canal was being constructed, Dr. J. H. White, president of the American Society of Tropical Medicine, articulated this fact almost like a religious prophet who had reached perfect enlightenment. In his address to the annual meeting of the Society in 1912, titled "The Eradication of Malaria," White called for his colleagues' undivided dedication. "The actual work of eradication of malaria is elemental in its simplicity, but calls for untiring watchfulness and absolute thoroughness." Four decades later, one of the nation's leading tropical medicine experts addressed the Society on the progress against malaria over the past half century. After reviewing all of the strides attempted in different parts of the world, he told the group that White's remark of 1912 remains "an amazing understatement."[111]

Finding Roles in the "Atom City" South

By the 1950s, scientific modernization had become the guidepost for agriculture, manufacturing, government, and higher education throughout the South. The mobilization for the Second World War had greatly increased the spread of industry in the South. Much of the new industrial activity spawned by the wartime demands for supplies provided groundwork for permanent industrialization. The construction of factories and warehouses and the manufacturing of electrical machines and transportation equipment for the war effort left in place industrial resources that served the South's industrial growth after the war. Gradually, new manufacturing thrusts took hold, especially metal manufacturing, food processing, and apparel manufacturing.[112]

Also during the late 1930s through the 1940s mechanized methods for cotton production expanded rapidly. These automated processes changed the internal structure of the plantations in the Black Belt. A mechanical cotton harvester had been developed in the 1930s. It led the way for wider use of new machinery that planted, thinned, cultivated, and harvested cotton crops. Advances in chemical production also generated synthetic fibers that, in turn, were bringing down the price of cotton. International competition was also stiffening from cotton-producing countries in Africa and Latin America.[113]

In this context of modernization in farming and textiles, the malaria crisis that had vexed the South waned and all but disappeared by the 1960s. But the poverty of blacks in the rural South persisted. As the traditional plantation's reliance on manual workers declined, black sharecroppers began to lose their function and, hence, their employment. The paternal supervision by local landlords of black sharecroppers diminished as short-term "operators" replaced the provincial landowners. Hundreds of thousands of black (and white) tenant farmers became displaced, moving to a different farm each year. As one southern sociologist wrote, "[t]hose tenants who have not succeeded in locating upon good land or with a fair landlord are continually searching for better conditions."[114] Steadily, black southerners were drawn away from the Black Belt plantations and small farms to nearby and then distant industrial sites and cities—areas that promised new jobs in industry or service.

At the same time, industrialization further advanced in the South and grew by leaps and bounds throughout the rest of the nation. During the 1940s southern farming made large strides in technical modernization. Use of machinery, fertilizer, and pesticides increased, and so did output per acre. Synthetic fabrics such as rayon also shrunk the national and foreign cotton markets.[115] The tougher conditions facing traditional sharecroppers, combined with these technical and market changes, produced huge waves of blacks leaving agriculture or the southern region altogether. By 1945 the FEPC estimated that one hundred thousand blacks a year were migrating from the southern states.[116]

Swept up in the frenzy of World War II, the South's economic, political, and educational leadership had joined fully the federal government's quest to lead the world by pursing technological progress and scientific universalism. The post–World War II momentum to reshape American society through scientific innovation was captured in a special report to the president by the government's leading scientific advisers. It was titled *Science, the Endless Frontier.* The distinguished scientists called attention to the great responsibility facing the United States in world politics and its current "deficit" in scientific and engineering personnel. "All patriotic citizens," the report exclaimed, "well-informed on these matters, agree that for military security, good public health, full employment and a higher standard of living, these deficits are very serious." The panel appealed for an increase in funds and other resources from both government and industry to expand the production of the U.S. scientific workforce. An increase in the number of doctorates and professionals in the sciences and medicine was drastically needed. Also, there was a need to convert the military personnel returning from abroad into scientists and technical personnel of all sorts.[117]

In the eyes of the nation's black political leaders and educators, black southerners were not benefiting from the buildup in American technology and scientific personnel. In the modern technical systems cropping up throughout the South their absence was particularly disturbing because most black Americans still resided in this region. Increasingly, this pattern of racial exclusion was fueling public and political agitation among prominent black American activists and educators and their allies in labor and politics. By the late 1940s giant-scale atomic energy projects were under construction throughout the South and the rest of the nation. The federal agencies in charge of these projects became targets for the outrage of black activists and their civil rights allies.

The journalism of Enoc Waters, a war correspondent for the *Chicago Defender,* captured black leadership's mood of frustration. His 1946 article, "Atom City Black Belt," circulated widely in the black press. It blasted segregation at the TVA's prime site in Oak Ridge, Tennessee. Waters called Oak Ridge "home of the awesome atom bomb," but also a "city of paradoxes" because many blacks there lived in poverty. Oak Ridge was an emblem of the most sophisticated and most seamy fruits of the United States' emergence as a superpower:

> Here at [Oak Ridge,] the secret birthplace of the atomic age, some of the nation's greatest brains are engaged in exploring the unknown future of atomic power but Negro children are denied the right to learn their ABC's. Here, cloaked in mystery, inanimate gadgets of steel and glass are housed in concrete palaces where temperatures are controlled to a fraction of a degree but Negro workers must live in flimsy packing box structures set flush on the muddy earth. Here millions of dollars are marshaled to exploit atomic theory but not enough pennies can be corralled to provide for the welfare and comfort of a few thousand Negro workers.[118]

As for the planning that went into the Oak Ridge community and its projects, Waters had more biting words. He called Oak Ridge unique because it is "the first community with slums that were deliberately planned. The concept

[in] back of the planning and operation of this small city is as backward socio-logically as the atomic bomb is advanced scientifically."[119]

The popular black magazine *Negro Digest* carried many pieces that con-demned America's leadership for having the capacity to win the war against Ger-many and Japan, but not against southern segregation. Fannie Cook, a popular liberal novelist, called for an end to segregation in her article "An Atomic Ap-proach to Racism." She applauded the patience of American blacks in the face of brazen discrimination. "If I were a Negro," she wrote, "I should ask my coun-try to search out the secret of wholesome interracial living from a group of spe-cialists as able and earnest as the group from which it sought and gained the secret of atomic energy." But in the meantime the United States continued to "stumble down the interracial highway guided only by the whims of the blind, sullen, confused Southern reactionary."[120]

Walter White, executive secretary of the NAACP, wrote a somber essay in *Negro Digest* called "Race in the Atom World." He was astonished that south-ern congressmen were threatening "everything from political secession to mob violence" to obstruct President Truman's civil rights initiatives. Although the United States was building ever more destructive weapons to fight totalitarian threats, White decried that "we have not yet been able to devise a formula through which men of diverse political, economic, religious and racial groups can live together in justice to each other." He pointed out that the inability of the nation to defeat Jim Crow segregation lowered America's moral leadership of the free world. "Frightened by the incredible power which the splitting of the atom has loosed," White exclaimed, Americans still continue to "set race against race, creed against creed, class against class in our fear and ignorance. We do this despite the fact that we know that another world war could mean nothing else but the complete destruction of civilization."[121]

The significant growth of mixed industry throughout the South was ac-companied by increased urbanization. By 1960 about 36 percent of blacks re-siding in the South lived in central cities, and another 12 percent in other parts of the South's metropolitan regions. Between 1940 and 1960 the nation's black population and number of black workers grew markedly. During this same pe-riod the number of black workers employed in the South dropped by 160,000. Some 923,000 of the South's blacks left farm work, 660,000 of whom entered nonagricultural industries.[122] Outside the South, job avenues appeared to open, prompting the remaining tens of thousands of employable blacks to leave per-manently the harsh life of the Black Belt South. In short, in the South at the opening of the twentieth century, technical organization had created the Black Belt. Now in the post–World War II decades, this same force destroyed it.

CHAPTER 6

Alliances and Utopias

LIBERIA'S SEARCH FOR DEVELOPMENT

*O*ver the course of the nineteenth and early twentieth centuries, U.S. advances in large-scale technology and public health programs developed mostly in its northeast states. In relationship to North America's neighboring Atlantic and African regions, technological spin-offs first reached the black populations of the plantation South, then Panama, Haiti, and, lastly and only faintly, Liberia. However, the outbreak of World War II marked an abrupt surge in America's interest in sending technical expertise and programs to Liberia. Compared to earlier periods, U.S. technical activities in Liberia differed markedly after World War II.

During the start of the twentieth century, U.S. missions abroad to improve public health and commercial and public infrastructures had been orchestrated mostly by individual industrialists, military administrators, government public health leaders, philanthropists, and religious missionaries. However, after successes such as the Panama Canal, intercoastal railroads, telephone expansion, and the U.S. military's performance in World War I, the nation's political leaders and "captains of industry" were even more bent on pointing the economy, indeed society, toward a science-derived utopia. A new class or strata of scientists, engineers, technicians, public health specialists, and managers was spawned inside the companies, government, and academe.

When the United States became involved in the World War II conflict, two immediate conditions drew some of the U.S. technical strata onto Liberia's shores. First, the U.S. military needed access to North and West Africa to further its intercontinental military strategies. Second, American medical scientists needed to intensify their search for means to control and treat malaria and other tropical diseases that threatened Allied military forces worldwide. As World War II raged, the U.S. military and contractors funded from foreign aid built airfields and related transportation links in Liberia to facilitate the movement of army personnel and supplies. Furthermore, the U.S. Public Health Service sent small teams of medical workers to Liberia to work in and around U.S. installations. Sanitary measures were implemented to minimize risks to U.S. soldiers from such diseases as malaria, yellow fever, and venereal disease. Some of these military health provisions were expanded to include local Liberians. Also, U.S. tropical medicine specialists established a research center in Liberia for studying tropical diseases endemic to Liberia and other parts of the tropical world in Africa and Asia.

At the same time that the United States began small levels of technical assistance in Liberia, Liberia itself embarked on its own mission of scientific

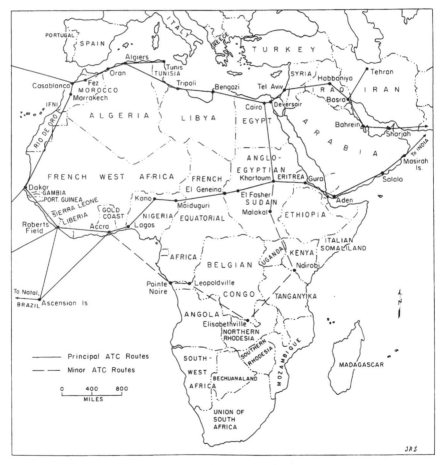

MAP 6.1. Air Transport Command Routes, U.S. Armed Forces, Africa and Middle East, 1941–45. Reprinted from U.S. Army, Medical Department, *Preventive Medicine in World War II,* vol. 6, *Communicable Diseases: Malaria* (Washington, D.C.: Office of the Surgeon General, 1963).

progress. Its new president, William V. S. Tubman, set out on a campaign to modernize Liberia, a campaign that would follow America's lead.

"A Little Place Called Liberia:" Assistance under FDR

Shortly after the U.S. military entered the World War II conflict, it established a treaty with Liberia to facilitate transportation activity in West and North Africa. A growing number of U.S. military personnel worked in or were transported through Liberia and other nations in North and West Africa. Liberian airfields became important to the Allies' air transport command. In order for the United States to deliver airplanes to the Allies in the Middle and Far East, Africa had to be crossed (see map 6.1). American forces had to man air stations across West and North Africa to relay these planes and troops. Finally, Africa was rich with resources that the United States and the Allies planned to protect from the hands of Germany, Italy, and Japan. Moreover, Japan had captured control of rubber-producing nations in the Far East. Thus, Liberian rubber, in particular, became even more vital to the military and civilian sectors of the United States and the other Allies. By the middle of the war, about five thousand U.S. soldiers were stationed in Liberia.[1]

The public health work in Liberia centered on U.S. air bases in West Africa and began soon after France succumbed to Germany. In the autumn of 1941, Germany encroached on Liberia. President Franklin Roosevelt informed the American public at a New York press conference: "Dispatches this morning showed a very definite attempt on the part of Germany to establish itself, by the infiltration method, in a little place called Liberia. . . . Liberia, of course, is awfully close to South America." He stressed that the United States had to develop a means of "resisting" such maneuvers "for the purpose of our own defense, and hemisphere defense."[2] Fearing that Dakar would also fall under the control of the Axis powers, in 1942 the United States entered into another accord with Liberia, the Defense Areas Agreement. This treaty gave the United States the sole right to construct and defend military and commercial airports and seaports in Liberia as well as to construct roads necessary for this purpose.[3]

The following year President Roosevelt visited Liberia. He stopped over in Liberia as he headed home from the Casablanca conference with Winston Churchill. Roosevelt became the first American president in office to visit Africa. In Liberia, Roosevelt met with Liberian president Edwin Barclay in Monrovia, visited the Firestone Rubber Plantation, and met black American troops stationed in Liberia. Shortly, another agreement was signed stipulating that the United States would provide more defense-related supplies, services, and information.[4] Among the projects was an airfield constructed near the Firestone operations. Also, lend-lease funds were provided for the construction of the Free Port of Monrovia. By 1950 this port would grow to a site that handled nearly fifty-one thousand tons of incoming cargo, while shipping out sixty-four thousand tons. Nearly one thousand ships called at the Free Port annually.[5]

Public health aid to Liberia accompanied the buildup of U.S. military installations and personnel in Liberia. The medical resources the United States began to bring into Liberia were also linked to the larger United States and other Allies' need to find ways to successfully manage diseases in tropical-zone nations worldwide. The Allied forces fighting in the Near East and in East Africa during World War I had grisly experiences with tropical diseases. Liberia offered a relatively close geographic setting for intensive study and demonstrations with preventive treatments and public health measures. Liberia had a perfect disease ecology for pursuing this new research. Malaria was so prevalent in Liberia that in many sections it was considered holoendemic. This is a condition in which transmission rates are very high, malaria is present year round, and adults have developed significant immunity to the disease.[6] However, compared to the massive public health campaigns in the Panama Canal Zone and in Haiti during the interwar period, the U.S. initiatives in Liberia during the World War II decade were sparse and delimited. American technical experts were looking for answers to scientific puzzles both in the microbiological realm for disease control and in the geopolitical realm of military strategy. The needs of U.S. soldiers and scientists came first and foremost, not Liberian communities.

The U.S. military's effort to control disease outbreaks in Liberia started precisely with its troop movements through the region. In August 1941 the United States sent personnel to Liberia for the construction of airfields and related transport facilities. Military officials had to establish public health protections in Liberia for these troops and other oncoming personnel. The climate and environment of Liberia were ideal for the breeding of the *Anopheles gambiae* mosquito. This is one of the most efficient and deadly vectors for malaria.[7] Moreover, Liberia had a tropical environment and endemic diseases similar to those of nations across northern and western Africa and the Far East. Medical authorities involved with the United States and Allied military forces were anxious to use Liberia as a site to study the geographical and epidemiological aspects of Liberia's environment and endemic diseases. Thus, alongside American troops, public health personnel and malariologists were also among those landing in Liberia.[8]

Initially, the U.S. military in Liberia focused efforts to control malaria in the vicinity of the Roberts Field air base. In 1942 a malaria control detachment was sent there. In February of 1943 it was replaced by two units of black personnel, one to conduct surveys, the other to do control work such as clearing brush and draining swamps.[9] The general policy of the U.S. military was to provide health care only to those native populations working directly on military projects. However, since the medical needs throughout Liberia were so serious, American medical units administered services in addition to malaria control. From May 1942 to March 1943, an army-run hospital served both military personnel and the local workers at Roberts Field. Also, a free clinic was established at Roberts Field for the treatment of local women with venereal disease.[10]

Tubman Takes up the Mission: Education and Medicine

William V. S. Tubman, Liberia's president from 1944 to 1971, cherished assistance in public health and medical training as much as foreign economic investments. Tubman intended to somehow build on the American projects and demonstrate to the Liberian people that his political program was employing the power of modern science for national development. What to U.S. political officials were public health measures based purely on military necessities, to Tubman were grand gifts of medical knowledge to the Liberian nation as a whole. Tubman imagined that somehow his leadership would translate the U.S. health activities in the Roberts Field area into medical benefits for the entire Liberian nation.

From early in his career, Tubman began emphasizing a mission to modernize the minds of the Liberian people. He would do this by spreading modern education, public services, and capitalist economics to local communities. The incorporation of the native peoples into government and modern life had been the biggest obstacle to Liberia's political and economic unification. Tubman ardently believed these fragmented ethnic populations could be unified. The unity would be realized by expanding modern education and health care.

At his inaugural in 1944, Tubman laid out his plan to unify and modernize Liberia. He emphasized he would work for improvements in all facets of national life, including foreign policy, education, agriculture, public construction and commerce, women's suffrage, and public health. Of the last area, Tubman remarked, "No pains will be spared in disseminating in every possible manner, throughout the Republic, information that will help our population in understanding and practicing sanitation." The nation's small community of health professionals was largely foreign nationals and was overburdened with massive numbers of sick people. Tubman urged Liberians to be patient. In the meantime, he said, "every encouragement will be given to qualified physicians and hospitals to carry on the good work they are doing, pending the time, and that soon, when [our] Government will be able to augment these by hospitals and physicians of our own."[11]

A deep pragmatism lay beneath Tubman's approach to politics, and positivism and scientific idealism lay beneath his overall philosophy. In his public discourse, Tubman frequently preached the theme of self-motivation backed by dutiful performance. Speaking at a college commencement in Monrovia early in his presidency, Tubman stressed that "thought without action has no force." He criticized those who imagined change but could not or would not take concrete steps to realize that change: "To think and not put into action such of our thoughts as are laudable and possible for execution, is like faith without work—dead. We become mere theorists. We may soar high in our thinking, formulate great ideas and ideals, but unless we can activate these ideas and ideals which are resultant of our thoughts, we do nothing; we live in the realm of thought and talk and build only air castles which are intangible."[12]

In Tubman's view, for any domestic policy to be effective for Liberians, the overriding ethos of public policy had to be scientific and secular. Tubman believed that Liberia could eventually take its place among the developed democracies of the world only through modern technology. From the first year of his election, President Tubman set about carefully bargaining with tribal leaders to broaden his base of loyalty nationwide. In his inaugural speech of 1944, Tubman stressed, "As long as the vast majority of our citizenry is alienated from full social and political participation in the government, we would be perpetuating the germ of dissatisfaction and eventual disunity within our brother." According to a Liberian political biographer of Tubman, Robert A. Smith, the president followed his ideas with action. Smith points out that Tubman believed "roads, schools, and hospitals must be built, not only because the people wanted these things, but because without them there can be no national reconstruction, no replacement of local feelings by wider allegiances."[13]

The dedication felt by Tubman to the mission of Western-style modernization was immediately noticed and endorsed not only by Western politicians and military officials, but also by philanthropies sponsoring programs in Liberia. U.S. religious missionaries commended Tubman's pledge to establish higher education and schools for the medical fields in Liberia. In 1945, Rev. C. C. Adams wrote an essay in the black American press, "Christianity and Education Bringing Light to Liberia." Adams, the secretary of the Baptist Foreign Mission Board, recounted that his denomination and eight others, including the Lutherans, Catholics, and Episcopalians, were operating mission schools throughout Liberia. Now, with the Liberian government working in direct cooperation with the religious schools, a new "educational saga" was unfolding. The "thousands of boys and girls . . . together make a mighty army engaged in the solemn task of conquering an empire of ignorance and superstition."[14]

Reverend Adams also stressed that although the church hospitals were currently training nurses, Liberia needed a medical school so that it could train its own doctors. He explained, "The witch doctor stays in Africa because the Christian doctor remains away." The minister stressed that "hospitals, clinics, doctors, and nurses are the only answer. . . . The witch doctor, like darkness, is only present by the absence of light—bring in the light and darkness cannot stay."[15]

Tubman's education program stressed a substantial increase in the number of educated Liberians, whether through public or mission schools. To his thinking, the minds of the ethnic peoples comprising Liberia's major population had to be pried open to functional, secular education at any cost. His government spurred as much public funds as possible to increase schools throughout the nation and gave support and educational guidelines to mission schools.[16] Between 1948 and 1959 the number of government schools rose from 105 with 203 teachers to 678 schools with 3,468 teachers. As for mission schools, during the same period they grew from 103 with 363 teachers to 151 with 576 teachers. The total number of students increased from about 20,000 to 72,000 out of about 225,000 school-age children in Liberia. Tubman also was not concerned with stressing one type of education for Liberia's citizenry. Early in his presidency he called for the "removal of illiteracy from within our borders," a goal that is "vitally essential to our very national existence." Thus, he supported "education in its general and broadest aspects, vocational as well as agricultural."[17]

However, Tubman never envisioned a serious attempt to incorporate into his modernity movement the traditional agrarian institutions involving health, economy, and education. Intellectually, Tubman knew that the ethnic rural peoples making up the bulk of Liberia's society had their own philosophy and values about education and their own indigenous technologies for food and market production. But he either refused or failed to understand that these same people had their own visions of what community development projects and welfare agencies should be or do.

The mission for science burnt deeply inside Tubman's political soul. Specifically, he was astonished at the progress in Western medical science and doctoring, and this progress became the paradigm for his movement to bring secular education and modern development to Liberia. Throughout his presidency, Tubman believed it was vital for government to provide medical services to all of Liberia's people. To him, modern medical care sponsored by government was vital for nurturing the Liberian populace and a healthy nation-state.

Inside Liberia, effective medical care would win loyalty to the president and government workers from the poverty-ridden and ethnically divided populace. Tubman embarked on not only increasing schools, but also spreading awareness about public hygiene and the benefits of the modern doctor. Tubman believed instinctively that by providing modern medical services for his fellow Liberians, he could blunt popular opposition to social problems linked to the large U.S. firms extracting Liberia's natural resources. He stressed to his political peers that Liberia needed to build a hospital network and medical profession patterned after those of the United States and European nations. Indeed, Tubman's view that modern U.S.-styled hospitals and doctors were vehicles for social progress, not neocolonialism, offers an interesting parallel to prominent leaders of the other regions we are investigating, especially Booker T. Washington and Francois Duvalier.

Medical Utopianism and Nationhood

In their social and professional activism, Tubman, Washington, and Duvalier all voiced strongly the "medical police" ideal. This was a political philosophy that extended great political importance to medicine for civic life and collective security. The medical-police ideal emerged in early modern Europe and America. During the eighteenth and nineteenth centuries, the working class throughout the West experienced food and housing shortages and epidemics caused by industrialization, mass warfare, and poorly organized governance. The central governments of European nations and the United States were forced to confront public health problems to stave off labor revolts and sustain and augment the power of the national state. Thus, monarchs and presidents in nations such as France, Germany, Austria, and the United States created and expanded administrative resources for controlling epidemic diseases and for sanitation. European governmental heads expanded the national state apparatus to include administering hospitals for the general public as well as for military garrisons. Public health agencies that could monitor and improve public sanitation, dispensaries for the poor, and services for maternity care and child protection also were greatly enlarged.[18]

Similarly, Liberia's President Tubman strove to expand the legitimacy and popularity of his government by espousing the gospel that public health and modern medicine must be high national priorities. Similarly, Booker T. Washington established the National Negro Health Week movement not just to educate the black lay public about effective public hygiene practices. The hygiene campaign was also intended to increase civic cohesion and personal loyalty to the Washington program throughout the communities of the Black Belt South.

Tubman's medical and technical utopianism was central to his broader political program, which he called "National Unification." This policy was the first serious effort to eliminate the century-old barrier between the ruling Americo-Liberian elite, centered in Monrovia, and the many distinct ethnolinguistic groups of Liberia's rural hinterland. Tubman envisioned somehow bringing the entire Liberian populace aboard his train of secularization, technological development, and medicalization. His emphasis would be on physical and cultural modernization and the institutions he considered reflective of this modernization. Modern schools, hospitals, and government agencies would be the connecting fibers between the Americo-Liberians and the so-called natives. Leopold Senghor praised the apparent pragmatism of Tubman's nationalism. "In 1943," Senghor recalled, "William Tubman . . . was going to revolutionize Liberia, and give it not only a modern ideal but modern means to achieve it."[19]

Throughout his presidency, Tubman expressed recurrent interest in waging national war against disease and ill health. His interest was not only in reaping the benefits of medical policing for his administration, but in going even further by eliminating the premodern cultural world views of Liberia's rural masses—world views based on religion and superstition.[20]

As Liberian nationals were absorbed into foreign-owned mining operations and plantations as workers, and the Americo-Liberians into the nation's small public sector, Tubman envisioned his government medical programs ensuring the healthiness of this "human capital." His administration would provide medical resources based on advanced Western medicine, thereby providing healthy, trainable personnel for foreign investors and projects. In his public speeches throughout his presidency, Tubman emphasized the idea that the "hospital will come to wield a great influence in . . . the nation" and will be served by "a corps of dedicated Liberian and foreign workers imbued with the tradition of the medical profession—service to God and humanity."[21] The modern hospital, to Tubman, was a virtual sacred institution. In his view the hospital was the heart of a medical utopia in which people were administered to with the most refined forms of scientific healing and expertise. In the mid–1960s Tubman spoke at the dedication of the new St. Joseph's Hospital and Pre-Medical College. The facilities had been built from support received by the Vatican, the University of Torino, the Italian government, and the Catholic Church in Monrovia. Tubman laid out his utmost faith in these facilities: "This hospital and its adjuncts, like all other hospitals, should be citadels waging war against the enemy of our commonality—Death. Sometimes its Captains, Lieutenants and Non-coms, the doctors, nurses and attendants, succeed in compelling him to retreat but at last he wins the battle and takes men away as his prey: but the battle goes on and those engaged in the medical army continue their fight against death in what I consider the most glorious warfare on earth."[22] As for the medical training the hospital would provide, Tubman called this a gallant service. He drew a distinction between armies and medical centers: "The mighty armies of great military powers . . . are armed with weapons of destruction, and which induce fighting and battling on the side of death, engaging in mass destruction and the carrying away of human beings like a flood." By contrast, "the hospital and clinic stand as a formidable fortification in the fight between life and death." According to Tubman, the role of doctors in all fields of the medical profession including "graduate, student and practical nurses, midwives, medical aides and hospitals attendants is beyond compare, for they are rendering a service to humanity that is more noble and more grandiose than all the formidable military armies of the world."[23]

However, in the context of Liberia's extreme poverty, disease, and ethnic diversity, Tubman's idea that the pursuit of clinical medicine would somehow power national modernization had fundamental flaws. To many non-Western intellectuals and political heads, Tubman's idea overlooked serious philosophical and political dangers inherent in the industrial West's science mission. To the generation of leaders of the less-developed world, convening meetings such as the Bandung Conference (1955), Western science was one that privileged and segregated the goals of the scientists' specialties over the immediate needs of society's poor for the basic elements of life and healthy environment. Dependence on Western scientists and technicians also perpetuated the condition in which sophisticated technical instruments and personnel—complex machinery, utilities, medical clinics, and the like—remained in foreign hands. Finally, Tubman's idea of national development systematically avoided attempting to build modern technologies in Liberia from the bottom up by using indigenous technologies and ethnic institutions.[24]

The actual state of medical resources for Liberia as a whole reflected the horrendously difficult road President Tubman was walking in trying to create a nationwide network of modern, Western-style medical services. In the early 1940s Liberia had a patchwork of a few government facilities, local health projects sponsored by foreign religious missions, a smattering of private physicians, largely of German, French, Hungarian, and American descent, and the medical services operated by foreign corporations at their work sites, primarily for their employees. Hospital facilities were sparse in quantity and gravely lacking in quality. The largest, most modern medical facility in Liberia remained the hospital (now having one hundred beds) operated by Firestone Plantations Company at Harbel, near Roberts Field. Firestone also had a medical facility at Gedetabo (Maryland County), close to Cape Palmas. There were two small hospitals in Monrovia—the Liberian Government Hospital, with about thirty-five beds, and the Baptist Foreign Mission's twenty-five-bed Carrie V. Dyer Memorial Hospital. Both of these facilities had inadequate equipment, staffing, and supplies. The government also operated four general clinics, one each in Tappeta (Central Province), Voinjama (Western Province) at Cape Palmas (Maryland County), Marshall (Marshall Territory), and Sinoe (Sinoe County).[25]

Also, some small hospitals and health stations were run by foreign religious missions: the Phoebe Lutheran Hospital, which lacked an attending physician, in Harrisburg (twenty-eight miles from Monrovia); a Methodist Mission Board clinic in Ganta (Central Province); and a clinic in Suoccoco, operated by the Assemblies of God Missions. But medical care resources designed to reach the general Liberian population outside of Monrovia and directly funded and administered by either the Liberian government or the United States were virtually nonexistent.[26]

Technical Projects without Democracy

Similar to medical utopianism, technological utopianism has inherent limits in the building of a popular democracy and viable economic system for a nation of rural poor. The execution of technological projects does not require parliamentary or popular democracy in the social context to succeed. Indeed, as we have seen in the case of the use of tenant farmers and convict labor on harsh southern plantations, as well as the U.S. military's corvée labor in occupied Haiti, democratic protections for workers were of little importance in these situations. Likewise, Tubman and his fellow government officials organized an elaborate patronage network between government, private investors, and tribal chiefs. This arrangement forged rural Liberians into cheap labor for the large foreign-owned rubber plantations. Another means by which the Liberian government bypassed basic needs and protections for Liberian workers was to let contracts to foreign interests. These expatriate firms then set up enclave technical projects—for example, construction of a government building, a hospital, or a road—employing only expatriate managers and workers.[27]

Throughout his administration, Tubman was hostile toward labor organizations and trade unions that attempted to represent large groups of rank-and-file workers. But this just fertilized recurrent strikes during his administration. Unionization was one of the few avenues of recourse available to Liberian workers amassing at the foreign-owned industrial sites and plantations. The first major strike of Liberian nationals occurred in December of 1945. American construction managers from the U.S. Navy and the main private contractor (the

Raymond Concrete Pile Company) were the key employers of dock workers at the Monrovia port. However, the workers at the port walked out because they were underpaid. They were also protesting other problems involving overtime pay and health benefits. Wages for foreign employees, including truck drivers, clerks, and technicians, were nearly double those of Liberians in these same occupations. As for overtime pay, foreign employees received this rate when their weeks' work exceeded forty-eight hours, while local Liberian workers labored sometimes as long as eighty-four hours without receiving the overtime wage.[28]

Inadequate medical care, lack of accident insurance, and prohibition from using the canteen were also key issues raised by the strikers. The Liberian workers only received medical care if injured on the job, while expatriate workers received medical services regardless of the source of their illness or injury. Moreover, Jim Crow segregation was practiced inside the camp hospital run by the construction company. The canteen was stocked with duty-free imported goods sold at cheap prices. Liberian workers wanted to benefit from the substantial savings that the purchase of these goods provided to the imported workers. The strike was settled when the construction contractors agreed to make improvements in all of the contested areas. They also hired one of the strike leaders to serve as a public relations officer to improve employee relations at the company.[29] Nonetheless, serious strikes continued to break out, including a strike by mechanics in 1949 followed by eleven to fourteen more strikes through 1961.[30]

The national revenues and foreign aid received by Liberia Tubman doled out in two directions. He funded specific technical projects such as the construction of new public hospitals. Also, Tubman used these funds as well as jobs in development projects as payoffs to political supporters, family members, and loyal tribal leaders. By producing concrete sites with modern facilities and medical services, no matter how limited the size of the populations that these facilities served, Tubman accomplished two ends. He was able to extend his dream of modern development while cultivating political stability with his politics of patronage and personal authority.[31]

The Public Health Mission in Liberia

Given the Tubman administration's preoccupation with developing modern health care and technical works modeled on those of Western nations, the appearance of any U.S. military and medical personnel in Liberia for assistance projects during World War II was most welcomed. President-elect Tubman and President Roosevelt had begun corresponding in the opening years of World War II. In early 1944 Tubman took office. Liberia broke from its policy of neutrality and officially declared war against Germany and the Axis powers. The doors for U.S. assistance to Liberia to further the Allied war effort were open. One of the first items Tubman addressed was the need for the United States to send a medical team to Liberia to assist in eliminating the severe health problems.

Likewise, American military officials were also interested in gaining more precise knowledge about the types of sanitation problems and diseases that American troops would face once in Liberia. In March of 1944 the U.S. State Department requested that the Public Health Service dispatch a medical and public health team to Liberia.[32]

As observed earlier, the United States' initial role was limited strictly to curtailing local diseases and poor sanitation threatening U.S. military personnel. U.S. soldiers returning from foreign service were bringing their tropical infections back to U.S. civilian populations. Some five hundred thousand cases of malaria had been recorded among U.S. soldiers during the early years of World War II. Consequently, the other main function of the Liberia medical team, according to its head, was "to protect other parts of the world, particularly the United States, from introduction of exotic vectors of disease."[33]

In November 1944 a small number of black medical personnel from the U.S. Public Health Service arrived in Liberia. Known as the U.S. Public Health Mission in Liberia (PHML), the group of eleven black American physicians, engineers, entomologists, and nurses was directed by the surgeon John B. West. Hildrus A. Poindexter, a leading bacteriologist, conducted the laboratory work. This mission had originated at the request of President Tubman, through Dr. West, who was stationed in Liberia with the U.S. Army medical corps.[34]

The PHML fulfilled the interests of both the U.S. military and Liberia's officialdom. By sending these African-descent medics, U.S. political officials were assisting symbolically President Tubman's intention to somehow build a nationwide medical service network that could reach all sectors of Liberian society. To the U.S. military and tropical medicine community, the PHML served two important functions. First, the medical mission contributed to the U.S. Army's efforts to control malaria mosquitoes "about the airports and port development." The U.S. acting secretary of war described this rationale for the mission to Dean Acheson, secretary of state: "The War Department has, at the present time, a military interest in Liberia. It is my opinion that the assignment of the [PHML] mission by the United States Public Health Service to Liberia would be complementary to our military interests . . . if limited to a type of extra military sanitation and disease control."[35] Second, field research by the PHML would contribute to the effort by the U.S. tropical medicine community to learn more about the parasites and treatments for infectious, warm-climate diseases, especially malaria. As for Tubman, he imagined the U.S. medical team was a large-scale assistance project. President Tubman saw to it personally that Dr. West would observe firsthand the severe medical needs of communities he was sent to visit. But usually all West and his teams ended up doing in the communities outside of Monrovia was lend a sympathetic ear.

The PHML first surveyed local health conditions and medical resources and consulted with physicians, some of whom had long years of practice in Liberia. General health conditions throughout Liberia, they discovered, were exceedingly poor. Infectious diseases were widespread, especially malaria, yaws, enteric diseases, and helminthiasis (intestinal parasites or worms). The immediate root for this high concentration of these diseases in Liberia was unhealthy environmental conditions. Unsanitary water supplies and crude sewage-disposal conditions provided a fertile breeding environment for flies, mosquitoes, infected animals, and other disease vectors.[36]

As for health care resources in Liberia such as doctors, health centers, and sanitation measures, the PHML found conditions shocking. Dr. West described the dire situation in Liberia in an article in *Public Health Reports,* the key research organ of the U.S. Public Health Service. Writing about the PHML, West stated, "Upon our arrival, we found that to the best of the available knowledge there were six physicians, two dentists and an indeterminate number of nurses practicing in Liberia, which has a population estimated at 2 million." Among all the church and government clinics throughout Liberia, only three had a physician with an M.D. degree in attendance.[37]

In a typical month of activity during its early phase of operation, the PHML team made improvements to some rooms of the government hospital in Monrovia. They also inspected nearby wells, ordering those that were unsanitary closed. The mission also conducted a small entomological project, identifying hundreds of malarious mosquitoes *(Anopheles gambiae)*. Finally, they oversaw construction of their public health compound.[38]

Given Liberia's acute medical care needs, Liberian and U.S. officials needed to devise health campaigns for any number of serious diseases and unhealthy living conditions. These problems included malnutrition, poor prenatal care causing high infant and maternal mortality, and diseases, such as tuberculosis, yaws, and venereal disease, that incubated primarily within immediate communal or interpersonal household surroundings. Posing little threat to U.S. soldiers in Liberia who were garrisoned in their own encampments, most of these conditions were only of secondary importance to U.S. officials and medical academics. To U.S. authorities, malaria was the most serious of these health problems and the disease most in need of frontier, high-level research. Its mosquito vectors could easily regenerate and then re-infect. Thus, malaria threatened and rethreatened U.S. military and engineering personnel throughout all parts of the tropical world. For this reason, malaria was seen as an impediment to U.S. globalism and a prime target for scientific mission work.

The PHML proceeded to survey malarial insects and conditions in Monrovia, Kakata, and Roberts Field, the site of the U.S. air base under construction. It began a reduction program with mass DDT spraying that covered over 90 percent of the inhabited structures of Monrovia and Roberts Field. The sprayings had to be repeated about every four weeks to be effective, and in Monrovia this involved about 1,290 houses in that city. Next, the now largely Liberian crews reduced breeding areas for the mosquitoes by laying drainage ditches and pumping casual waters. This program was repeated on Bushrod Island, another area dense with malarious mosquitoes and part of the site for the new dock project.[39]

This early anti-malaria program resulted in significant reductions of the disease in Monrovia and Roberts Field. However, these reductions were temporary and, by themselves, could never be lasting. The malaria mosquitoes were becoming resistant to insecticides like DDT. Moreover, such pilot projects did not reach the environmental conditions of the great masses of Liberians in the hinterland, daily facing rampant malaria infection. One small project that did prove successful involved a smallpox vaccination program in the hinterland by the PHML. In 1946 and 1947 one of the PHML doctors noted dozens of smallpox cases in hinterland villages. The PHML, now made up of twenty-seven professional and technical employees, started a fairly large smallpox eradication campaign. By 1948 there were locally trained vaccinators employed in this mass vaccination program.[40]

The Liberian Institute for Tropical Medicine

The PHML had helped to provide protection against epidemics for the U.S. military personnel in Liberia. But officials in the U.S. military and foreign affairs agencies had much broader goals in tropical disease and malaria research in mind. At the close of World War II, Liberia became the site of one of the world's most sophisticated malaria research centers. Throughout World War II the U.S. military had gained firsthand evidence of the devastating impact of tropical diseases. In World War I the military forces of the United States and the Allies had to contend primarily with malaria and dysentery diseases. But in World War II the United States fought on fronts in both Asia and Africa. Between 1942 and 1945 alone, the intensity and array of tropical diseases among the American armed forces were alarming. There were more than 750,000 cases of dysentery or diarrhea, over 500,000 cases of malaria, nearly 200,000 cases of infectious hepatitis, and about 122,000 cases of dengue. As the U.S. soldiers who had served in tropical nations were returning home, medical authorities saw signs that the tropical infections many of them harbored could threaten America's domestic population.[41]

Now that tropical diseases were crossing borders into the United States and Western nations, the mission to uncover the biochemical basis and cure for infectious diseases was as vigorous as ever among the medical researchers of these nations. Indeed, American medicine had entered its "golden age of antibiotics." Pharmaceuticals such as penicillin and streptomycin were proving astonishingly effective for treating diseases such as syphilis and tuberculosis. Throughout the United States, Europe, and Japan, jubilant expectations abounded that other drugs could be found to eliminate the world's ages-old plagues. Researchers of viral diseases began laboriously shifting focus from pathology to biochemistry, and bacteriophage agents were found to be the vital link to understanding viruses and host bacteria cells as well as the nature of genes.[42]

An offshoot of the energetic and growing agenda for U.S. biomedical research was a renewed focus on malaria. To American tropical disease researchers, the high prevalence of malaria as well as other major tropical diseases made Liberia a logical site for a malariology and tropical medicine research center.

Shortly after World War II, private philanthropic and scientific interest groups in the United States established the Liberian Institute for Tropical Medicine. Planning for this research center had begun in 1947 with a gift of $250,000 from Harvey S. Firestone Jr., supplemented by special grants from the U.S. government. The institute was operated by the American Foundation for Tropical Medicine. The foundation coordinated the support of about twenty U.S. universities and scientific societies interested in tropical medical research. In later years the institute would receive funds from the International Cooperation Administration, the National Institutes of Health, and the Rockefeller Foundation.[43]

Predictably, from its inception the Liberian Institute received hearty support from President Tubman. In 1947 he wrote to the foundation's president, Dr. Thomas T. Mackie: "We look forward to the benefits that we hope will be derived from the project envisaged by the Institute." Tubman arranged for the institute to receive a one-hundred-acre site, upon which were constructed a laboratory building with two two-story wings, an animal building, and staff houses.[44]

The researchers behind the Liberian Institute stressed the perfect coincidence between the facility and the health needs of tropical countries like Liberia. Mackie stated, "Liberia, the African Republic founded just 100 years ago by freed American slaves, offers ideal conditions to study the three factors which at present keep the peoples of the tropics in virtual bondage." The three conditions were poor agricultural yield, domestic animals' endemic diseases that ultimately reduced local food production, and infectious diseases that kept infant mortality rates high.[45]

The institute's agreement with the Liberian government allowed it tax exemption and permitted its personnel to conduct research projects and practice veterinary medicine in any location of the country. The institute was located three miles from Robertsfield (Marshall Territory), and the institute's staff averaged about eight professionals (three M.D.'s and five Ph.D.'s). Research projects at the institute covered a gamut of laboratory fields. These projects involved protozoology related to human and primate malaria, entomological studies connected with malaria and trypanosomiasis, and helminthological studies centered on snail hosts of schistosomiasis.[46]

Tubman overestimated the usefulness of the Liberian Institute for improving the immediate health problems facing Liberia. Actually, the institute had no mandates, either from its U.S. sponsors or the Liberia government, requiring it to perform educational or public health services for Liberian communities or institutions. Instead, the Liberian Institute was a free-floating center conducting and funneling basic microbiological studies of tropical diseases back to major academic centers in America and Europe. Its officials emphasized, "The Institute is unique in that while it is located in Liberia, it is dedicated to the betterment of all mankind, through progress against the diseases which handicap people living in tropical regions throughout the world."[47]

Similarly, the medical services of the U.S. military were hoping that the basic research from centers like the institute would eventually lead to more effective treatments for tropical diseases. Thousands of World War I and II veterans lay in U.S. hospitals with chronic symptoms and infectivity caused by tropical diseases. Another high priority for the U.S. military was the discovery of anti-malaria compounds that could help those soldiers at risk for infections or already suffering and disabled by malaria. Even though the research of both the Liberian Institute and the U.S. military medical personnel in Liberia was useful to the U.S. medical community, neither these research teams nor the Liberian-government host managed to do much toward local disease eradication significant enough to improve health conditions of typical Liberian communities. Moreover, any idea of transferring the institute to the control of the resident Liberian medical professionals was out of the question given the wholly inadequate medical and public health operations existent in Liberia at this time.

A National Public Health System: Vision without Reality

While the Liberian Institute for Tropical Medicine quietly pursued its objectives, the PHML pushed forward along a more public path. Those in the U.S. public interested in Liberia's fate were impressed by the new reports that the mission generated. During 1947 and 1948 the PHML managed to gain support from some of the church missions as well as Liberian officials. The popular black U.S. magazine *Negro Digest* reported that in Liberia "a US Public Health Service mission is fighting a spectacular battle against ignorance and voodoo doctors." The *Digest* emphasized that the program consisted "entirely of American Negroes." It pointed out that when Dr. West's team arrived in 1944, "it was distinctly unpopular with the voodoo doctors, who use fires and incantations, and with the bush doctors, who use roots and herbs." "Go away; you don't belong here," West was first told. But on blocks in Monrovia, house-to-house sprayings of DDT eliminated mosquitoes, and malaria admissions in hospitals dropped. "'Now,' West chuckled, 'we catch the devil if the sprayers are a day late.'"[48]

Attempting to broaden its programs, the PHML managed to implement a health survey of disparate Liberian populations. Collecting clinical and laboratory data from some 8,846 persons sampled throughout the Liberian countryside, the PHML provided the most concrete assessment yet compiled of disease patterns and their implication for the overall development of Liberia.[49] The PHML discovered that specific tropical diseases were most responsible for high mortality, morbidity, and social and psychological disability among Liberia's people. These were intestinal helminthiasis, treponematosis, diarrheal dysenteries of both bacterial and endamoebic etiologies, and malaria. They also found that in some sections of Liberia, trypanosomiasis, schistosomiasis, mycotic infections, and chronic indolent ulcers were widespread. According to the head of the PHML's laboratory and medical research, many of these incipient medical problems were "notoriously neglected by physicians and the laity as non-fatal conditions of minor importance."[50]

The PHML team intended its reports from Liberia to provide a realistic picture of Liberia's sizable health problems to the U.S. Public Health Service. Their reports also aimed to stimulate public interest throughout the United States about the plight of Liberia. The PHML also devised a long-term health plan for Liberia. Working with other foreign health experts in Liberia under the Liberian government's auspices as well as with local missionaries, the PHML health plan recommended six major goals:

1. Annual physical examinations for all school children;
2. Premarital testing and free treatment for venereal disease cases;
3. Free treatment for students and tribal people treated in government hospitals and clinics;
4. Broader employment of foreign doctors;
5. Increase in clinics and hospital construction throughout the nation; and
6. Local institutions for medical training.[51]

"It is our belief," West wrote in 1948, "that the Public Health Service of Liberia will constantly improve."[52]

President Tubman's administration and other U.S. officials publicly endorsed these health goals for Liberia. But neither the U.S. planners nor the Liberian endorsers mentioned the prerequisite resources these objectives implied. These preconditions included, first, revenues and trained personnel to implement medical education institutions and hospital-based training programs on a mass scale. These institutions and programs function as pipelines for the lengthy and costly preparation of Western-style medical and nursing professors and professionals (goal numbers 5 and 6). Government agencies were also needed to provide the funds that would remunerate hospitals and private physicians to treat ordinary Liberians as well as to supply and monitor demographic data for health planning (goal numbers 1 through 3). State-sanctioned medical bodies and public health agencies also needed to be created or expanded. Finally, all the rungs of the national medical facilities and agencies would have to develop means to incorporate traditional folk health values, practices, practitioners, and midwives into their plans and operations.

Given these gross limitations, any permanent gains from the Tubman administration's health care initiatives were highly significant for Liberia's political persona throughout the West. By 1950 Liberia had a small but fully operational public health department staffed by a few physicians, trained technicians, and nurses. Over the next decade this department, the National Public Health Service (NPHS), expanded slowly to include most of Liberia's few medical facilities in addition to public health services. By 1962 it operated nearly one-half of all the medical centers in the nation; foreign missions and companies ran the rest. All told, the NPHS was overseeing eight general hospitals and five specialty facilities; the two largest were in Monrovia: the Liberian Government Hospital (180 beds and ten physicians) and the Maternity Center (120 beds and eight physicians). Although medical school education for Liberians still was obtained abroad, physicians or other health professionals were practicing in all of Liberia's provinces.[53]

In 1964 the Government Hospital expanded to 200 beds and two tuberculosis annexes caring for about one hundred inpatients. The C. V. Dyer Hospital had evolved into a 140-bed institution specializing in obstetrics and pediatrics. Construction was underway for a new 250-bed hospital at Sinkor in Monrovia. Financed mainly by the United States, this new center incorporated the Tubman National Institute of Medical Arts. This school was a residential facility that trained 230 students in nursing, sanitation, and other health fields.[54] Given the fact that Liberia's urban population amounted to only about one to two hundred thousand people out of its total population of two million people, these hospital beds were only a tiny portion of what Liberians in Monrovia actually needed by modern standards. However, to the Tubman administration and U.S. benefactors, strides were being made in Liberia's quest for modern development. The full-scale campaigns that U.S. administrators and public health experts had waged against diseases earlier in the century in Panama and Haiti, much less the precise technical approaches used in these campaigns, were all but forgotten by U.S. foreign aid officials during the World War II years.

When President Truman took office for his second term, the U.S. approach to international military affairs changed profoundly. Locked in a global military power struggle, first against Hitler's Germany, and now against the Soviet Union, a new phalanx of foreign-affairs agencies and military resources emerged to protect U.S. national security. Occupancy of land protected by conventional ground, air, and naval forces was no longer the ultimate basis for national sovereignty and security. In the minds of the Cold War era political leadership, power, instead, rested in a nation's capacity to protect itself from and to deliver nuclear destruction.[55]

Under the umbrella of its new superpower position in international Cold War politics, the U.S. government developed three means for foreign assistance: bilateral aid programs, multilateral aid programs, and military and fiscal assistance. The United States had gained experience coordinating and delivering massive bilateral aid through its Lend-Lease Program and the Marshall Plan. Second, the United States began to work through multilateral international assistance organizations, especially the United Nations and the World Health Organization (WHO). Finally, the third means for the United States to extend international aid became military and fiscal partnerships with the leading free-world powers, especially the North Atlantic Treaty Organization (NATO), the World Bank, and the International Monetary Fund (IMF).[56]

The U.S. leadership in the new international political system of free-world nations was a manifestation of the more deeply woven mission to realize utopia through science and technology. From the start of World War II, federal expenditures for research and development grew steadily by about 12 percent annually. In 1940 the nation allocated about $74 million for science and technology; by 1962 and 1963 this figure grew to $12.2 billion. By 1967 this figure stood at $16.5 billion.[57] Expansion in the organizational complexity of U.S. technological production, armed forces, and medical care was complemented by immense growth in corporate management and government administration. In 1899 the ratio of productive workers to administrative employees was ten to one, by 1947 this ratio was ten to two, and by 1957, ten to three.[58]

Changes in the priorities of U.S. medical and scientific research institutions also occurred during the World War II decade. U.S. public health agencies became dominantly focused on microbiological research and surveillance as the backbone for disease control programs. Thus, eradication campaigns in communities throughout the tropical world where epidemics were occurring became less important in U.S. assistance programs. More focus was placed on learning about the microbiology of the specific diseases found in these tropical settings—diseases which could, and occasionally did, strike U.S. military personnel and, increasingly, U.S. tourists and business personnel traveling by airlines through tropical nations. Consequently, by the late 1940s, academic and military medical experts from the United States were tracking tropical diseases worldwide—inside the geographic band that stretched from Panama to Liberia, from Libya to Greece, and from Yemen to Pakistan. In short, most medical activities by the U.S. government in Liberia were designed to serve, first, U.S. medical science and government agencies; and, only second, the public medical care programs and needs of the host Liberian state, medical profession, or populous ethnic communities.[59]

The New Aid Broadens

To the Liberian government, the PHML and the Liberian Institute were of great symbolic importance. But the Tubman administration's most urgent and realistic hope for health and technical improvements rested on expecting a substantial infusion of U.S. economic assistance. Tubman believed that with this philanthropic aid, his administration could build a nationwide public works and health-care delivery system. The Truman administration appeared to Tubman as a godsend.

In June of 1949 President Harry S. Truman unveiled what would become his famous Point Four Program. Truman stressed that this foreign policy initiative was "a bold, new program for making the benefits of our scientific advances and industrial progress available for the improvement and growth of underdeveloped areas." Under Truman's Point Four Program, the United States stepped up significantly its economic assistance to allied African and Asian countries.[60]

U.S. bilateral aid to Liberia during and immediately after World War II had stressed road development, finances for the deep-water port in Monrovia, and the public health research programs and improvements. Under the Point Four

Program the sights were much higher. This assistance was intended to improve such areas as communications media, industry and mining, and development banking. Civil engineering and sanitation projects also aimed at improving roads and ushering in the building of water supply facilities, hydroelectric utilities, and agricultural modernization.[61]

In 1950 the Joint Liberian-American Economic Commission was established. The commission was made up of seven Americans and six Liberians. The Americans were experts in such areas as education, economic development, and public health. The Liberian members were heads of Liberia's national agencies responsible for these domestic areas. Eighty percent of its funds were Point Four allocations, while Liberia contributed the remaining 20 percent. Its purpose was to review the implementation of the development program growing out of the U.S. aid.

Between 1949 and 1963 the United States provided Liberia with about $45.7 million in assistance and another $51 million (from 1951 and 1961) in loans. The bulk of the U.S. grants to Liberia during the 1950s were for education (22 percent), agriculture (20 percent), transportation (15 percent), health and sanitation (10 percent), and public administration and safety (8 percent). In 1961 about 160 foreign technical experts were employed in U.S. aid programs. The distribution of U.S. technical employees in Liberia was roughly similar to these funding allocations.[62]

A small but significant portion (about 9 percent) of this increased U.S. aid was targeted specifically for public health and became part of the budget for Liberia's national public health program. The agency also received grants and loans from other foreign sources (for instance, the governments of Israel, Germany, and Italy) and international health organizations such as WHO, FAO (the United Nations Food and Agricultural Organization), and UNESCO (the United Nations Educational, Scientific, and Cultural Organization) for specific health programs such as malaria control and health education. These external funds were coupled with growing Liberian revenues from the government's introduction of income taxes (1951) and iron-ore royalties (1953). Annual expenditures by the government for public health rose from about $385,000 in 1949 to about $1.15 million dollars in 1962, and $2.82 million in 1963. But this does not represent a proportional increase in Liberian expenditures for health-care operations. From 1955 through the 1960s, the percentage of the annual budget allocated for public health remained from 7 to 10 percent.[63]

U.S. financial assistance to Liberia also came in the form of loans to private enterprises for operating companies in Liberia. Between 1948 and 1960 the United States provided $40.8 million for this purpose. The bulk of these loans went to mining companies: the Liberian-American-Swedish Minerals Company (LAMCO) for iron-ore production ($30 million), the National Iron Ore Company ($6 million), and the Liberian Mining Company (LMC, $4 million), to construct railroads. Smaller loans, totaling less than $800,000, were provided to three companies involved with road machinery, lumber, and asphalt.[64]

Throughout the 1950s the United Nations agencies committed $4.2 million dollars to Liberia. In 1961 eighteen UN technicians were working in Liberia. Also, during this same period, the WHO implemented a number of health projects in local communities of Liberia and other underdeveloped nations throughout the Third World. In the mid–1950s the organization estimated that 1,472 million people lived in malarious regions like those found in Liberia. Within seven years, as a result of WHO efforts, this figure fell to 572 million. Among its African programs, WHO opened several projects in Liberia. As part of its efforts to bring "pre-eradication malaria programs" to least-developed nations, WHO helped set up some rural health services in Liberia. It initiated a pilot mosquito-control project in the Kpain zone of Liberia. It stressed spraying houses with residual insecticides as the means to break the malaria transmission cycle in that area.[65]

Iron-Ore Mining: Liberia's New Enclave Technology

At the same time that U.S. bilateral aid and international organizations gave Liberia increased financial and technical assistance, Tubman moved ahead with what he believed was an economic revolution. A centerpiece in Tubman's plans for national modernization was his Open Door Policy. He organized the governmental infrastructure to offer generous incentives for foreign investment in Liberia.[66] During the 1950s and 1960s these Liberian incentives, as well as large loans from the U.S. government, spurred Liberia into a period of unprecedented growth in its technological development and public revenues. The government prospered from income gained through concessions of its raw materials, concessions which, in turn, enabled Tubman to expand the public sector along the lines of his notion of scientific modernization.

Developing Liberia's human capital into a trained, functional public sector and developing the nation's rich rubber resources had been at the top of the Tubman administration's modernization plans. But during the 1950s iron-ore mining became the key new technology that the Open Door Policy encouraged. Liberia held some of the richest iron-ore deposits in the world. U.S.- and foreign-owned and managed mining companies in Liberia skyrocketed. The earliest and largest of these mining enterprises was the U.S.-owned LMC. This mining operation was a subsidiary of Republic Steel. The firm began its project in 1946, after a few years of exploratory surveys and negotiations, as part of President Roosevelt's involvement with Liberia. LAMCO (1953), the German-Liberian Mining Company (DELIMCO, 1958), and the National Iron Ore Company (1958, American with Liberian shareholders) followed the LMC.[67]

The Free Port also provided excellent facilities for the shipment of iron ore. By the mid–1950s it was the most modern port on the west coast of Africa. The port had modern warehouses, petroleum storage tanks, fire services, water tanks for ships, huge fenced-in open-storage spaces, and two electric power plants (one owned by the LMC). Through the 1950s, Firestone Rubber generated about

39 percent of Liberia's general revenues. However, by the mid–1960s the mining companies had exceeded the rubber plantations as Liberia's number one revenue producer [68] (see map 6.2).

The large extraction companies in Liberia were also responsible for introducing additional modern power and transportation technologies. The large extraction industries in Liberia had electricity plants built to service their operations throughout the country. In 1961 five major electric power installations were operating in Liberia: the Monrovia municipal plant, the Bomi Hills plant, operated by the LMC, the Firestone hydroelectricity plant, the Firestone Cavalla plant, and the LMC's plant at the Free Port. These installations produced 24,000 kilowatts of the nation's total 28,000 kilowatts of electric power. Through 1963 the municipal plant in Monrovia had to purchase power from two of the mining company's stations to meet the public's electricity demands.[69] LAMCO built a 165-mile railroad that ran from its mines in the Nimba iron ranges to Buchanan. By the mid–1960s, sixty- and ninety-car trains carrying up to 8,100 tons of ore were being dispatched from the Nimba mines each day.[70]

As of 1965 the iron-ore concessions of the four major mining companies were producing 25 tons of ore annually. Liberia had become Africa's leading iron-ore producer and was among the largest iron-ore producers in the world. Combined, Firestone and LMC were producing over 80 percent of Liberia's exports and more than 50 percent of the nation's total production. Ten years later

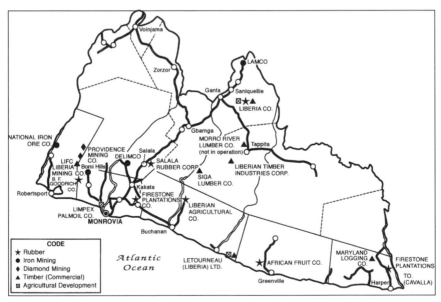

MAP 6.2.   Foreign-owned commercial development sites in Liberia active between 1961 and 1966. Adapted from Robert W. Clower et al., *Growth without Development: An Economic Survey of Liberia* (Evanston, Ill.: Northwestern University Press, 1966), map 1.

iron ore had greatly outdistanced rubber as Liberia's leading export (at 75 percent), and its portion of the gross domestic product was 28 percent.[71]

However, the influx of technologies related to extraction industries did not spell a diffusion of professional and technical employment sites that hired Liberians. These mining corporations remained operationally and financially "folded inward," bringing in their own managerial personnel. With the aid of Tubman's Open Door practices, the mining firms swiftly processed, removed, and exported Liberia's raw wealth. But in fact, Liberians gained only very few of the managerial, professional, and financial occupations the mining firms created. Liberians fortunate to be hired by the mining companies were restricted mostly to unskilled or semiskilled jobs. In 1962 the four large iron-ore companies employed 6,250 Africans and 430 non-Africans. Additionally, at least 5,000 Liberians and 750 expatriates were employed by subcontractors that supplied industrial and construction support to the mining companies.[72]

The employees in the managerial and administrative positions were overwhelming non-Liberians. Nonetheless, indigenous Liberian mine workers were still prosperous compared to typical Liberian workers and their families, who had to live on traditional subsistence farming. Furthermore, Liberian mine workers and their families benefited from some of the ancillary employee services operated by the mining companies. By 1962 the mining companies were operating five hospitals and dispensaries. Two years later, in Yekepa and Buchanan, LAMCO opened additional modern hospitals that grew to a total of 260 beds by 1970. About 40 percent of the patients at these LAMCO facilities were families of employees.[73]

Despite the growing number of health care resources that foreign companies were establishing in Liberia, work and living conditions for the unskilled workers were harsh compared to those of the foreign employees. It usually took the buildup of labor discord to force foreign-owned industries to improve work and housing conditions for Liberian nationals. For example, early in its operations, or at times when prices for iron-ore dropped on the world market, LMC presented especially bad conditions for its Liberian employees. A Liberian university researcher described them:

> The Company's housing facilities for its Liberian workers were both primitive and unhealthy. . . . They generally consisted of one bedroom units for workers with families and dependents. They lacked indoor toilet facilities and easily accessible water supply. Because they also lacked kitchens or cooking space, the employees and/or their wives were obliged to prepare their meals outside . . . amidst the persistent clouds of dust and other debris created by passing vehicles. And even though the Company paved the roads within the elegant section of the Camp in which its expatriate staff—mostly white Americans—lived, it never bothered to pave the roads in the overcrowded section of the Camp where most of its workers lived.[74]

As for providing revenues for its public sector, Liberia as a whole paid a steep price for hosting these mining companies. The mining firms enjoyed substantial tax exemptions for their agents and managers who were importing resources for their mining companies. Tolls of any kind for harbor, freight, wharf, or similar resources were usually not charged. Finally, royalties were low; for example, 4 percent of net sales for LAMCO.[75]

Moreover, indigenous women fared poorly in the modern "enclave industries" of rubber and iron-ore production. Neither provided an avenue for substantial numbers of women into the paid-employment sector. The large rubber and mining companies provided about one-half of the nation's wage employment. Yet in 1961 a survey of Liberian labor found that Firestone (Harbel) employed only seventeen women in unskilled work, LAMCO employed one, and Goodrich, eighty-nine. At Firestone, these female employees worked as nurses' aids or laundresses. Goodrich employed the women in tree nurseries under shaded conditions. However, increasingly, the women were requesting work in the open fields as tappers. Chiefs received larger fees for providing male laborers. But they opposed concession companies employing women from their clans. Women controlled the basic agricultural system and markets of most rural Liberian tribes, even among the traditional ethnic groups residing in Monrovia. Thus, chiefs of many indigenous peoples feared that if women entered employment for Western companies, the entire village, town, or kinship unit could eventually dissipate.[76]

A Dead End for Modernization

The expectation among U.S. and Liberian policy makers during the 1950s and early 1960s was that as public and foreign corporate infrastructures were constructed, the Liberian civilian sector would service and financially support them. The enthusiasm the U.S. State Department had for Liberia's development potential is reflected in America's AID (Agency for International Development) program. In 1965 alone the United States provided over $16 million in assistance to Liberia. Only three other African nations received larger funds: Nigeria ($27.6 million), Tunisia ($18.4 million), and the Congo ($16.5 million). Moreover, Liberia's AID funds were over ten times what Haiti received even though Haiti's population was about double that of Liberia (see fig. 10).[77]

By far, Tubman's development policy was most successful in expanding formal education throughout the nation. From the early 1950s to 1960, the number of elementary school students in Liberia grew from about 20,000 to over 58,000. By 1970 this figure reached 120,000. The number of secondary schools and students increased from nine schools and 870 students in 1951 to sixty schools and 5,574 students by 1962. The Booker Washington Institute also experienced growth. Between 1954 and 1961 Prairie View A & M College of Texas supplied technical teachers to the BWI as part of a contract. By 1961 BWI had 351 full-time students studying fourteen trades, and its graduates were employed in business, industry, and government. That same year, foreign concessions provided eighty scholarships for students attending BWI.[78]

Yet, even with Liberia's great strides in education, the U.S. foreign policy makers' assumption that Liberia would become a beacon of modern capitalism on the African continent proved dead wrong. The great majority of Liberians remained in subsistence agriculture, and the nation's government remained inept. Moreover, educated, experienced Liberians in the higher occupations of university, engineering, hospital, clinic, and corporate operations did not nearly reach the numbers these technical institutions required.

Despite the apparently large amounts of U.S. assistance it received, Liberia was wrestling with its ongoing inability to establish an autonomous, moderately healthy internal economy and public sector. Having accepted the large loans during the 1950s and 1960s, by the late 1960s Liberia was sliding into heavy debt to the United States. Between 1965 and 1970 U.S. aid fell drastically from about $22 million to $13 million. Loan payments were due from Liberia for the large technical projects the United States had implemented earlier, especially the Port of Monrovia (1944), smaller ports and roads (1951–1955), and electric and hydroelectric power facilities (1959–1962)[79] (see map 6.2).

By the early 1960s, to his credit, Tubman had put Liberia on track to assemble its own small-scale, professionalized medical and public health sector. He and other Liberian officials, with spirited support from the Kennedy administration, managed to build or expand several hospitals and clinics throughout their country as well as start a medical training center in Monrovia for attracting more physicians and nurses who would stay in Liberia. In 1961 about eighty-five physicians and eleven dentists were practicing in Liberia. Thirty-six were in private practice or employed by Christian missions, while the Liberian government employed forty-six.[80] The population census of Liberia for 1962 indicated even higher numbers of medical professionals (about 174). They were concentrated mostly in Montserrado County, which included Monrovia and the Marshall Territory. As of 1962 there were fifty-four Liberian medical students training abroad, mostly in the United States and Germany, about four of whom expected to return each year to practice in Liberia.[81]

During 1961 Tubman made an official visit to the United States and met with President Kennedy. Afterward, the Kennedy administration arranged a loan of $6.8 million to Liberia to begin construction of the medical training center. Ten years later the center was opened, its core facility named after its original benefactor, the John F. Kennedy Memorial Medical Center. The facility provided a 250-bed general hospital as well as a 140-bed maternity center. It boasted the full range of clinical services as well as a pathological laboratory, dental treatment, radiation therapy, and physical rehabilitation; and it performed nearly twenty-four hundred deliveries each year. At its opening dedication in June of 1971, the president of the black U.S. physicians' major organization, the National Medical Association, stood in attendance. Also at the ceremony were the mayor of Washington, D.C., Walter Washington, and his wife, as well as the U.S. ambassador to Liberia, Samuel Z. Westerfield (see fig. 11).[82]

In support of Tubman's progressivist surge, even the Firestone Company's health services, considered among the nation's best medical facilities, branched out to a substantial nonemployee segment. With a total of 510 beds in their Harbel and Cavalla facilities, the Firestone facilities administered over five hundred thousand patient visits per year. In addition, the Firestone medical services operated some fifty dispensaries that served outlying villages and work sites. Company officials estimated that about 25 percent of the medical patients were not connected with Firestone plantations. By 1973, when the company employed about fifteen thousand, Firestone medical facilities served these employees as well as an additional fifty-five thousand dependents and nonemployees.[83]

Other company medical facilities expanded the care for their Liberian employees beyond just injuries that occurred on the job. LMC maintained a 25-bed hospital that could provide emergency care for seventy-five workers. This company and others operating the extractive industries in Liberia found it necessary to offer preventive and primary care as well. One development specialist wrote in 1966, "Many new workers suffer from common tropical diseases which prevent them from being efficient workers. These maladies must be cured before the individual begins work." Thus, most companies "find that the easiest and surest way of meeting this responsibility is to provide competent medical care [for these diseases] themselves."[84]

Liberians benefited from a few new hospitals. In 1965 President Tubman dedicated the new Phoebe Hospital in Suacoco (Bong County), in the heart of rural central Liberia. Tubman praised the Lutheran mission and other church men and women who had worked to have the hospital constructed. Phoebe Hospital, Tubman remarked, "is the quintessence of faith, broad planning and splendid co-operation." Although it was a small hospital, Tubman envisioned that one day it would become a great medical center. "With the construction and dedication of this building, our tasks have just begun," he stressed. With his characteristic missionary verve, he urged, "There is need for more and yet more facilities to preserve the health of the people . . . and we assure you of the Government's cooperation and support in the great and rewarding program of education and evangelism. May God bless the labors of your hands and raise up many helpers in this noble enterprise."[85]

Finally, some of the hospitals and medical stations established earlier in the century by religious groups and international health organizations were being turned over to government control or Liberian professionals. For instance, in 1961, as one Liberian researcher put it, "the majority of WHO personnel turned over their offices to Liberians and said good-bye in a most cordial way, satisfied that a task assigned them had been successfully carried out."[86]

In the end, there were many obstacles that doomed the growth and national spread of U.S.-centered medical activities in Liberia. Combined, the African American PHML, the modern medical facilities President Tubman managed to have constructed, and U.S. and international agency funds for public health were nowhere near sufficient to develop into a truly nationwide medical care system for two million Liberians. These meager health care resources also did little to alter the vast environmental and economic conditions that fueled the spread of diseases such as malaria and schistosomiasis and high infant and maternal mortality rates.

Moreover, building a medical system based on the preventive and clinical models derived from modern Western biomedicine requires several resources that Liberia lacked. First and foremost, the nation needed a public and medical education infrastructure of sufficient size. All the Western industrial countries had public education systems that were financed by all levels of government and overseen by local school districts. Modern nations also had professional schools that were sanctioned and monitored by government agencies as well as professional organizations. But in Liberia, public education was still seriously deficient in both its quantity (availability) and quality, especially at the high school, college, and professional school levels. The education and research of medical professionals must be rooted in universities and academic hospitals. Liberia had no such higher education resources. Consequently, Liberia did not develop a supply of health professionals nearly appropriate for the size of its population or the nation's dire health circumstances. U.S. assistance programs downplayed the education problem and, therefore, the long-term viability of homegrown health and technical professionals that Liberia so desperately needed.

In the early 1960s U.S. development specialists sponsored by AID did a thorough assessment of Liberia's development effort and the U.S. role. They criticized the mass education initiatives of the Tubman administration as unproductive and wasteful. According to these experts, the Liberian government's effort to bring public education to all sectors of Liberian society was prematurely trying to copy the U.S. model of public education. According to the U.S. experts, "specifically, Liberia shows signs of having accepted some of the tenets of American educational philosophy which are likely to produce pernicious effects if applied uncritically to the Liberian situation [namely] emphasis on free, universal, compulsory education."[87] The development group recommended instead that Liberia focus on educating a much smaller segment of the Liberian population. They also recommended that Liberia rely on sending its most capable students to professional schools abroad. Their thinking was that by targeting the education of some Liberians, who would then become useful to the commercial enterprises that were generating foreign investments in Liberia, the nation's wealth would increase and, eventually, so would its public revenues for education and other domestic needs.

U.S. assistance to Liberia in general followed the approach of the development specialists. But in so doing, it neglected the social and economic gravity of the educational needs in Liberia. Even with compulsory education laws in effect for all children between the ages of six and sixteen, as of 1962 only 25 percent of these children were in school. Also, less than 10 percent of Liberia's population was literate. The traditional ethnic blocs comprising the bulk of Liberia's population were still deeply divided. The main ethnolinguistic groups—such as the Mandingo, the Gissi (or Kissi), the Gola, the Kpelle, the Kruh, and the Grebo—retained their cultural practices of educating their youth. In addition, in the western, northern, and northeastern parts of the country, largely Moslem regions, Koranic schools flourished. The best the Tubman education program was able to accomplish in key rural areas was to try to work with or around so-called bush schools. These traditional schools—Poro for boys and Sande for girls—taught children the discipline, conduct, and rituals of the various local ethnic peoples. It was hoped that while "these bush schools confer the status of membership in the tribe [eventually they] could be used in promoting economic and social development." However, conflict often set in between these local traditional schools and the government schools.[88]

As for the United States facilitating the building of a modern medical school in Liberia, American development specialists urged that such a project would be fruitless. They recommended that, to train physicians, Liberia keep converting its public education funds into scholarships and use these scholarships to continue to send its best medical-school candidates abroad for training.[89] But Liberia had been taking this approach to medical and technical education since 1944. Between 1944 and 1962 more than 2,550 Liberian students had been sent abroad to study engineering, medicine, auto mechanics, and plumbing on government scholarships or other scholarships provided by the United States and other foreign nations. The United States and Western nations had shouldered a modest number of the Liberian medical students on Liberian government scholarships. For example, during 1962 most of these 54 medical students were studying in Germany (20), followed by the United States (16), and other European nations (17). As of 1961 there were only one hundred doctors in Liberia, a mere fourteen of whom were Liberian. This study-abroad system was obviously not producing anywhere near the number of physicians Liberia needed. Instead, it was perpetuating a brain drain of medical talent.[90]

In addition, Liberia lacked ample public finances. Its national revenues were too low since the government's tax base and duties on the private sector were both paltry. The Tubman administration used the royalties, rents, and profits from land concessions for its own political ends.[91] The miniscule federal allocation that Liberia managed to exact from foreign assistance and firms was far from sufficient to pay for nationwide health care. A substantial wage-earning class and significant third-party health coverage plans (sponsored by government and industry) also did not exist to financially support the medical profession and hospital network.

Overall, medical care resources had to be sizable and flexible enough if they were to serve adequately the rural poor, the majority of Liberia's population. Yet not only did Liberia lack hospitals, it had never developed a national medical profession. This physician sector would have provided the backbone for a national network of health professionals at the various levels of medical care as well as a Liberian-operated medical school around which these professionals could sustain themselves. In the context of these barriers (barriers that also prevailed in Haiti, as we see in chapter 7), the hospital-physician approach, thinly distributed, had not worked for Liberia in the three decades following World War II. This long period of laying a cracked foundation for modern medical institutions precluded hope for the development of a modern medical system in Liberia as this nation entered the last quarter of the twentieth century.

The Lost Republic

We have seen that even before World War II, in the Deep South and Haiti, U.S. federal authorities directly administered public health campaigns. But in Liberia, U.S. interests were primarily concerned with transferring medical resources or research projects to establish means for protecting U.S. military personnel in Liberia. Rather than become directly involved with implementing large-scale public works and health campaigns in Liberia, the United States left Liberians to fend largely for themselves. In terms of public health initiatives, U.S. military and international health authorities chose instead to mete out a small number of technical consultants or small projects to protect U.S. personnel or conduct tropical research inside Liberia. The United States also provided small financial grants or loans to the Liberian government so that Liberia could operate its own public health unit and partner with international relief agencies on projects in Liberia.

By the unfolding of the Nixon administration, the U.S. government and corporations were firmly entrenched in a laissez-faire, "developmentalist" world view with respect to the Third World. They expected that in less-developed nations like Liberia, capitalist modernization would kick in gear sooner or later, with or without U.S. support, and as long as the communist-bloc alliances for aid of any sort were kept out of Liberia's political rule. But rather than a beacon for democratic modernization under U.S. stewardship, Liberia had become just another "overpopulated" poor nation in the troubled Third World. Carrying heavy debts, government heads of these nations stared in distress at the United States and the other nations of the affluent West. One Liberian foreign-affairs specialist remarked in the late 1970s, "The hope that infrastructure will be internally stimulated is dim, especially when one considers that one-third of all government revenues goes toward repayment of existing debts."[92]

As for assessing U.S. technological flows to modern Liberia on the broader scale of modern Atlantic history, a comparative look at the Canal Zone, Haiti, and the black South is instructive. In both the Canal Zone and Haiti, the transfusion of U.S. technology into these societies had been centralized. The transfusion in Panama had two vehicles. First, the United States established the Zone as an autonomous administrative entity. This provided the means for U.S. managerial, public health, and fiscal authority over the canal project as well as canal employees, their families, and local communities. Second, civil engineering, sanitation, and public health projects were designed, implemented, and operated directly by U.S. engineering firms and public health authorities. During the occupation, U.S. technical resources in Haiti flowed primarily from the U.S. military's engineering and public health units, in cooperation with philanthropic foundations interested in tropical medicine and public health.

In the U.S. South, public health improvements were largely the result of federal and philanthropic campaigns aimed at reducing some of the most pervasive and costly communicable diseases. During the 1930s and 1940s, federal agricultural programs reinforced economic segregation and, thus, sped the out-migration of black southerners. Black Americans had fashioned their own medical profession and health care network. Until World War II, the black medical profession played a major role in training black medical personnel as well as delivering health care to the black communities of the South.

But other federal initiatives reaching the South expanded public health resources, the availability of nutritious foods, and hospital construction resources. The changes during World War II and through the 1950s brought large segments of the South's lower and middle class into the region's modernizing infrastructure and health care institutions. These segments included veterans and civilian populations of all ages covered by government health insurance programs such as public welfare and Social Security, and later Medicaid and Medicare. In addition, federal antidiscrimination regulations and agencies enabled a large part of the South's black population to obtain a share of the new public and professional schools, hospital resources, and government-linked employment. With greater access to hospitals, education, and jobs, the standard of living for many black southerners rose significantly. In communities throughout the old Black Belt South, the working-and middle-class clientele of the black medical sector waned, and, consequently, so did most of the South's black hospitals and traditional midwife sector. As for the poorest rural blacks most hurt by the heritage of economic discrimination, they remained in poverty throughout the old Black Belt region or other parts of the rural and urban South. Also, tens of thousands of blacks each year simply left the South altogether. Some three million did so between 1940 and 1960.[93]

The path of U.S. technical missions that emerged in Liberia throughout the twentieth century was nothing like the centralized industrial buildup that had occurred in Panama, in the military administration of interwar Haiti, or in the peripheral industrialization that eventually soaked up the Black Belt. U.S. industrial and health projects in Liberia were erratic and scattershot. It was not Liberia's geographic remoteness, compared to the other three regions, that accounted for the miniscule and ineffective technical aid Liberia received. U.S. military installations and businesses had accelerated worldwide during the Cold War era. The spread of these institutions was hand in hand with the modernization of shipping, air travel, and communications. In fact, after World War II, huge exports by the United States of its military and technical personnel and equipment as well as military resources across the Atlantic and Pacific Oceans successfully created permanent, multilateral alliances. NATO served as the chief hub for this administrative and military network that the United States forged with the other nations along the northern Atlantic Ocean.[94]

But for post–World War II Liberia, such a strong link with the United States never emerged. The American technicians and public health experts that went to Liberia were seeking knowledge, produce, and military advantages of little practical benefit or relevance to the majority of Liberians. Moreover, by attempting to put Liberia onto its own course of scientific progress while leaving engineering technologies, medical planning, and disease control agendas in foreign hands, President Tubman's administration also left out the Liberian masses. Worse, the Tubman government had paid for the drive for modern development with expensive loans from the United States and Western interests, eventually plunging the nation further into economic and civic stress. In 1979 massive riots broke out in Monrovia when the government doubled the price of rice. The president, William Tolbert, was assassinated, and this set in motion nearly two decades of brutal military coups and political anarchy. By the late 1970s, the United States' mission in pursuit of leadership in world science was as vigorous as ever, but the Liberian nation lay dying in its wake.

CHAPTER 7

Technology of Reaction

HAITI AND THE NEW DISPERSAL

$\mathcal{P}$rior to the 1930s, discourse in the United States regarding Haiti and other tropical nations emphasized the need for these societies to pursue a wholesale leap in politics, economics, and health. To leaders such as Theodore Roosevelt and Woodrow Wilson, the only hope for these nations was for them to become productive economic partners of the civilized West. U.S. and European leaders expected that the tropical nations could improve their governments and economies immensely after a period of colonial rule. By sending energetic administrators, military units, engineers, and public health experts, U.S. foreign policy makers believed societies such as Haiti could become orderly, healthy nations.

Through the New Deal and World War II, U.S. political and corporate leaders grew even more fervent in their belief that social problems could be solved by the nation's technical and medical disciplines. However, the Haiti occupation experience had shown U.S. policy makers that individual technical professionals, taken out of their institutional frameworks in the United States and sent to nations like Haiti, could not single-handedly overhaul public infrastructures and health conditions. Moreover, installing U.S. administrators and technicians to act as temporary direct rulers in politically unstable societies like Haiti exacted a political and financial price the United States was not prepared to shoulder.

During the 1940s technical modernization was deepening in the United States and the rest of the industrial West and Japan. All of these nations expanded their educational, production, research, and military infrastructures—the various institutional building blocks for their technological modernization. This multidimensional growth of their scientific communities and technologies was vital for these nations' capacity to wage and recover from World War II. However,

Haiti's national development was moving in the opposite direction. Health, economic, and social welfare problems abounded, but homegrown technical resources were miniscule. Malaria, for example, was intensifying as a social problem throughout Haiti. Contrary to popular myths carried over from the nineteenth century, widespread epidemic malaria in mid-twentieth-century Haiti was neither natural nor irreversible. But during and after the occupation, specific, human-made conditions were stoking malaria's omnipresence. Cities throughout Haiti had been built on flat coastal plains, making drainage difficult and swamps abundant. In post-occupation Haiti, poorly or partially completed road construction, uncontrolled irrigation, and borrow pits that were not drained also compounded Haiti's malaria menace. Under these conditions malaria remained endemic in Haiti. According to Rockefeller Foundation investigations in the early 1940s, prevalence rates for malaria were as high as 70 to 80 percent in some sections of the nation.[1]

In addition to the serious malaria problem, local doctors estimated that over one-fourth of rural Haitians were infected with hookworm and over three-fourths were still infected with yaws. Tuberculosis presented less publicly visible, nonetheless severe health and social problems, causing about one-quarter of the deaths registered during the 1940s in Haiti's one dozen hospitals.[2] The public health and medical services implemented by the U.S. occupation had not strengthened the local medical profession that attended the small, mostly urban upper and middle classes. Nor had the U.S. health program raised the standards of the informal folk healers and midwives, the sole sources of care for the vast poor and working-class populations of Haiti. With the lack of national health institutions and environmental improvements for a growing population, the epidemic diseases remained deeply entrenched. In the meantime, the high levels of disease drained family and community resources, hurting productivity in both Haiti's traditional agrarian economy and its small business sector.

By the outbreak of World War II, Haiti was still overwhelmingly agrarian and poor. Of its estimated 3 million people in 1940, nearly 2.6 million were rural dwellers, and 90 percent of the nation's people were illiterate. With most Haitians living in dire poverty and health circumstances, the nation's supply of three hundred or so physicians was seriously inadequate. During the 1940s this amounted to about one physician per ten thousand residents. Malaria, hookworm, and yaws were still widely prevalent and considered the nation's most deadly triad by Haiti's medical community. Malnutrition and enteric diseases were even more deadly to the nation as a whole.[3]

Educational resources in Haiti were also in dreadful shape. At the onset of World War II, Haiti had a total of only 236 students in its haphazard, mostly private medical, law, agriculture, dental, and engineering schools. An international survey of education in 1942 reported that Haitians "who desire to pursue advanced studies in education, liberal arts, pure science, and philosophy, must go abroad since no provision has ever been made in these fields in Haiti."[4]

During World War II, Haiti became eclipsed as a target for substantial U.S.

bilateral assistance. By contrast, nations hosting American military facilities and personnel, including Panama and Liberia, also acquired substantial U.S. economic links. Panama and Liberia received significant U.S. military aid for construction of transportation facilities and public health services. By the height of the war, the Canal Zone and Panama were sites for forty landing fields. About seventy thousand troops were on duty in the Canal Zone. U.S. military resources in and around the Zone were so large that the Panamanian government and private sector catered the entire national economy to the Canal Zone. At the same time, these Panamanian leaders ignored investing in and developing their nation's own natural resources, port cities, industries, food and housing production, and export links. Furthermore, local merchants and producers lost income because they could not compete with the goods sold by the commissaries run by the Canal Zone and the Panama Railroad Company. As one Panama leader said, "Panama exists by and for the Canal." A 1946 report compiled by the U.S. Embassy in Panama stated candidly: "All major economic pursuits in Panama are related to some degree to demands arising in the Canal Zone. . . . Commercial and financial relations with the Canal Zone . . . lie at the heart of Panama's economic system."[5]

Liberia also benefited from the U.S. involvement in World War II. As we observed in chapter 4, Liberia attracted U.S. military operations and health missions, mining ventures, and tropical medicine researchers. In contrast, the U.S. military established no bases in Haiti during World War II. Occasionally it sent patrol boats and artillery to Haitian ports and landing strips. However, this military aid was more to guard Haiti's harbor or maintain internal political stability than to propagate U.S. military operations against Axis maneuvers.[6]

While post-occupation Haiti drifted to the remote periphery of U.S. foreign affairs, and Haiti's infrastructure remained highly inadequate, U.S. society was surging ahead in its quest to bring its institutions under the guidance of networks of scientists. With international problems in Europe looming, government, industry, and universities in the United States pursued even more vigorously technological innovation and institutions. Early in the twentieth century, heroic individual scientists and inventors led innovation in U.S. industry, medicine, and scientific research. They were workbench scientists, commercial inventors, military doctors, and government agents. Many were self-supporting entrepreneurs, medical school professors, and lone bacteriologists or malariologists heading small project staffs. However, from the New Deal through the Cold War, the number, specialties, and funding of scientists, engineers, and technicians in the United States skyrocketed.

During the 1930s the National Cancer Institute and other National Institutes of Health were established in the Public Health Service. When World War II broke out, these resources were greatly supplemented by the Committee on Medical Research of the Office of Scientific Research and Development. Established in 1941, the committee functioned to "initiate and support scientific research on medical problems affecting the national defense." By 1944, the office had granted 496 research contracts to 120 institutions. The National Research Council organized some thirteen committees and forty-three subcommittees to guide these projects.[7] As part of this research upswing, the nation's leading physical scientists worked feverishly to design weapons for America's military supremacy. The investment in the research of these scientists resulted in the Manhattan Project. In a few short years after its start, the atomic bombs developed by the project leveled Japanese cities, delivering Japan's surrender.

Following the war, major research foundations emphasized that the advances in the basic sciences were vital for future success of America's medical institutions. One leading medical research foundation reported: "Between the microscopic and the atomic levels lie the macromolecules, huge protein complexes believed to hold the wheels and the driving power of the living machine. . . . These giant conjugated proteins, some now visible by electron microscopy, some measurable by other physical techniques, are the chief focus of investigation in biology as the basis of medical progress." The influence of this new focus of molecular biology in medical education and research in the United States was clear: "Medical research is less and less concerned with symptoms and manifestations, more and more concerned with comprehension of molecular mechanisms underlying biological activities."[8]

Following World War II, a major medical foundation hailed the successes that it and many institutional leaders in the United States saw emerging from molecular science. "The evidence of the successes of medical science to prevent and reverse the course of disease cannot be questioned," the report read. Antibiotic drugs like sulfonamides and penicillin had been discovered and had improved immensely the treatment of individual patients. These pharmaceuticals were proving profoundly effective in controlling an infected person's tuberculosis, venereal disease, meningitis, or pneumonia. By the close of the World War II decade, American leadership was fully preoccupied with a "romance with medical science."[9] Steadfastly, the United States assumed its political and military leadership of the free world. U.S. leaders envisioned that America's technical capabilities—in public administration, engineering, food production, medicine—now more than ever had to be exported to struggling governments and communities of the less-developed world.

Politics and the Limits of Eradication

In the wake of World War II, U.S. technoscience was broadening its presence throughout the international community through military installations, communication networks, and foreign projects of major corporations. Given the high level of technological resources that America was providing to postwar Europe through programs like the Marshall Plan, it seems ironic that the transfer of U.S. technical aid to Haiti immediately following the war moved exceedingly slow. This situation was partially the result of the United States having little military use for Haiti. Haiti's lack of large commercial establishments and stable civilian political institutions, as well as U.S. preoccupation with political and military affairs in Europe and the Soviet bloc, limited any flow of substantial U.S. technical assistance into Haiti's civilian sectors. Instead, most U.S. aid would be limited specifically to Haiti's military rulers.

At the start of Word War II, an elite of mostly mulattoes ruled Haiti. It was composed of large property owners, government bureaucrats, military and police heads, and professionals. Both Haiti's elite and middle class combined were but a tiny segment of the nation's overall population—estimated at about 1 and 7 percent, respectively. With poverty and illiteracy so rampant, the Haitian elite served as a buffer between U.S. and European politicians and financiers, on one hand, and Haiti's general population, on the other. The elite itself was deeply divided. One sociologist studying Haiti in 1942 wrote bluntly: "The elite . . . is anything but a solidly united ruling class. . . . Factions and cliques are based upon color differences, regional animosities, family connections, income, education, cultural interest, and personal rivalries."[10] Lacking a strong military interest in Haiti during the war years, U.S. authorities gave minimal support to Haiti's political and economic elite, and only at times when that support would prevent outbreaks of extreme civic turmoil.

In 1941 Elie Lescot, Haiti's ambassador to the United States, became Haiti's president. He replaced the ultra-nationalist and authoritarian Stenio Vincent, who had presided since 1930. Lescot proclaimed that he would be responsive to the social and economic problems of the potentially explosive population at large. However, in line with Vincent's regime, Lescot's four years of leadership relied on aggressive centralization of power within the executive branch.[11] Lescot's stance toward Haiti's intellectuals, professionals, and artists initially was relatively tolerant. But gradually it deteriorated into repression. Promising to bring the Haitian people into the fold of modern education and development, Lescot took a hard line against the Voodoo religion and its community influence. Lescot cooperated fully with the wartime strategy of the United States. In exchange, he managed to have a small dry dock constructed in Haiti and received provisions of light military weapons.[12]

During Lescot's rule, health experts from U.S. agencies and foundations joined with international organizations to undertake eradication campaigns in Haiti. The most notable U.S. health program in Haiti during World War II was a campaign against yaws. In 1942 the anti-yaws operations of the Rockefeller Foundation were incorporated into a branch of the American Sanitary Mission. Working with the Haitian public health department, the Sanitary Mission greatly expanded clinical screening and treatment. Between March and December of 1943 some 141,000 cases were treated, receiving 208,000 injections; and from January to September 1944, 175,000 cases (217,000 injections). In addition to the yaws-control program, the Sanitary Mission and Haiti's public health department laid drainage works in the area of Port-au-Prince to reduce malarial conditions and conducted some health education publicity for lay populations.[13]

In 1944 a critical point in this campaign was reached. That year American health experts began dispensing penicillin, resulting in drastic reductions in syphilis and yaws cases. In the past, it had been almost impossible to get rural patients to return for second and third arsenical injections. But the superior performance of penicillin, compared to arsenicals, in eliminating severe lesions and other illnesses in yaws patients greatly impressed the Haitians throughout the countryside. By the mid–1940s penicillin could be administered in oil with beeswax with just two injections at a ten- to twelve-hour interval. A few years later, improvements in pharmaceutical manufacturing resulted in inexpensive penicillin preparations especially effective against endemic treponematoses. These new drugs could be administered with one intramuscular injection.[14]

Strangely, the curative power of penicillin helped fertilize the career of one of the Western world's most notorious despots, Haiti's prime minister, Francois Duvalier. A physician who specialized in infectious diseases and public health, Duvalier had served as a medical worker and researcher in the yaws-control programs set up by the United States in his country. In 1946 he became a section chief in the malaria program for the American Sanitary Mission. Prominent in Haiti as a public health officer, Duvalier also became well known in medical circles in North America and Europe. He provided some of the first medical studies on penicillin for treatment of yaws and syphilis and the drug's dissemination in large-scale campaigns. Duvalier's professional career as a physician instilled in him a high personal esteem for modern science. But as in other historical instances when doctors turn into administrators of violence and torture, it also helped to instill in him the "distanced" emotions and exposure to death that blended well with his later autocratic behavior and politics.[15]

Before the advent of penicillin, the search for more effective syphilis treatment had made grindingly slow progress. However, as with other tropical medical specialists throughout the West, Duvalier's professional confidence swelled upon observing and using the curative power of penicillin.[16] In an account of public health efforts against yaws in Haiti, he wrote in a Canadian medical journal, "1944 marked a turning point in the history of the yaws in Haiti." Two new clinics were opened, in Grand-Goâve and La Vallée. For patients who managed to get to one of the treatment centers, penicillin cleared up virtually all the serious symptoms of yaws. Duvalier advocated that special attention be given to Haitians who lived in the rural provinces and suffered from yaws. In his medical articles based on case studies from patients of these early clinics, Duvalier wrote of their remarkable recoveries with just one or two days of penicillin treatment. Duvalier saw winning the battle against yaws as a precondition for attacking other public health problems in peasant communities. Gradually, the prevalence of the dreadful disease began to fall.[17]

The mass treatment campaigns in Haiti generated enthusiasm among the lay population. The rapidness with which the penicillin treatments had been dispensed and the immediacy of penicillin's effects were perfect for winning lay admiration for public health personnel such as Duvalier. But despite this success, Haiti's larger public health crisis and medical care needs persisted. For example, just to permanently reduce yaws cases, serological surveys had to be continuous to track relapses and reinfections occurring in communities in which mass therapy drives had been conducted.[18] But Haiti had no such laboratory resources. Even more serious, Haiti lacked health care personnel and facilities needed to address its other major health problems, such as widespread tuberculosis, tetanus, malnutrition, and infant mortality.

Pulled deeper and deeper into medical specialization at home, U.S. health experts in the less-developed nations brought with them what Rene Dubos called a focus primarily limited to "specific etiology." In this approach, medical and public health professionals give highest priority to understanding the microbial causes for clinical disease and its treatment in individual patients. Discovering insecticides and ways to use them in eradication campaigns also were high priorities for these specialists. Eliminating environmental conditions known to make people vulnerable to these diseases was only of secondary importance to U.S. aid officials and health care personnel.[19] While the U.S. medical community narrowed its focus toward explaining and solving clinical and entomological problems, Haiti still was experiencing some of the highest mortality rates of any nation in the Western world. Less dramatic, largely preventable health problems and diseases, especially malnutrition, tuberculosis, and kwashiorkor (caused by protein deficiency), were major contributors to Haiti's egregiously high death rates. Moreover, injuries and deaths due to work accidents, crime, and political suppression were not calculated, much less considered, by U.S. aid programs.[20]

In 1950 the Haitian ministry of health, assisted by the World Health Organizations (WHO) and the United Nations International Children's Emergency Fund (UNICEF), initiated yet another major campaign against yaws. Health workers found that the disease was still a virtual plague and managed to provide penicillin to about one-third of the nation's estimated three million residents. At the end of the program, the health personnel cautioned that a sizable population had eluded their first screening efforts. They pointed out the need for new permanent clinics for screening and treatment in communities with high endemicity.[21]

The following year the Pan American Sanitary Bureau discovered that much more work was still needed. In spring of that year one of the bureau's medical officers traveled to Haiti and reviewed the results of the earlier programs that had used the "ambulatory clinic method." He found that "the population coverage of the treated communities was very far from 100 percent. In fact . . . it became clear that in certain areas no more than 15 to 25 percent of the population had been treated." In response, a new campaign was initiated, one that used a house-to-house approach. In the fall of 1951, after a three-week sweep of houses throughout Haiti, virtually all of the nation's three million people had been reached, and yaws was largely eliminated.[22]

The emphasis on projects intended to control one specific tropical disease, by the United States and other Western donor nations, had other negative consequences. The U.S. projects did little to remedy the stunted supply and alienation of Haiti's local doctors. Nor did the emphasis on eradicating one disease improve the practice of Haiti's vast array of lay healers and midwives. For Haiti, there were neither sufficient health professionals nor an organized network of health care facilities so that modern medical procedures and drugs could be made available to most Haitians. Only then could one expect health care interventions on a scale capable of lowering national morbidity and mortality rates. Indeed, nationwide, Haiti's government agencies and public health, education, and public works institutions were barely functioning. At the same time, Haiti's presidents grew more totalitarian. These two negative conditions led to a brain drain of Haiti's talented professionals. After the start of the Duvalier regime in 1957, this outward stream becomes a flood.

The Social Promise of Technical Assistance

Following the Allied victory, U.S. political and economic leaders believed that America's greatest product and gift for fostering allies was its scientific and technical expertise. In the aftermath of World War II, the United States' Marshall Plan became one of the largest direct-aid programs in modern history. The program focused almost completely on rebuilding the public works and industrial structures that had been productive prior to the war in European nations. Building on the experience and momentum of the Marshall Plan, the United States began experimenting with technical aid packages for the newly emerging less-developed nations. U.S. foreign assistance to these nations became focused on development projects. The United States donated funds (credit and grants) but more frequently contributed technical experts, equipment, and supplies to spread the ethos and influence of the free-world international alliance.

U.S. officials preferred to provide assistance directly to the needy nations. At the same time, they also gave foreign assistance in cooperation with the growing number of international organizations, especially the United Nations and WHO, or hemispheric bodies such as the Pan American Health Organization and the Alliance for Progress. Working with these international aid organizations, the U.S. government and private interests dispensed experts to provide technical assistance, or know-how, and relief resources.[23]

In Haiti, U.S. technical assistance was funneled to its central government. Even though Haiti's government became more authoritarian in the post–World War II decades, and civil order repeatedly broke down, U.S. assistance programs continued to flow. U.S. officials believed that their developmental aid would help these poorer nations accomplish several ends. First, aid programs could assist the poor countries to begin to link their economies with the larger, innovative free-market nations. With this link, living standards would rise, breaking the cycle of poverty and political instability in these weaker countries. Finally, with open-market economies as well as wider availability of education and political participation, nations like Haiti would thwart communist aggression.[24]

President Truman was a major prophet of the idea that American science held limitless promise for the world community. In one of his most famous presidential addresses on U.S. foreign policy, Truman spelled out the Point Four Program. Delivering it in January 1949, Truman called for the world's poor countries to follow the United States' lead. The Point Four Program was America's bold effort to make "the benefits of our scientific advances and industrial progress" available to the less-developed nations of the world. Truman urged that while the United States could not give direct resources to all the needy nations of the world, "our imponderable resources in technical knowledge are certainly growing and inexhaustible." The Point Four Program would enable the United States to share "our store of technical knowledge" with all "peace-loving peoples" and help them "realize their aspirations for a better life."[25]

Buoyed by advances in everything from antibiotics to the Manhattan Project, national leaders and institutions greatly expanded the scientific sector. Networks of scientists were recruited and compartmentalized inside government, business, and academia. Collectively, the nation's science-led government, management, and medical care comprised what John Kenneth Galbraith called in the 1960s America's new "technostructure."[26]

For the U.S. politicians and philanthropists involved in technical and medical aid abroad in the late 1940s through the 1960s, the question was not How could the United States, using its vast technostructure, help its suffering neighbor Haiti? but Which specific Haitian political segment and communities should U.S. officials select to aid with funds, technical experts, or supplies? The Cold War was intensifying and Communist revolution in Cuba was gaining ground. Thus, in the United States the incentive grew for sending technical assistance programs and resources into Haiti regardless of Haiti's domestic political climate. U.S. officials justified aid to Haiti as necessary to keep Haiti from falling either into "chaos or Castroism."[27]

Transferring the New Deal: IDASH

At the same time the United States along with international organizations implemented programs in Haiti such as yaws-control projects, the Haitian government's own plan for national development was stagnating under the weight of internal political turmoil. Facing pressures from popular strikes, in 1946 Lescot fell from power. Later that year Dumarsais Estimé replaced him. Early in his administration, the Estimé government entered the political pincer that would face subsequent rulers. On the one hand, there stood the nation's peasants, urban poor, students, teachers, religious Voodoo adherents in the hundreds of thousands, and labor, community, and newspaper activists. These were always on the brink of massive unrest and desperate political action.

On the other hand, there were Haitian elite elements—the president, the military and national police, business owners, Catholic Church leaders, and large landholders. Contrary to popular assumptions, peasant revolts in modern history have not always been efforts to establish popular democracy. They frequently have been reactionary, totalitarian, and politically directionless. Thus, Estimé had to exercise care to prevent any revolt of the masses. He attempted to follow a middle-of-the-road approach, making some reform promises a reality. Drawing support from the ruling circle as well as from the popular segments, Estimé slowly structured a dictatorship of *pouvoir noir.* He supported the expansion of trade unions and rural cooperatives, enlarged the public school system, upgraded work conditions for government employees and contract agricultural workers, and started bringing non-elite blacks into civil service employment.[28] In short, the Haitian masses were receiving political concessions—a form of symbolic freedom from mulatto rule—by Haiti's ruling clique.[29]

One of the Estimé regime's most important liberal measures passed in the winter of 1948–1949: a law creating social and health insurance. In 1946 Maurice Dartigue, Haiti's minister of agriculture, had informed the Institute for Inter-American Affairs that health insurance, beyond just "humanitarian consider-ations," would benefit clerks, teachers, retailers, and government workers. These workers were the "class . . . which reacts most quickly to subversive propaganda, being the most vulnerable to the repercussions of economic fluctuations." Estimé's insurance law established an autonomous administrative program, the Institut d'Assurances Sociales d'Haïti (IDASH).[30]

U.S. foreign aid officials believed erecting a national social insurance agency for Haiti was the perfect means to generate a modern system of health care delivery for Haiti. If Haiti could use this new IDASH program to organize its own social insurance agency, the agency could in turn greatly expand access to health care services throughout Haiti. This agency also could improve the ad-ministrative and financial self-sufficiency of the overall medical sector for the Haitian government. Moreover, politically, efforts to organize a social insurance program would please both conservative and liberal political elements at home. The government officials who were proponents of this assistance could easily justify the effort and resources they invested. The social insurance program, they argued, was a means to fortify living standards for Haiti's white-collar segment, a small sector, but one critical to the functioning of the Haitian government overall.

For these reasons, in the fall of 1950, the U.S. Social Security Adminis-tration (SSA) sent I. S. Falk to Haiti to provide the long-awaited technical as-sistance for the IDASH program. Falk was one of the nation's premier experts on medical care policy and financing. Trained in bacteriology and public health at Yale, Falk later served on the faculties of Yale and the University of Chicago and was a key researcher with the Chicago Department of Health through the 1930s. He became a leading architect of the federal government's New Deal health policies, serving as associate director of the Committee on Costs of Medi-cal Care and, between 1940 and 1954, as director of the Division of Research and Statistics in the Social Security Administration.[31]

The Haitian social insurance law was modest in scope. Its wording covered only work accidents and sickness (including maternity) but not other common risks such as unemployment or non-work-connected disabilities or death. As writ-ten, the law provided work-accident insurance for public employees as well as those in industry, commerce, private teaching institutions, domestic service, and agriculture. The cost of the plan was to be borne entirely by the employer contributions at an initial rate of 1 percent of the basic wage of workers. However, before the IDASH could be set up, the Estimé government collapsed. The arrival of U.S. experts to design and coach implementation of the insurance program was held up until late 1950.[32]

Falk visited Haitian officials who had drafted the insurance law and president-elect Colonel Paul E. Magloire. He collected statistical data and cost figures relating to social insurance coverage programs. Falk also visited some hospitals and clinics, as well as groups of employers and physicians, and amassed information about Haiti's overall medical resources.[33] For the next year, policy materials and correspondence between Falk and U.S. and Haitian officials increased, but still the IDASH law remained far from implementation. The essential question to Falk and the SSA became not what form the IDASH agency and plan would take once started, but whether any agency or plan could be implemented at all.

The Plan Stalls

U.S. officials were hoping a consultant such as Falk could provide the practical expertise to help the Haitian government personnel put in place a self-regulating program—a program designed to be run and funded by the Haitian government itself. Yet early after his visits began, Falk was shocked by the "almost unbelievable poverty." He learned that the Haitian government and employers were "*completely* lacking experience in s.i. [i.e., social insurance] of any kind (not even having had any background in voluntary insurance) and having a *total* lack of trained personnel or anybody equipped with substantial knowledge of the subject (to say nothing of administrative experience)." Moreover, Falk found that Haiti lacked sufficient hospital and medical services to function as part of a social insurance network.[34]

Another major obstacle Falk and the SSA ran into was the ambivalence toward the IDASH plan among Haiti's physician community. When the IDASH law passed, the Haitian Medical Association expressed deep suspicion about the impact of the social insurance plan on physicians. To this organization, the social insurance plan posed a serious economic threat to Haitian doctors. The physicians feared the plan would trigger competition from Haiti's Public Health Service. This agency provided health care free of charge or for small donations. The Haitian physician community harbored a long-standing mistrust of the Public Health Service. During the occupation, the Public Health Service had been a key tool of U.S. domination and had disrupted Haiti's traditional physician community. "[I]n many respects we do not share the opinion of IDASH's creators," the doctors wrote in one of their political pamphlets. They pointed out that the social insurance program could mean merely an expansion of government facilities run by the Public Health Service. Such an expansion "would risk transforming social insurance (health or life, or overall comprehensive insurance) into unlimited welfare." The doctors believed this would amount to their losing virtually all of their patients to the Public Health Service.[35]

In addition to the dearth of experienced administrative personnel in both the government and commercial sectors and the opposition of the nation's physicians toward IDASH, the IDASH plan had to overcome the severe limitations in medical resources throughout Haiti. The IDASH agency was to have broad authority, providing not only the medical-benefit payments to insured persons but also new revenue sources for expanding medical facilities and services for these beneficiaries. But Falk and other IDASH supporters had to worry that the availability of social insurance benefits would trigger such an initial burst of new demands on Haiti's already meager medical facilities that work conditions at the facilities would become unmanageable. Overburdened health centers, for example, would intensify the emigration of Haitian physicians, already an ongoing problem. Since in Haiti "hospitals, clinics and medical personnel are quite inadequate to meet current needs," the only alternative they could suggest was that "the insurance beneficiaries must be given priority in receipt of care from existing facilities and personnel."[36] This measure would not be, of course, a genuine solution for the general inadequate health care resources in Haiti. It merely meant that insurance recipients would bump the indigent from access to the already extremely scarce medical services.

Confronted with these circumstances, Falk's expectations for IDASH's success became progressively lowered. In his final report to the SSA, a report that was supposed to be a blueprint for the implementation of IDASH, Falk was extremely pessimistic. He advised that the maternity and sickness insurance components of the plan be scrapped because they were "wholly impractical unless and until the resources for medical, clinic and hospital services are substantially expanded." Also, most farm workers lacked income to pay into the plan's funds. Thus, farm workers had to be left out of the plan. A UN report in 1949 indicated that the vast majority of Haitian people still worked in subsistence agriculture and, to a minor degree, fisheries, charcoal making, lime burning, and village handicrafts. Even the townspeople were linked to agriculture since they lived primarily from distributive trades tied to farm produce and handicrafts.[37] Paying workers' benefits to such a mass of laborers would instantly bankrupt a fledgling social insurance system in Haiti, no matter what form that system took. On the other hand, by cutting out the Haitian agricultural sector, the U.S.-backed insurance program could expect to benefit only a miniscule portion of Haiti's overall workforce and population.[38]

In the face of Haiti's intractable political and financial problems, Falk emphasized that no social insurance program in Haiti was possible without first addressing a multitude of practical, start-up conditions. Most importantly, competent administrative personnel had to be selected, and trained abroad if necessary; Haiti's medical resources had to be expanded by the Haitian government, even if this was an unpopular measure to the private physicians; and public education programs about the benefits of social insurance had to be initiated and aimed at the general public, employers and employees. Falk also recommended that the IDASH not be organized as an independent agency, given the extremely politicized nature of Haiti's executive government. Instead, it must somehow be organized as a civil service unit and part of the Haiti Department of Labor.[39]

In the end, the social insurance plan (including IDASH) was not implemented in the early 1950s. Instead, Haiti fell into the throes of political turmoil involving general strikes, student protests, assassinations of journalists and community populists, juntas, and rigged presidential elections. Indeed, political order in Haiti was not established until several years following Francois Duvalier's rise to the presidency in 1957. As for IDASH, it managed to develop an orthopedic hospital. Beginning in the early 1960s, IDASH provided coverage for a small number of workers (less than two thousand yearly) and elderly retirees.[40]

International Organizations: The UNESCO Example

The U.S. effort to install a health insurance apparatus had almost no immediate impact in Haiti. However, new community development aid continued to flow from international health and relief agencies. Assistance programs from these agencies and other multilateral initiatives providing disease control and health care did have some impact. Progress had been made in eliminating yaws. Also, the United States continued to provide financial and technical assistance for community development projects under its Point Four and Agency for International Development (AID) programs and other inter-American initiatives, especially the Alliance for Progress.

In 1948 UNESCO started one of the first multilateral aid programs in Marbial Valley. It was called the Fundamental Education Pilot Project. Located south of Port-au-Prince, Marbial had a population of about twenty-eight thousand peasants. Most were illiterate peasants who lived scattered along the floor and steep slopes of the valley. The region was cut off from governmental resources such as roads, schools, public housing, or political institutions. Since Marbial Valley had become overpopulated and over-farmed, extreme conditions of poverty had set in throughout this community (see figs. 12 and 13).

The staff of the UNESCO program had to struggle with developing a means of standardizing reading instruction for the region's peasantry. The program first used printed material in Creole as a way to build literacy in formal French. Also, UNESCO experts operated a small health center as well as a program of technical instruction to try to improve agricultural production. The project's officials admitted their health center was only a "modest beginning." But nonetheless, a positive message was still sent throughout the local communities. The UNESCO workers, according to the project's official report, were "not initiating any large scale campaign of preventive medicine or sanitary engineering but [were] concentrating on making all possible progress by changing attitudes and eliciting the active co-operation of the people themselves."[41]

In 1951 the Marbial project was phased out. The results of the Marbial education and health program had been meager. Some text material in Creole had been developed for the project's schools. The Marbial literacy initiative fell far short of the numbers it planned to reach in Marbial Valley because of a long list of factors. These included a constant shortage of funds in the face of extreme rural poverty, opposition by local leaders, difficulty in changing the folk behaviors of many of the peasants, and corruption or mismanagement on the part of some Haitian program administrators.[42]

Yet the Marbial project still held significance to the United States and other foreign governments interested in providing aid to Haiti. The photographs of UNESCO teachers and technicians working in peasant settings circulated widely throughout the free world. Such publicity impressed world opinion about Haiti's ongoing poverty and development needs, especially needs for educational, agricultural, and medical assistance throughout the countryside. The Marbial project furnished fieldwork data and a concrete program model that inspired Western foreign-aid interests to attempt community-level projects in Haiti and similar settings.[43] The Marbial effort also gave confidence to black American intellectuals and U.S. foreign policy reformists about the importance of such aid to Haiti's future. Education had raised the living standards and political stature of U.S. blacks. They believed it could do likewise for the peasants of Haiti. Rayford Logan, a prominent Howard University scholar, singled out the Marbial program in his study of U.S.-Haitian links, published in 1953. Writing in a leading U.S.–Latin American policy journal, Logan emphasized that the Marbial project indicated that, with the proper international assistance and local community involvement, Haiti could develop "mutual self-respect" from the leaders and people of the U.S. mainland.[44]

The Marbial project had demonstrated to United States and international aid agencies that the community (not Haiti's national government) could be an effective starting point for assistance programs for Haiti's people. Donor agencies could mobilize community programs and at the same time largely bypass or minimize constraints by Haiti's horrendous central government. Other international bodies, such as WHO and the hemispheric Alliance for Progress, became encouraged to set up community development initiatives in other locales throughout Haiti. These organizations pressed forward in Haiti in spite of the terrible government and public-sector conditions with which these projects had to contend. Hundreds of small, private relief efforts continued to crop up in Haiti in the 1960s and 1970s. These programs provided food, medicine, shelter, and other emergency resources directly to the nation's most needy populations.[45]

A Missionary Gift: Schweitzer Hospital

Since their very beginning, independent religious groups and philanthropic projects centered in the United States have flowed to these Atlantic regions we have been analyzing. These missions managed to operate small, local projects in spite of the larger political and economic conflicts of the times. Whether in the communities of destitute black sharecroppers in the Jim Crow South or in the interior of Liberia with indigenous people toiling for the rubber plantations— churches, schools, and health stations organized by religious denominations and philanthropies appeared. In post–World War II Haiti, one of the most successful such projects was the Albert Schweitzer Hospital of Deschappelles, in the Artibonite Valley. A physician, William Larimer Mellon, and his wife, Gwen Grant Mellon, founded this hospital in 1956. Leaving well-to-do lives in Arizona, the two set out to emulate the work of Albert Schweitzer, whom they met in 1949. They spent seven years planning and raising the funds necessary to build and operate the hospital.

When the hospital opened, the surrounding district had an estimated seventy-three thousand residents. The district had no towns with more than three thousand residents and no industries. Valley inhabitants depended on subsistence agriculture, with only about two hundred families earning salaries. Of the salaried workers, most worked for the hospital, while others were domestic workers or small merchants. Approximately 85 percent of deliveries occurred at home with the mother alone, or attended by traditional midwives, or the mother's relatives. High levels of contamination during the birthing process resulted in infant death rates of 270 per 1,000 in 1952, and 60 deaths per 1,000 in 1966.[46]

Initially, the medical staff had to be drawn from the United States and Europe. The majority of the hospital's nursing corps and lab technicians in its early years also were from Switzerland, Germany, Canada, the Netherlands, and the United States. Poorly paid and motivated by missionary zeal, the medical teams at this institution quickly became besieged with patients. The 150-bed hospital became especially adept at treating infants and children in various stages of malnutrition. The hospital provided hosts of other services, from dental care to isolation for tuberculosis. By the late 1960s, Schweitzer Hospital had more than three thousand admissions each year as well as seventy thousand to one hundred thousand consultations.[47]

The hospital continued its operations through the 1990s, relying mostly on foundation grants, unsolicited donations, and modest fees from patients. A key to the Schweitzer Hospital's longevity has been its independence from the Haitian and U.S. governments. In 1978, Dr. Mellon wrote about his practice of keeping his institution free of any political strings: "I feel strongly opposed to accepting funds from any government." As for Haiti's dictators, the view he maintained for nearly four decades was, "I'm only a visitor here and have no right to criticize long established customs and traditions."[48]

Reactionary Technology

International aid agencies and philanthropists such as the Mellons managed to improve conditions in many local communities throughout Haiti. In the meantime, in the mid–1950s the American government became interested in financing larger, bilateral links with Haiti's central government. These links were intended to support the politico-strategic priorities of the United States. American officials were determined to prevent Soviet influence and socialist political movements from triumphing in Haiti, as was happening in Cuba. The emergence of the Francois Duvalier regime (lasting from 1956 to 1971) was attractive to U.S. foreign policy makers. Duvalier had a reputation throughout Haiti as an admirable country doctor and public servant, a reputation that had been three decades in the making. He became widely respected especially because as the minister of health he had taken part in the popular 1946 social movement.[49]

Yet behind Duvalier's public persona was an emerging autocrat bent on turning the biopower of modern medicine and the annihilative power of modern weapons to his political advantage. The political aspects of Duvalier's dictatorship have been covered extensively by scholars and journalists specializing in Haiti and Latin America.[50] But the connection that Duvalier developed to the social gospel and instrumentalities of technoscience has been totally overlooked. This connection began with Duvalier's participation in the 1930s in a nationalist intellectual movement that stressed the Haitian people's African heritage. As one of the founders of the literary journal *Les Griots,* Duvalier, along with fellow nationalist writers, identified African retentions as the heart of Haitian popular culture and Negritude as the nation's latent revolutionary ideology. Next, throughout his long years as a leader in public health and government agencies, Duvalier nurtured ties with Voodoo priests and temples as a means for working within peasant communities.[51]

Through his research in anthropology, stressing Haiti's black African folk culture and Voodoo, and his leadership in health programs throughout the countryside, Duvalier gained important political insights and contacts. He became sophisticated at sensing the power of peasant and village culture and institutions, including the pivotal importance of Voodoo. Although to the modernist outsider peasant life seemed backward and superstitious, political leaders in such societies undergoing revolutionary pressures could sponsor and manipulate these folk institutions to obtain tyrannical ends.[52] Duvalier became especially skilled at this tactic since, as a public health physician (trained in part in the United States) Duvalier projected a public image outside of Haiti of highly competent technical ability and quiet confidence.

His combination of professional expertise, public service, community links, and black racial nationalism enamored Duvalier to Haiti's ruling circles as well as U.S. and European foreign affairs officials. To U.S. and European politicians and government aid authorities, he seemed a potentially even-handed, populist leader with an understanding of Western technical modernization. Instead, in the opening years of the 1950s, as Haiti's minister of labor, Duvalier began building a self-serving and self-enclosed network inside Haiti's governing elite and military. This network would enable him to seize power in 1956.

Duvalier knew that, ultimately, political rule in Haiti rested on a leader's ability to assuage the peasant masses or pay the price of political turmoil, mass violence, and coup attempts. Throughout his political career Duvalier kept ties to the Voodoo leadership throughout Haiti's city slums and villages. Once in the presidency, Duvalier went further and developed a security force of persons of both Voodoo and other social origins throughout Haiti's communities; it acted as a secret police force.[53] He then used his presidential authority to implement wave upon wave of political persecutions. He tightly controlled foreign assistance, building a program of empty, that is, symbolic, development. Mats Lundahl, a leading authority on Haiti, described the tragic path that Duvalier's politics took: "It had been generally hoped that Duvalier would become a modernizer of Haitian politics. He possessed a comparatively clean record: for having taken part in the cultural renaissance connected with the *négritude* movement, for having been a reform [public health] minister under Estimé, and for presenting what was interpreted as an enlightened attitude towards the peasants. Immediately after assuming power, however, he began a violent eradication of every group that could possibly have constituted a future threat to his position."[54] How could the highly publicized U.S. policy to spread democratic and scientific development proceed in Haiti in such a context of obvious repression? To understand the technical links made by the United States inside Haiti through the 1960s and 1970s, we must look more closely at how Haiti's leaders themselves wrestled with scientific and technological modernization.

Symbolic Development

Throughout the nearly two decades of Duvalier's rule, newspaper accounts, popular books, and studies about the dictatorship appeared in the United States and Europe. Curiously, these materials did not consider the relationship between President Duvalier's political megalomania, on the one hand, and his professional skills as a public health physician and administrator, on the other. Research on the atrocities committed by physicians in Germany and occupied areas during World War II is most applicable. Physicians for the Nationalist Socialist government, backed by scientists and anthropologists, conducted intentionally deadly experiments on Jewish populations and other non-Aryan social segments as well as planned the techniques of death for the concentration camps of the Holocaust. Their actions were considered so horrendously unethical that they resulted in the Nuremberg Codes. These are the first international guidelines for protecting human subjects from inhumane medical experiments and procedures.

The war crimes of German physicians reveal that under conditions of frenzied political and racial nationalism, apparently normal professionals can quickly transform into inhumane planners and operatives. Indeed, the submissive professionalization of doctors, scientists, and other professionals whose work involves human care make these professionals easily prone to dissociate or numb themselves from the pain or death they may inflict on powerless clients. In Germany under Hitler and in other fascist societies of Europe, medical professionals quietly and dutifully used their medical techniques and treatment facilities to inflict egregious suffering on civilians considered enemies of the state.[55]

Duvalier's advocacy of modern medicine and mass education for national development evinced a deep belief in the social, transformative power of science. But his scientism had the form of what the historian Jeffrey Herf describes as reactionary modernism. In his analysis of Weimar Germany, Herf shows that leaders in politically troubled states undergoing rapid industrialization and economic depression often fervently embrace modern technology. Through their speeches and politics, they romanticize the capacity of science and technology to ameliorate social problems, enshrouding this vision with fanatical cultural and ethno-racial nationalism. At the same time they ignore, indeed suppress, the intermediate institutions necessary for open-ended public discourse and scientific scholarship. These institutions include widely available public schools, democratically run universities, protections for popular freedom of public opinion, and professions that develop and follow ethical guidelines. As Herf stated about Germany in the era of national socialist politics, the "paradox of reactionary modernism is that it rejects reason but embraces technology."[56]

U.S. policy makers and other U.S. policy watchers who were concerned with Haiti and the Caribbean overlooked European parallels to Duvalier's authoritarianism. These policy experts and advocates clashed over the degree of aid that should be sent to Haiti. But most all believed Duvalier to be a competent administrative professional, so technical aid flowed nonetheless. Inside Haiti, this aid (funds, equipment, weapons) was commandeered by this political leader/physician. Duvalier was determined to use both the cultural power of modern medicine and piecemeal technologies and heavily armed military and paramilitary forces. Blending the two forces that is, what cultural historians call *violence douce* and *violence directe*—was the only means for Duvalier to reign personally over a nation still so overwhelmingly rooted in African rural cultures.[57]

The merger of Duvalier's medical vision and political vision crystallized throughout his career. In his early medical articles of the 1940s, Duvalier described his findings concerning the tremendous effectiveness of penicillin for treating his patients afflicted with yaws and syphilis. But in the same reports he also set forth his philosophy for the national development of Haiti. In his 1945 medical article "Contribution to the Study of Yaws in Haiti," Duvalier glorified the anti-yaws campaign. He drew political conclusions, calling for the need to organize the rural areas into administrative units. These local sections, or "cellules," would serve as the basis for making Haiti a great society, "un vrai centre de civilisation." In his medical papers, Duvalier also stressed the need to eliminate illiteracy among the Haitian peasants. But this education would not be for improving the political self-development of the Haitian masses. Instead, it was intended to expand their awareness about the demands and rewards of modern medical treatment. In describing the yaws campaign, Duvalier criticized popular ignorance that, linked with diseases like yaws and malaria, "darkens the future of the Negro community" of Haiti.[58] He wrote, "Medicine, in its medical social activities, finds itself to be handicapped by the fact that even the slightly cured peasant [suffering from yaws] thinks he is completely cured. Therefore, he stays away from the clinic. This way of viewing his future is caused by a behavior linked to the problem of mass education."[59]

By 1961 Duvalier's political march (although not his psycho-cultural drives) toward a predatory oligarchy was widely known throughout U.S. foreign policy, military, and journalism circles. Highly respected news and academic sources, especially the *New York Times* and the *Hispanic American Report,* regularly featured reportage on the gruesome martial state that Duvalier was throwing over Haitian society.[60] Yet, despite occasional suspensions of aid, U.S. funds and technical development projects continued to flow into Duvalier's government. U.S. policy makers repeatedly expressed the concern that outright withdrawal of all aid to Haiti would result in a collapse of the Haitian government altogether.[61]

Key liberal U.S. politicians and civil rights activists criticized the continuation of American assistance to Haiti's government. They saw it as an immense contradiction to U.S. political values and public morality. Senator J. William Fulbright, the head of the U.S. Senate Foreign Relations Committee, condemned U.S. aid to Haiti. Upon Duvalier's "re-election" in 1961, Fulbright called Haiti "one of the worst dictatorships that remains in Latin America." In response, the standard explanation for the continued support sent to the Duvalier government became a diplomatic, strategic one: the United States feared, as a threat to its national security, the spread of Castro-led communism into the Dominican Republic and then Haiti. Thus, the United States supported Duvalier at any cost even though U.S. presidents and many in Congress publicly detested Duvalier's repressive leadership. Secretary of State Dean Rusk called the aid to Haiti an incentive for the Duvalier government to democratize the country: "We made all types of efforts to bring about changes in Haiti. We used persuasion, aid, pressure and almost all techniques short of the landing of outside forces, but President Duvalier was extraordinarily resistant."[62]

A deeper impulse underlay U.S. aid to Haiti during the Duvalier decades. It was the belief of U.S. officials that technical aid was a catalyst from which technologies and democratic, laissez-faire economic institutions in Haiti, and in the undeveloped world in general, would somehow take root. We have seen that long before Duvalier took office, indeed, before "less-developed nations" became a popular political term in Western diplomacy, U.S. political and corporate culture had become enthralled with the promise of technoscience. This preoccupation became manifested in the industrial activism of the early northern states and the modernization of the Union military, in the public health campaigns of the Spanish-American War, in the building of the Panama Canal, as well as in the TVA and the New Deal projects.

Now in the post–World War II era, U.S. leaders believed that technical assistance projects were natural spin-offs of their nation's continuing journey toward becoming a scientific republic. The science ideal gained strength in the U.S. government and its international relations each time U.S. doctors and medical scientists subdued a perilous disease or engineers designed more effective weapons to use against a totalitarian foe or took another step forward in space exploration. Against this backdrop, it is understandable that U.S. foreign policy makers extended technical aid to Haiti even while the nation became a police state ruled by an imperturbable physician-statesman.

The preoccupation in the postwar United States with scientific progress was reflected in the Alliance for Progress of the Kennedy administration. The alliance initiative followed the Point Four Program of the Truman and Eisenhower administrations. President Kennedy commenced the Alliance for Progress in 1961 to realize what he called the "Decade of Development." The alliance aimed to bring a fundamental change in U.S. relations with the other nations of the Americas.

As one alliance official stated, the policy intended to merge "strategic and security interests with economic and social measures considered vital for sustained democratic growth." Working through the U.S. Agency for International Development (AID) and the Inter-American Development Bank (IDB), Alliance for Progress planners set very specific goals for each aid program abroad. Lowering infant mortality rates, improving sanitation, expanding potable water supplies, increasing farm production—these were the types of quantifiable objectives that alliance administrators built into AID and IDB programs.[63]

Yet this aid policy had inherent limits stemming from its over reliance on the idea that once planted on a small-scale, scientific enlightenment and technical networks sprout spontaneously. Levins and Lewontin, leading analysts of modern American science, have described the limits of restricting approaches to social problems to scientific, technical solutions: "The working scientists usually do not deal with . . . global [society-wide] objectives but, rather, see themselves as pursuing humanitarian, nonpolitical objectives." A technical specialist believes, above all, "precise scientific information" and is "adamant in refusing to pursue a problem beyond the narrowest possible boundaries of his or her specialty and in refusing to allow considerations from the broader areas to inform his or her own work." The results of over-compartmentalized science in development assistance, according to another leading scholar of development issues, "means not only disciplinary specialization and the differentiation of the scientists among themselves, but also the impossibility of a connected grasp of reality and a critical judgment of it."[64]

In the Alliance for Progress initiatives, the technology transferred was typically homegrown (i.e., in the United States) and conveyed in a unilateral process. The project specialists used U.S. equipment and supplies to fill their aid packages. Over 90 percent of the commodities financed by AID in 1965 had been brought from U.S. firms. As AID officials wrote, these resources, "not dollars, are sent overseas by AID. The list includes iron and steel, industrial machinery, chemicals, motor vehicles, fertilizers, construction equipment, electrical apparatus, copper, and petroleum products." By the mid–1960s the alliance assistance programs—technicians, equipment, funds, and military expertise—were reaching seventy nations in Latin America, Africa, and Asia. These programs emphasized long-term economic development under the guidance of U.S. technical experts and guaranteed physical security for pro-U.S. political heads by providing specific weaponry or military training.[65]

As the Cold War intensified, so did U.S. involvement in foreign lands. Consequently, even during the Duvalier dictatorship, U.S. technicians, supplies, and funds arrived in Haiti. The United States, in its new global strategy of containing the Soviet bloc on every front, teamed up with allied industrial nations to tactfully spread technical aid projects. In 1961, under the U.S. Mutual Security Program, Haiti received assistance amounting to $5.47 million. These resources were provided for "such functions as rural credit, watershed management, agricultural education, hydrological investigations, access roads, small irrigation and drainage works, rural public health, and the development of the Artibonite Valley."[66] The U.S.-led Organization of American States picked Haiti for its first large-scale program under the Alliance for Progress. In 1962 Haiti received $6 million in aid under the alliance, largely for jetport and road construction.[67]

The sheer variety of the U.S. aid programs of the 1960s—agricultural, engineering, medical—is amazing compared to the almost total neglect of Haiti by the United States early in the twentieth century. But each project was largely a black box, separate from other projects as well as separate from relevant institutions and people that were not part of the Duvalier government in Haiti. In the Haitian political and socioeconomic context, this meant that over the long haul failure was almost inevitable. In the meantime, informal peasant technologies in Haiti moved forward with the same inevitability as nature's seasons.

Peasants and Technologies of Survival

While the Duvalier regime tightened its hold, the best the Haitian poor could do was to continue their traditional subsistence institutions and folk culture. Relying on food production, family, and social networks carved out and passed down for generations, peasants moved on with their lives and dreams. In 1960 Haiti's population was estimated at 3.5 million, with about 88 percent rural. The majority of these rural residents farmed for subsistence and local sales on small plots of three acres or so. Agriculture accounted for almost all of the national income, with coffee as the largest export product (12 percent of national income) followed by small crops of cocoa, bananas, sugar, sisal, and rubber.[68]

Modern farm techniques for large plots were of no importance to this population of small farmers. Instead, traditional links between food production and family life held together the genders and generations of each rural community. The vast majority of rural Haitians lived divorced from their nation's formal politics and foreign assistance. But strong ties between women, religion, family farming, and local markets kept social and economic life viable. Gender, education, local economy, and travel customs operated interactively. A leading anthropologist described the socialization of children in these local communities in a 1959 field report: "Girl children learn how to trade from their mothers, aunts, older sisters and other female relatives and ritual kinfolk. They customarily accompany some older woman, first to the local market places and market-like gatherings. . . . Later they are taken to the large regional market places, the town market places, and even the great market places of the capital. They are taught to buy and sell, to calculate value and to recognize currency, to measure and to judge quantity and quality, to assess various products, as boys are taught to plant, cultivate and harvest."[69] The division between market skills and farming were not fixed, but varied by households and communities. Most women and men had skills for both farming and marketing, with men traditionally doing most of the former and women the latter. Even though both men and women could inherit land, both genders preferred this customary division of labor.[70]

The complex rural life and geographical landscape of Haiti and the extremism of Haiti's centralized government were major impediments to any foreign development aid program that did not take them into account. Households and communities had little cash resources but, we have seen, were complexly organized around numerous traditional functions, farming, and markets. Clusters of rural households and farms were scattered geographically and separated by natural geographic barriers.

Even though these rural Haitians were fundamentally excluded from government and persecuted by Haiti's ruling and financial elites, age-old institutions of self-help and community cooperation carried the peasant population on from generation to generation. A cooperative work system known as a "coumbite" and adopted from African social life proliferated among peasant farmers. Coumbites could involve a few farmers coming together for just a few hours to perform a strenuous task on behalf of a coumbite member. The coumbite also could assemble large groups working together for days on a project for a member's farm. Collective meals and recreation followed work.[71]

One UNESCO worker described the typical life and world view of the Haitian peasants. "The people have never felt themselves to be part of any well defined community, large or small," he wrote. "The strongest ties are between members of the family, and the largest co-operative groups are the neighbors, relatives and friends who band themselves loosely together in coumbites for work." The UNESCO worker emphasized that most peasants owned a small plot that they cultivated with the help of their family. Although politically alienated, the peasants maintained a collective life of social and economic continuity: "Isolated as the people are, they are nevertheless much on the move. Women, for instance, must constantly go to the town or around the countryside to attend the markets and to sell their wares."[72]

Cotton, Dams, and Immunizations

U.S. aid programs for Haiti sometimes tried to influence the peasant economy, but with little results. We have seen that the United States attempted to build a miniature TVA project in Haiti's Artibonite Valley. The hope was to produce energy and irrigation resources for that region. U.S. aid authorities also tried to redirect the Haitian farm sector toward cultivating cotton. Large plantations had been key to the economy of the Black Belt South and Liberia's rubber production. However, in Haiti the wealthy class that owned large plantations (several thousand estates in number) and controlled the economy during the centuries of slavery had been smashed by the Haitian Revolution; the land was redistributed in tiny parcels. Tens of thousands of families subsisted on their small farms. Thus, any rapid drive for large-scale farming was essentially pointless.

During the 1930s Haitian farmers had managed to produce enough cotton that it became a significant export crop. However, damage caused by the Mexican boll weevil destroyed this crop. But by the 1960s some U.S. aid officials were attempting to reverse this condition. Since the United States was a world leader in modernized, large-scale farm production, U.S. aid officials tried to bring to Haiti a technically sophisticated form of cotton agriculture—farming that was centrally managed and could produce large amounts of cotton for export.[73]

In 1961 IDB officials estimated that in Haiti over 93 percent of the property holdings were less than three acres. They emphasized, "This extreme example of dwarf holdings or 'pulverization' of farm property is the fundamental obstacle to the socio-economic progress of agriculture and requires a very special approach." The following year, AID provided technical and financial resources to Haiti's Ministry of Agriculture for setting up a credit plan for farmers in the Artibonite Valley. This unit, the Institute of Agricultural and Industrial Development (IDAI), encouraged the cultivation of cotton. A new type of cotton plant was used, more efficient planting methods were taught, and insecticides were applied widely.[74]

In the meantime, exploiting this incoming assistance, Duvalier and his aides steered the Haitian state into a disjointed array of development activities. With the assistance of what came to be known as the OAS-IDB-ECLA Tripartite Mission, the Duvalier government established the National Development and Planning Council. In 1963 the council, under the tight grip of Duvalier officials, devised a short-range development plan known as "Le Démarrage," or "the starting." The plan called for a relatively huge amount of financial aid from the United States and its Western allies (some $50 million), yet actually garnered only a few million dollars annually from the United States.[75]

Duvalier wanted total administrative control of the foreign assistance projects. The ongoing conflicts over this issue between the U.S. State Department and the Duvalier administration made U.S. aid officials give much smaller amounts of development aid than Duvalier requested and, moreover, pinpoint

this assistance for specific projects in Haiti. The cotton production initiative received continuous IDB support. The program included the start-up of textile
workshops in Port-au-Prince to train workers, and a few mills were established
in various towns. In theory, the aim was to produce enough cotton and textiles
to meet the demands of Haiti's domestic market. However, in practice, by 1964
results were modest. Twelve workshops had been established, but with tiny production capacities (there were only three to six looms in each center). Over three
hundred farmers were receiving loans for their participation in the cotton program, but their total acreage was only about 1,310 acres.[76]

The irrigation and hydroelectric power plan for the Artibonite Valley became awash with delays and conflicts between Haitian government officials, under the dictates of Duvalier, and U.S. project officials. Throughout the late 1950s
and 1960s, Haiti had the lowest electrical resources by far of any nation in the
Caribbean and Latin America. Most available electricity, four-fifths of which
was owned by foreign companies, was in Port-au-Prince. Blackouts, unmetered
consumption (of as much as 50 percent of the electricity produced), and daily
rationing were common in Haiti. Rural electrification was nonexistent. Prior to
the Duvalier government, Haitian president Magloire had planned to issue a contract with the Westinghouse Electric Company to install the Artibonite hydroelectric plant. However, as the Duvalier presidency dug in, the construction of
the hydroelectric plant at Peligré dragged on. U.S. technical planners expected
the Peligré plant would boost Haiti's electrical resources up some 150 percent.
But Duvalier insisted on controlling key technical personnel constructing
the power plant, appointing political cronies of questionable competence and
character.[77]

As early as 1960, Duvalier appointed a personal ally to a chief engineering inspection post on the dam project, triggering U.S. displeasure. This incident was one of many similar disagreements about control over personnel for
U.S. projects in Haiti, occasionally causing suspensions of U.S. funds. U.S. officials demanded they oversee hiring, firing, and wages of Haitians employed
on U.S. aid projects, a boundary Duvalier repeatedly overstepped. The work on
the Artibonite project proceeded tumultuously and was fraught with delays.
Moreover, multitudes of poor farmers were dislocated from the Artibonite Valley to make way for irrigation sections or local farmers favored by the
Duvalierists. These uprooted farmers ended up as seasonal workers in sugar fields
of the Dominican Republic or Cuba, exiled, or facing starvation. It was not until 1971 and 1972 that two turbines were installed at Peligré, doubling the nation's
total resources for generating electricity to sixty thousand kilowatts.[78]

Even with this improvement in electrical power, most of Haiti's poor, living both in and outside cities, relied on hand-gathered fuels and ingenuity for
their cooking and heating needs. Most used kerosene lamps for lighting. Urban
dwellers fortunate to live near electric lines fashioned makeshift circuits. A U.S.
government-sponsored survey of Haitian life conducted in the late 1960s and

early 1970s found that "After dusk, in unlighted neighborhoods, bamboo poles with wires attached are hoisted by householders to power lines in order to make an electrical contact and provide light."[79]

Throughout the 1950s and 1960s, U.S. technical assistance in Haiti generated modest, although sorely needed benefits. On average, between 1950 and 1970 Haiti received about $2.5 million annually from the United States, about half of its total annual foreign aid. The bulk of this U.S. aid was used for agricultural development, road construction, emergency relief, disease control, and government budget subsidies.[80] On the health care front, U.S. funds were used for "crash" disease-control programs. In cooperation with technical and funding support from the Pan American Health Bureau, UNICEF (United Nations Children's Fund), and WHO, between 1962 and 1965 U.S. funds fueled several malaria and smallpox campaigns. An estimated two million houses were fumigated, and anti-malarial drugs were given to about fifty thousand patients. Also, an estimated 850,000 Haitians received smallpox vaccinations. Finally, U.S. assistance helped improve the water supply in sections of Port-au-Prince.[81]

Many international health experts and advocates criticized these "mass impact" eradication programs in Haiti run by the U.S. government and international organizations. They pointed out that the preoccupation with measures to quickly snuff out epidemic diseases ignored the more difficult challenge of attacking high mortality that was caused by malnutrition and intestinal disorders. To stamp out these less dramatic causes of death would require the elimination of social conditions such as the poverty and lack of primary care facing the general Haitian population. Moreover, drops in disease rates caused by rapid eradication campaigns involving DDT were proving short-lived. Infectious mosquitoes were proving resistant to DDT. Other health advocates pointed out that the eradication projects short-circuited resources for more diverse approaches in the research of malaria.[82]

The public health triumphs in the Canal Zone over yellow fever and malaria had resulted from two primary measures: closed-system engineering for water and sewerage supplies in conjunction with comprehensive public health surveillance and medical treatment. But such broad steps were never tried in post–World War II Haiti. In fact, these steps were never mentioned, much less considered for practical implementation, by either the eradicationists or the anti-eradicationists running U.S. and multilateral foreign health-aid programs in Haiti.

A Haitian Diaspora

The impact of the Duvalier government on Haiti's potential to modernize technically was profound. Technical assistance programs in vital areas of national life—agriculture, health care, education, and the like—failed one after another. In 1962 only eighty-six thousand of the nation's seven hundred thousand primary-school-age children had schools. There were only about one thousand university students (one-fifth female) and eighty-five hundred school teachers, most untrained. Worse, Haitians with professional training and commitment to improving Haiti left the country or were exiled.[83]

The Duvalier regime, followed by that of his son and subsequent military governments, generated a "sub-African Diaspora" of Afro-Haitians who scattered throughout North America and other parts of the world. Fleeing black dictatorial oppression, this Diaspora was composed of the steady wave of immigrants, exiles, and refugees from Haiti's preciously small professional and skilled segments. In 1963 Leslie Manigat, a former professor at the University of Haiti and former member of the Haitian Department of Foreign Affairs, was a participant-observer of the accelerating brain drain. Situated anew in a Washington (D.C.) university, he called Haiti's outflow of talent the "sorriest symptom" of his nation's political repression. A "general exodus" was underway "of Haitian professionals and technicians to the United States, Canada, Latin America, Europe, North Africa, and, particularly, black Africa."[84]

Nearly 70 percent of Haiti's physicians who graduated between the mid–1950s and early 1960s had exited the country, as did large numbers of nurses. "International organizations have been assaulted by [Haitian] engineers, teachers, economists, physicians, and lawyers seeking jobs in the newly emerging nations," Manigat wrote. He estimated that over one thousand Haitian professionals and technicians were on the staffs of agencies and development services in Africa. "We are witnessing a kind of new Haitian diaspora," Manigat stated, "which spreads its youngest and most dynamic elements throughout the world."[85]

Haitians trained and experienced in the various aspects of legal institutions also left in growing streams. The wholesale political repression in Haiti under the Duvaliers and the ensuing military heads all but destroyed the Haitian court system. According to a leading specialist on Haitian legal institutions, during and after the 1960s "Haiti's best judges and lawyers joined the growing diaspora and often ended up running the judiciaries of the newly independent Francophone African states." Haitian experts approximate that 80 percent of the nation's lawyers and other professionals (i.e., physicians, engineers, teachers, and public administrators) were living in the United States, Canada, or African nations. Many of these Haitian professionals went on to distinguished careers in their new homelands. Many, about three hundred, held positions in UN programs in the Congo, Rwanda, Burundi, Togo, and Guinea. The UN Technical Assistance Program had more Haitian professional employees from Haiti than from any other nation of the Caribbean and Latin America.[86]

By 1970 approximately 35,000 Haitians had managed to gain residence in the Bahamas, even though many such sojourners were frequently deported back to Haiti. By the mid–1970s New York City had 150,000 Haitian emigrés, and thousands more were residing in Chicago, Washington, and Montreal. The number of legal aliens from Haiti who settled in the United States by 1992 stood at 250,000. During these decades, the Haitian immigrant community maintained strong political, cultural, and charitable ties with the Haitian homeland.[87] However, Haiti remains among the most impoverished and politically fragmented states in the Western Atlantic world.

The dispersal of tens of thousands of Haitians from Haiti in recent decades has been exceeded by the ongoing debacle in Liberia. Periods of political anarchy and violent civil war have been recurrent in Liberia since 1980. By 1990 nearly one-fourth of Liberia's 2.3 million people had fled the country. Most took refuge in Sierra Leone, Guinea, and Côte d'Ivoire. In Monrovia that year, eyewitnesses reported, "Most of Liberia . . . has been without electricity, water or telephone for more than two months, and food is scarce everywhere." Moreover, cholera had broken out throughout the city and refugee areas. At the close of the 1990s, about one-half of Liberia's people were internally displaced due to political violence, and cholera remained a problem.[88] Cholera is an ancient bacterial killer that brings a quick, excruciating death. Ironically, this disease had been eliminated from U.S. and European cities in the nineteenth century by the application of simple public health and sanitation measures.

Conclusion

SCIENCE AND HOPE

The transfer of U.S. technology and medicine throughout the Atlantic world was shaped by the nation's most cherished vision. From its colonial origins through its rise to worldwide leadership in the twentieth century, the idea most constant in the U.S. political imagination has been that the nation's destiny depends on pursuing science and its widespread application. With railroads finally connecting the Atlantic and Pacific coasts, at the end of the nineteenth century U.S. leaders moved forward with the ideology that wherever in the Atlantic world they brought their modern science knowledge and technical expertise, improvements would ensue.

From the late nineteenth through the mid-twentieth century, in the Black Belt South, northern industrial finance and the South's economic leaders geared the regional economy around large-scale plantations, mechanized agriculture, and, finally, mixed industry. Next, national political leaders and industrialists made Panama the site for one of world's largest, most modern canals. Building on their Spanish-American War experience with tropical disease, the U.S. public health authorities in the Panama Canal Zone carried out one of world's most effective sanitation programs. During the interwar decades, national public health leaders and business philanthropies implemented medical research and public health campaigns to control major infectious diseases in the southern states. Most notable were the Rockefeller Foundation's campaign against hookworm and, somewhat later, the federal government's malaria projects throughout the Black Belt to advance a kind of public-health medical science.

Having had success in the Canal Zone, as well as ongoing experience in the southern communities of the U.S., America's political, military, and medical leadership took on a much bigger modernization challenge: the implementation of an entire public health and public works system for Haiti. Federal government and military leaders were confident that this small nation could be rid of its major diseases and that a modern infrastructure and administration could be put in place. Subsequently, Liberia became the site of less-focused, but significant U.S. medical (tropical) research as well as the site for corporate investments and plantations for raw materials production.

With the outbreak of World War II, U.S. technical links with Haiti and Liberia deepened. Euphoric following the Allied victory in the war, U.S. government, corporations, and academe invested in expanding their echelons of scientists and technicians. America's leaders believed that the new technostructure of scientists, engineers, and medical researchers was vital for national prosperity and leadership of the free world. Research and development (so-called "R&D") in science, technology, and medicine received historically high levels of support from government and industry. A presidential panel on biomedical research stated in 1971, "Federal support for biomedical research during the past two decades has brought us to the threshold of an era [of] unparalleled national capability." As years passed, U.S. presidents trumpeted the cause of science even more loudly. Richard Nixon, in his 1974 state of the union address, emphasized that his administration "recognizes that the need for progress in every major area of American life requires technological input." A few years later, Jimmy Carter, too, exclaimed that "our vision of the future [is] largely defined by the bounty that we anticipate science and technology will bring."[1]

The progression of the Cold War caused U.S. leaders to grow even more unswerving about the imagined future benefits of science. To counter the military threat of the Soviet bloc, the United States used its scientific prowess to research and erect the world's most advanced military and communication technologies. As the commander-in-chief of the nation's military and sole holder of the authority to release a nuclear holocaust, the U.S. president now commands the world's most powerful military defense. In its struggle against worldwide communism, the U.S. also made itself a hub of international aid and promoted political alliances—especially NATO—uniting national governments and militaries of the Atlantic world.

At the same time that the United States rose to international superpower, black activists and civil rights leaders of the twentieth century internationalized their world views. They imagined that there was a moral link between their struggles in the U.S. and the plights of "Negro" nations such as Liberia and Haiti. After World War II, one by one, African nations gained independence. Prominent scholars and civil rights leaders believed that somehow tight U.S. international policy and economic bonds would broaden to embrace this new family of multicultural and multiracial nations. Awed by the United States' international political and economic power, the civil rights community believed that black Americans and Africans would somehow be brought onto equal footing with the prosperous citizens of the United States and other nations of the industrial world.

Harold Isaacs, one of the United States' leading foreign policy journalists during the Kennedy-Johnson and Nixon years, wrote about this vision of global civil rights and racial equality. He warned that the United States faced unprecedented challenges in foreign relations. Isaacs attributed these challenges to the end of colonialism, the new communication technology, and rising expectations throughout newly independent African and Asian nations. The U.S. foreign-policy-making community, politically conscious black citizens, and Africa's leaders and youth were experiencing "a great jumble of new impulses, sensations, moods, and experiences." The emergence of Independent Africa, in particular, was creating "new leverage in world affairs working on the affairs of the Negro in America."[2]

However, the rising hopes that America's black poor and the less-developed nations of the world would be swept to progress by the United States' technological advances went unfulfilled. The high expectations of the 1960s and 1970s obscured the historical reality that U.S. technical involvement in the black Atlantic had come at a high cost to black labor and communities. In the Deep South, the plantation set generations of black sharecroppers and their communities far behind other Americans in the race for middle-class assimilation. In Panama, thousands of West Indian workers, European work crews, and indigenous people experienced bare wages and preventable ill-health and deaths from disease while building the Panama Canal. And in Haiti and Liberia, the U.S. technical transfer benefited, first and foremost, the U.S. military and economic interests in raw materials, then, second, the civilian populations of these nations.

Indeed, by the 1970s there were glaring social problems unaffected by the technical and medical resources the United States sent into Atlantic nations such as Haiti, Panama, and Liberia. In post–World War II Haiti and Liberia, the leadership of these nations became either inept or autocratic and enforced its autocracies using modern weapons or technical projects. But U.S. leaders claimed neither responsibility nor the power to help these nations' majorities. Instead, U.S. military interventions (or evacuations) became reoccurring in these countries. The harsh political and social realities of these regions dashed the expectations of the U.S. civil rights community that scientific progress, modern standards of living, and civil rights could be conveniently exported.

Can larger foreign policy frameworks explain the difficulties in technological transfer from the United States to Black Atlantic nations' contexts? In recent decades historians, political scientists, and development economists have striven to identify the nature of U.S. global expansionism—the roots and permutations of U.S. political and economic influence throughout the Atlantic and Pacific worlds. They deem wholly inapplicable the notion of formal empire, that is, "government from abroad." The traditional empires established by Britain and other European nations used military forces to maintain conquest and then established permanent settler-nations. In these empires' colonies in which permanent settlement was the aim, plundering of raw materials ensued and sometimes the expulsion of indigenous populations.[3]

In assessing the United States' rise to global superpower, these scholars emphasize that no such direct domination occurred. Instead, they have come to emphasize approaches such as "informal" empire, Wilsonian internationalism, mutual dependency, and balance-of-power frameworks. Others, leading historians, stress the United States' international influence as spiritual, political, and philosophical in nature. They write as celebrants of the United States as the world's one true Republic of Technology, the leader by virtue of it having obtained humanity's highest level of intellectual, scientific, and moral development.[4]

Even though there is no general agreement on an exact explanatory paradigm for U.S. expansionism, all agree that the U.S. global network has involved using a variety of foreign aid and international military security links. These links have been in the form of large private investments, military installations or interventions, and indirect political maneuvers to generate pro-U.S. officials and democratic political discourse in these foreign nations.[5] In the case of the African Diaspora nations we have examined, varying degrees of such links did emerge. But that has only been part of the story.

From our research on African Diaspora societies, there has been a deeper, less-visible technological relationship between the United States and its weaker Atlantic neighbors. This relationship involves the liberating (mostly in the United States) and hegemonic (mostly in the African nations) impacts of U.S. technology. It has had three dimensions. The first concerns the cultural (not strictly intellectual) and sociological evolution of technologies and basic and medical sciences inside the United States. The second dimension of this relationship has been the spin-off and exportation of specific technical projects from the U.S. domestic technostructure to these foreign societies. The third dimension has been the impact of this U.S. technical aid on the collective social, cultural, and political life of the recipient regions. Overall, the development of agricultural technology and tropical medicine inside the United States has historically adapted to the nation's racial and cultural heritage—a heritage rooted in slavery, racial caste, and the poverty of the southern Black Belt. Furthermore, the contradiction between the benefits of technology for its sponsors and makers in the United States and the detriments experienced in African-Atlantic societies has all too often been obscured in standard histories of American technology.

For the United States, its centuries-old science and technology project raised standards of living and built its government, military, and businesses into a global force. In this sense, the mission for science has been a rousing success. However, for many regions of the African Diaspora, the history of U.S. technology transfer bears out no grand theory or gospel of science utopianism. Overall, for these regions it has been a history of hegemony and social problems. Hegemony involves the exercise of domination through local consent and cultural encroachment. The scientific idealism of the post-Sputnik United States— the belief that scientific and technological superiority was the only sure avenue to prosperity and international power—spread to the leaders of Haiti, Liberia, and other less-developed nations. Yet in the end, U.S.-led technical endeavors in Haiti and Liberia through the 1970s left behind little that promoted the large-scale institutions and resources necessary for these nations to modernize themselves.

The United States contributed technical experts, funds, and equipment for small, multilateral aid projects to Haiti and Liberia. Large U.S. companies also sponsored production firms to process and market these two nations' raw materials. Finally, the U.S. military extended experts, weapons, and equipment to bolster pro-U.S. officials and political activists in Haiti and Liberia, and in Panama as well. But all together these technical links did little to equip the Haitians or Liberians to eliminate their nations' intransigent political suppression, poverty, major diseases, and public health problems. During the 1960s and 1970s, Haitians remained paralyzed under a stagnant dictatorship. In the meantime, Liberia's half-baked gains in national public works and industry, planned according to U.S. development schemes and technical tutorship, unraveled.

Back in the United States, following the *Brown v. Board of Education* decision and the start of the modern civil rights movement, black Americans of the traditional plantation South splintered into three segments in relationship to the national economy's technological transformation. The first segment comprised hundreds of thousands of black workers and their families who had to leave the farm areas of the South or the South altogether. A second segment, the majority of black southerners, remained in the South, becoming lower-level or semiskilled workers in the service and manufacturing sectors. These southern black wage earners managed to move into income brackets close to the region's lower middle class. The small professional and business circle within this black population became the nation's much maligned "black bourgeoisie."[6] Finally, a large population of black sharecroppers and other unskilled rural blacks in the South remained in poverty. Generation upon generation still struggled to survive unemployment, poor health, inadequate housing and schools, and racial discrimination by courts and politicians. This was especially true in traditional Black Belt areas, such as the Mississippi Delta communities.

The southern black Americans who migrated to northern and western cities during the 1950s and 1960s were later joined by thousands of Haitian immigrants. Many of the Haitians coming to the United States were professionals and skilled personnel forced to flee political repression or poverty in their homeland. In Panama, the United States terminated the Canal Zone as a legal entity (in 1976) but maintained control of the Panama Canal. Panama as a nation has had intense political and economic problems. As in nearby Haiti, U.S. interventions to maintain civil order have been recurrent, coming and going like the seasons.

The persistent poverty of blacks in the United States and other parts of the less developed world engendered deep dissatisfaction among activists and intellectuals, with promises by U.S. policy leaders that technological advances would somehow save the day for all of humanity. Frederick Douglass, as ambassador to Haiti in the late nineteenth century, espoused optimism about the power of science. Nearly a century later, Rev. Martin Luther King, Jr. was much more troubled by this power. To King there was something missing, a void in the scientific advances of America's Space Age. Throughout his public career, King emphasized that the high-speed transportation and communication technologies were shrinking the world, but also enabling U.S. political and economic leaders to leave behind old social attachments. King believed that moral insensitivity created by the expanding global technology was producing social alienation between the haves and the have-nots around the world. The technological advances in the militaries of the superpower nations were also causing an ever-growing threat of nuclear annihilation.

In an academic essay on this issue, "Facing the Challenge of a New Age," King pointed out that modern society's "scientific and technological genius" had enabled humankind to "dwarf distance and place time in chains; [and] carve highways through the stratosphere." The nations and individuals of the globe were now linked into a "new world of geographical togetherness." But while modern technology was creating speedier air, land, sea, and telephonic communication, it was also compounding alienation and conflict between societies and social groups that had previously existed physically isolated from each other. Although through science "we have made of the world a neighborhood," King emphasized, "now through our moral and spiritual genius we must make of it a brotherhood."[7]

King's challenge to the United States not to place boundless faith in science remains unmet. Instead, many of America's leaders still believe more than ever in promoting the mission of science. Assessing the unparalleled success of the United States and Europe in modern world history, recently a leading American political philosopher gave credit to the primary role of science in Western civilization. He wrote that the "unfolding of modern natural science has had a uniform effect on all societies that have experienced it." This science, in the form of technology, "makes possible the limitless accumulation of wealth, and thus the satisfaction of an ever-expanding set of human desires."[8] Our glimpse into the histories of the black South, Haiti, Liberia, and other Atlantic-African populations says otherwise. It has shown that for these places and people such rewards from quests for modern science are still either a hope or a prayer.

NOTES

Introduction

1. B.L.R. Smith, *American Science Policy since World War II* (Washington, D.C.: Brookings Institution, 1990); C. E. Rosenberg, *No Other Gods: On Science and American Social Thought,* rev. and exp. ed. (Baltimore: Johns Hopkins University Press, 1997); Ronald L. Numbers, ed., *Scientific Authority and Twentieth-Century America* (Baltimore: Johns Hopkins University Press, 1997). Developments in, for example, genetics, the World Wide Web, space exploration, robotics, and nuclear energy are ongoing political and social concerns. The writings about these fields are an ever-growing mountain, underscoring the central value and role that science continues to hold in modern America.

2. Y. M. Rabkin, *Science between the Superpowers* (New York: Priority Press, 1988); Robert Anderson et al., eds., *Innovative Systems in a Global Context: The North American Experience* (Montreal: McGill-Queen's University, 1998); Jessica Wang, *American Science in the Age of Anxiety: Scientists, Anticommunism, and the Cold War* (Chapel Hill: University of North Carolina Press, 1999).

3. Historical studies of American technology and non-European societies focus mostly on culture and the diffusion of technology from the West (as a whole) to the non-West. Most important are Daniel R. Headrick, *The Tentacles of Progress: Technology Transfer in the Age of Imperialism, 1850–1940* (New York: Oxford University Press, 1988); Michael Adas, *Machines and the Measure of Men: Science, Technology, and Ideologies of Western Domination* (Ithaca, N.Y.: Cornell University Press, 1989); Ronald T. Takaki, *Iron Cages: Race and Culture in Nineteenth-Century America* (New York: Knopf, 1979); and Arnold Pacey's two works, *Technology in World Civilization: A Thousand-Year History* (Cambridge, Mass.: MIT Press, 1990), and *The Maze of Ingenuity: Ideas and Idealism in the Development of Technology,* 2nd ed. (Cambridge, Mass.: MIT Press, 1992). For views of Western technology transfer from the angle of the modern African and Asian recipient nations, see Jacques Gaillard, V. V. Krishna, and Roland Waast, eds., *Scientific Communities in the Developing World* (New Delhi: Sage, 1997). A small body of diplomatic and political history looks into the less-visible strands of U.S. relations with Third World nations. Most notable are Emily S. Rosenberg's works: *Spreading the American Dream: American Economic and Cultural Expansion, 1890–1945* (New York: Hill and Wang, 1982); "The Invisible Protectorate: The United States, Liberia, and the

Evolution of Neocolonialism, 1909–40," *Diplomatic History* 9 (summer 1985): 191–214; and "Revisiting Dollar Diplomacy: Narratives of Money and Manliness," *Diplomatic History* 22 (spring 1998): 154–176.

4. Frank Tannenbaum largely initiated comparative history of slavery and post-Emancipation race relations with his classic *Slave and Citizen* (1946; reprint, Boston: Beacon, 1992). For an overview of comparative slave studies through the 1980s, see David Turley, "Slavery in the Americas: Resistance, Liberation, Emancipation," *Slavery and Abolition* 14 (August 1991): 109–116. Key comparative studies of slave societies and early agrarian capitalism include Peter Kolchin, *Unfree Labor: American Slavery and Russian Serfdom* (Cambridge, Mass.: Harvard University Press, 1987); Shearer D. Bowman, *Masters and Lords: Mid-Nineteenth-Century U.S. Planters and Prussian Junkers* (New York: Oxford University Press, 1993). The historiography of comparative disease experiences during the slavery era is assessed in Kenneth F. Kiple, *The African Exchange: Toward a Biological History of Black People* (Durham, N.C.: Duke University Press, 1988). The few histories that venture into the twentieth century and compare African diaspora societies and African nations are restricted to political or cultural themes. These valuable works include Joseph E. Harris, ed., *Global Dimensions of the African Diaspora* (Washington, D.C.: Howard University Press, 1993); Michael L. Conniff and Thomas J. Davis, eds., *Africans in the Americas: A History of the Black Diaspora* (New York: St. Martin's Press, 1994); Stanley B. Greenberg, *Race and State in Capitalist Development: Comparative Perspectives* (New Haven, Conn.: Yale University Press, 1980); Ronald Walters, *Pan Africanism in the African Diaspora: An Analysis of Modern Afrocentric Political Movements* (Detroit, Mich.: Wayne State University Press, 1993); the two major contributions by George M. Fredrickson, *White Supremacy: A Comparative Study in American and South African History* (New York: Oxford University Press, 1981) and *Black Liberation: A Comparative History of Black Ideologies in the United States and South Africa* (New York: Oxford University Press, 1995); and Anthony W. Marx, *Making Race and Nation: A Comparison of South Africa, the United States, and Brazil* (New York: Cambridge University Press, 1999).

5. James T. Campbell, *Songs of Zion: The African Methodist Episcopal Church in the United States and South Africa* (New York: Oxford University Press, 1995); Brenda Gayle Plummer, *Rising Wind: Black Americans and U.S. Foreign Policy, 1935–1960* (Chapel Hill: University of North Carolina Press, 1996); Penny M. Von Eschan, *Race against Empire: Black Americans and Anticolonialism, 1937–1957* (Ithaca, N.Y.: Cornell University Press, 1997).

6. Steven Yearley, *Science, Technology, and Social Change* (London: Unwin Hyman, 1988). For social constructionist studies of epidemiology and medical discoveries, see, for example, François Delaporte, *The History of Yellow Fever: An Essay on the Birth of Tropical Medicine* (Cambridge, Mass.: MIT Press, 1991); and Bruno Latour, *The Pasteurization of France* (Cambridge, Mass.: Harvard University Press, 1988).

7. G. Parayil, "Models of Technological Change: A Critical Review of Current Knowledge," *History and Technology* 10 (1993): 105–126; Thomas P. Hughes, *Networks of Power: Electrification in the West, 1880–1930* (Baltimore: Johns Hopkins University Press, 1993); W. E. Bijker, T. P. Hughes, and T. J. Pinch, eds., *The Social*

Construction of Technological Systems: New Directions in the Sociology and History of Technology (Cambridge, Mass.: MIT Press, 1987); R. Macleod and M. Lewis, eds., *Disease, Medicine, and Empire: Perspectives on Western Medicine and the Experience of European Expansion* (London: Routledge, 1988).

8. Langdon Winner, *Autonomous Technology: Technics-out-of-Control as a Theme in Political Thought* (Cambridge, Mass.: MIT Press, 1992).

9. Robert S. Johnson, "Science, Technology, and Black Community Development," *Black Scholar* 15 (1984): 32–44; Anthony Walton, "Technology versus African-Americans," *Atlantic Monthly* 283 (January 1999): 14, 16–18.

10. Two important exceptions to this tendency are Kenneth R. Manning, *Black Apollo of Science: The Life of Ernest Everett Just* (New York: Oxford University Press, 1983); and Linda O. McMurry, *George Washington Carver: Scientist and Symbol* (New York: Oxford University Press, 1981).

CHAPTER 1   *Machines and Plantations*

1. Alfred W. Cosby, *The Measure of Reality: Quantification and Western Society, 1250 1600* (New York: Cambridge University Press, 1997); Bernard Barber, *Science and the Social Order* (New York: Collier, 1970).

2. Thomas Paine, *The Age of Reason: Being an Investigation of True and Fabulous Theology* [1794], in *Collected Writings/Thomas Paine* (New York: Library of America, 1995), 326.

3. Ibid., 328.

4. Richard Delgado and D. R. Millen, "God, Galileo, and Government: Toward Constitutional Protection for Scientific Inquiry," *Washington Law Review* 53 (1978): 349 361, Jefferson quote (made in 1785) on 358; J. L. Heilbron, "Introductory Essay," in *The Quantifying Spirit in the Eighteenth Century*, ed. T. Frangsmyr et al. (Berkeley: University of California Press, 1990); W. D. Pattison, *Beginnings of the American Rectangular Land Survey System, 1784–1800* (Chicago: University of Chicago Press, 1957).

5. Thomas Jefferson to John Adams, letter, October 28, 1813, in *The Writings of Thomas Jefferson,* vol. 9, *1807–1815*, ed. P. L. Ford (New York: Putnam's, 1898), 425, 429.

6. Carl Sagan, *The Demon-Haunted World: Science as a Candle in the Dark* (New York: Ballantine, 1996), 405–406, 424–431, 433–434, quote on 424.

7. Abraham Lincoln, "Annual Message to Congress, December 3, 1861," in *The Collected Works of Abraham Lincoln*, vol. 5, ed. Roy P. Basler (New Brunswick, N.J.: Rutgers University Press, 1953), 39.

8. Arthur F. Raper, *Preface to Peasantry: A Tale of Two Black Belt Counties* (1936; reprint, New York: Atheneum, 1968); Charles S. Johnson, *Shadow of the Plantation* (Chicago: University of Chicago Press, 1934).

9. Johnson, *Shadow of the Plantation*, 103. Johnson cites, in particular, U. B. Phillips's essay, "The Decadence of the Plantation," *Annals of the American Academy of Political and Social Science* 35 (1910): 37–41.

10. William Barrett, *The Illusion of Technique: A Search for Meaning in a Technological Civilization* (Garden City, N.Y.: Anchor Press/Doubleday, 1978), 17–19; Arnold Pacey, *Technology in World Civilization* (Cambridge, Mass.: MIT Press, 1990).

11. Sir A. Grenfell Price, *White Settlers and Native Peoples* (1950; reprint, Westport, Conn.: Greenwood, 1972).

12. Edwin R. Embree, *Indians of the Americas* (1939; reprint, New York: Macmillan, 1970), 9.

13. P. A. Gilje, "The Rise of Capitalism in the Early Republic," *Journal of the Early Republic* 16 (1996): 159–181.

14. Louis C. Hunter, *A History of Industrial Power in the United States, 1780–1930,* vol. 2, *Steam Power* (Charlottesville: University Press of Virginia, 1986); Robert W. Fogel, *Without Consent or Contract: The Rise and Fall of American Slavery* (New York: W. W. Norton, 1989), 102–111.

15. Roger Burlingham, *Engines of Democracy: Inventions and Society in Mature America* (New York: Charles Scribner, 1940), 33.

16. W.E.B. Du Bois, *Black Reconstruction: An Essay toward a History of the Part Which Black Folk Played in an Attempt to Reconstruct Democracy in America, 1860–1880* (New York: Harcourt, Brace, 1935).

17. D. Gontar, "A Version of the South, or Deconstructing Reconstruction: V. S. Naipaul's *A Turn in the South*," *Plantation Society in the Americas* 3 (1993): 93–113.

18. Howard W. Odum, *Southern Regions of the United States* (Chapel Hill: University of North Carolina Press, 1936), 355–359.

19. Lawrence Shore, *Southern Capitalists: The Ideological Leadership of an Elite, 1832–1885* (Chapel Hill: University of North Carolina Press, 1986); Shearer D. Bowman, *Masters and Lords: Mid-Nineteenth-Century U.S. Planters and Prussian Junkers* (New York: Oxford University Press, 1993).

20. Herbert S. Klein, *Slavery in the Americas: A Comparative Study of Virginia and Cuba* (Chicago: University of Chicago Press, 1967), 179–180, quote on 179; B. H. Barrow cited by Fogel, *Without Consent or Contract,* 26; C. S. David, *The Cotton Kingdom in Alabama* (Montgomery: Alabama State Department of Archives and History, 1939), 57–65.

21. Phillips, "Decadence of the Plantation," 37.

22. Du Bois, *Black Reconstruction in America*; Robert W. Fogel and Stanley Engerman, *Time on the Cross: The Economics of American Negro Slavery* (Boston: Little, Brown, 1974), 204; Fogel, *Without Consent or Contract,* 26–28.

23. Rupert B. Vance, *Human Factors in Cotton Culture: A Study in the Social Geography of the American South* (Chapel Hill: University of North Carolina Press, 1929), 43.

24. N. T. Wilcox, "The Overseer Problem: A New Data Set and Method," in *Without Consent or Contract: The Rise and Fall of American Slavery: Evidence and Methods,* ed. R. W. Fogel, R. A. Galantine, and R. L. Manning (New York: W. W. Norton, 1992), 84–109; and R. L. Manning, "The Gang System and the Structure of Slave Employment," ibid., 109–119. Also, sometimes black slaves served as drivers alongside overseers to supervise the gang workers. See W. K. Scarborough, *The Overseer: Plantation Management in the Old South* (Athens: University of Georgia Press, 1984), 16–18, 82–84.

25. W. T. Howard Jr., *Public Health Administration and the Natural History of Disease in Baltimore, Maryland, 1797–1920* (Washington, D.C.: Carnegie Institution, 1924).

26. L. P. Jackson, *Free Negro Labor and Property Holding in Virginia, 1830–1860* (New

York: Russell and Russell, 1942), 76–88; J. H. Moore, *The Emergence of the Cotton Kingdom in the Old Southwest: Mississippi, 1770–1860* (Baton Rouge: Louisiana State University, 1988), 268–274.

27. Fogel and Engerman, *Time on the Cross*, 235; E. D. Genovese, *The Political Economy of Slavery: Studies in the Economy and Society of the Slave South* (New York: Vintage, 1965), 204, 225.

28. Fogel and Engerman, *Time on the Cross*, 234–235; M. F. Berry and J. W. Blassingame, *Long Memory: The Black Experience in America* (New York: Oxford University Press, 1982), 14.

29. J. B. Sellers, *Slavery in Alabama* (Tuscaloosa: University of Alabama Press, 1950), 71.

30. H. M. Morais, *The History of the Negro in Medicine* (New York: Publishers Co./ Association for the Study of Negro Life and History, 1968), 11–18; David McBride, "Medicine and Medical Care," in *Macmillan Encyclopedia of World Slavery*, vol. 1, ed. Paul Finkelman and J. C. Miller (New York: Macmillan, 1998), 383–387.

31. Jackson, *Free Negro Labor*, 76–88; Sellers, *Slavery in Alabama*, 195–196, includes quote.

32. J. J. Farley, *Making Arms in the Machine Age: Philadelphia's Frankford Arsenal, 1816–1870* (University Park: Penn State University Press, 1994); M. R. Smith, *Harpers Ferry Armory and the New Technology: The Challenge of Change* (Ithaca, N.Y.: Cornell University Press, 1977).

33. John Ellis, *The Social History of the Machine Gun* (Baltimore: Johns Hopkins University Press, 1975), 23–25; Charles Singer et al., *A History of Technology* (New York: Oxford University Press, 1957), 5: 819, includes quote; W. H. McNeill, *The Pursuit of Power: Technology, Armed Force, and Society since A. D. 1000* (Chicago: University of Chicago, 1982), 242–243.

34. B. Perret, *A Country Made by War: From the Revolution to Vietnam—The Story of America's Rise to Power* (New York: Vintage, 1989), 185–186, quote on 185.

35. C. H. Wesley, "The Employment of Negroes as Soldiers in the Confederate Army," *Journal of Negro History* 4 (1919): 239–253; B. I. Wiley, *Southern Negroes, 1861–1865* (1938; reprint, New Haven, Conn.: Yale University Press, 1965), 146–152; Sellers, *Slavery in Alabama*.

36. J. D. Smith (statement), cited in *The Negro Artisan: Report of a Social Study Made under the Direction of Atlanta University,* ed. W.E.B. Du Bois (Atlanta, Ga.: Atlanta University Press, 1902), 17.

37. U.S. Department of Commerce, Bureau of the Census, *Negro Population: 1790–1915* (Washington, D.C.: GPO, 1918), 33, 36; quote on 33.

38. Ibid., 108–143; E. Franklin Frazier, *The Negro in the United States* (1949; reprint, New York: Macmillan, 1957), 188–187.

39. G. K. Lewis, *The Growth of the Modern West Indies* (London: MacGibbon and Kee, 1968), 50, includes quote; L. E. Fisher, *Colonial Madness: Mental Health in the Barbadian Social Order* (New Brunswick, N.J.: Rutgers University Press, 1985) 32–34.

40. E. T. Thompson, introduction to *The Plantation: An International Bibliography* (Boston: G. K. Hall, 1983), xv, includes quote. On the persistence of plantations in other regions, see D.P.S. Ahluwalia, *Plantations and the Politics of Sugar in Uganda* (Kampala, Uganda: Fountain, 1995), 1–9; Edgar Graham, *The Modern Plantation*

in the Third World (New York: St. Martin's Press, 1984); Marc Edelman, *The Logic of the Latifundio: The Large Estates of Northwestern Costa Rica Since the Late Nineteenth Century* (Stanford: Stanford University Press, 1992); A. L. Stoler, *Capitalism and Confrontation in Sumatra's Plantation Belt, 1870–1979,* 2nd ed. (Ann Arbor: University of Michigan Press, 1995); J. M. Fortt and D. A. Hougham, "Environment, Population, and Economic History," in *Subsistence to Commercial Farming in Present-Day Buganda*, ed. A. I. Richards et al. (Cambridge, U.K.: Cambridge University Press, 1973), 24.

41. Vance, *Human Factors in Cotton Culture*, 38.

42. Ibid., 39.

43. T. J. Woofter Jr., *Landlord and Tenant on the Cotton Plantation,* Research Monograph 5 (Washington, D.C.: Works Progress Administration, 1936), 1, 3.

44. Stanley Lieberson, *A Piece of the Pie: Blacks and White Immigrants since 1880* (Berkeley: University of California Press, 1980), 20–24; G. G. Eggert, *Harrisburg Industrializes: The Coming of Factories to an American Community* (University Park: Penn State Press, 1993), 261–262.

45. A. H. Stone, "Census Statistics of the Negro," *Yale Review*, November 1904, reprinted in *Studies in the American Race Problem,* by A. H. Stone (New York: Doubleday, Page & Co., 1908), 493–494, includes quote. On Stone generally, see R. L. Brandfon, *Cotton Kingdom of the New South: A History of the Yazoo Mississippi Delta from Reconstruction to the Twentieth Century* (Cambridge, Mass.: Harvard University Press, 1967), 123–124, 143–145.

46. On rural peonage see Carter G. Woodson, *The Rural Negro* (Washington, D.C.: Association for the Study of Negro Life and History, 1930), 67–88. Also see sources in note 47.

47. M. T. Carleton, *Politics and Punishment: The History of the Louisiana State Penal System* (Baton Rouge: Louisiana State University Press, 1971), 88–90; D. M. Oshinsky, *"Worse Than Slavery": Parchman Farm and the Ordeal of Jim Crow Justice* (New York: Free Press, 1996); M. J. Mancini, *One Dies, Get Another: Convict Leasing in the American South, 1866–1928* (Columbia: University of South Carolina Press, 1996), 20–21; Alex Lichtenstein, *Twice the Work of Free Labor: The Political Economy of Convict Labor in the New South* (London: Verso, 1996); M. C. Fierce, *Slavery Revisited: Blacks and the Southern Convict Lease System, 1865–1933* ([New York]: Africana Studies Research Center, Brooklyn College, City University of New York, 1994).

48. J. A. Hobson, "The Negro Problem in the United States," *The Nineteenth Century and After* 59 (July–December 1903): 592–593, quote on 593.

49. Oshinsky, *"Worse Than Slavery."*

50. T. O. Powell, "The Increase of Insanity and Turberculosis in the Southern Negro since 1860, and Its Alliance and Some of the Supposed Causes," *Journal of the American Medical Association* 27 (1896): 1185, cited in Julian H. Lewis, *The Biology of the Negro* (Chicago: University of Chicago Press, 1942), 266–267.

51. J. A. LePrince, "Historical Review of Development of Control of Disease-Bearing Mosquitos," *Transactions of the American Society of Civil Engineers* 92 (1928): 1259.

52. G. H. Bradley, "A Review of Malaria Control and Eradication in the United States," *Mosquito News* 26 (December 1966): 462; V. Heister, *An American Doctor's Odyssey: Adventures in Forty-Five Countries* (New York: W. W. Norton, 1939), 446.

53. A. M. Kraut, *Silent Travelers: Germs, Genes, and the "Immigrant Menace"* (Baltimore: Johns Hopkins University Press, 1995).

54. LePrince, "Control of Disease-Bearing Mosquitoes," 1260.

55. G. E. Bushnell, *A Study in the Epidemiology of Tuberculosis with Especial Reference to Tuberculosis of the Tropics and of the Negro Race* (New York: William Wood, 1920), 149–150; J. M. Richardson, *The Negro in the Reconstruction of Florida, 1865–1877* (Tallahassee, Fla.: FSU Research Council, 1965), 27–29.

56. LePrince, "Control of Disease-Bearing Mosquitos," 1260, includes quote; A. I. Marcus, "The South's Native Foreigners: Hookworm as a Factor in Southern Distinctiveness," in *Disease and Distinctiveness in the American South*, ed. T. L. Savitt and J. H. Young (Knoxville: University of Tennessee Press, 1988), 79–99; John Duffy, "The Impact of Malaria on the South," ibid., 29–54.

57. Lewis, *The Biology of the Negro*, 211; Margaret Humphreys, *Yellow Fever and the South* (New Brunswick, N.J.: Rutgers University Press, 1992).

58. Charles W. Stiles, "The Industrial Conditions of the Tenant Class (White and Black) as Influenced by the Medical Conditions," in *The South in the Building of the Nation*, vol. 6 (Richmond, Va.: Southern Historical Publication Society, 1909), 596–597.

59. E. H. Beardsley, *A History of Neglect: Health Care for Blacks and Mill Workers in the Twentieth-Century South* (Knoxville: University of Tennessee Press, 1987); David McBride, *From TB to AIDS: Epidemics among Urban Blacks since 1900* (Albany: SUNY Press, 1991); Dalila de Sousa Kiple, "Darwin and Medical Perceptions of the Black: A Comparative Study of the United States and Brazil, 1871–1918" (Ph.D. diss., Bowling Green State University, 1987), 62–85.

60. Jacqueline Jones, *The Dispossessed: America's Underclasses from the Civil War to the Present* (New York: Basic Books, 1992), 104–126.

61. M. Wayne, *The Reshaping of Plantation Society: The Natchez District, 1860–1880* (Baton Rouge: Louisiana State University Press, 1983), 197–204.

62. Charles S. Johnson, *Statistical Atlas of Southern Counties: Listing and Analysis of Socio-Economic Indices of 1,104 Southern Counties* (Chapel Hill: University of North Carolina Press, 1941), 17–20; Jaqueline Jones, *Labor of Love, Labor of Sorrow: Black Women, Work, and the Family from Slavery to the Present* (New York: Vintage, 1985), 95, includes quote.

63. Stone, *Studies in the American Race Problem*, 493; J. Jones, "Work Now, Get Paid Much Later: 'Free Labor' in the Postbellum South," *Reviews in American History* 15 (1987): 265–271.

64. Du Bois, *The Negro Artisan*, 21–24; Lieberson, *A Piece of the Pie*, 304–313.

65. S. H. Preston, *Fatal Years: Child Mortality in Late Nineteenth-Century America* (Princeton, N.J.: Princeton University Press, 1991).

66. Stone, *Studies in the American Race Problem*, 492.

67. Ibid., 490, 492, quote on 490.

68. *New York Age*, August 24, 1905, 2, cited in Stone, *Studies in the American Race Problem*, 154.

69. H. R. Northrup, *Negro Employment in Basic Industry: A Study in Racial Policies in Six Industries*, vol. 1, *Studies of Negro Employment* (Philadelphia: Wharton School of Finance and Commerce, 1970), 5.

70. Joel Williamson, *A Rage for Order: Black-White Relations in the American South since Emancipation* (New York: Oxford University Press, 1986), 219.

71. The convergence of science, the professions, and management in the early twentieth-century United States has been called the "control revolution" by historians. See A. H. Teich, ed., *Technology and the Future* (New York: St. Martin's Press, 1993).

72. B. T. Washington, *[Tuskegee Institute,] Annual Report, May 30, 1901*, 8, cited in Stone, *Studies in the American Race Problem*, 171.

73. B. T. Washington, *Working with the Hands: Being a Sequel to "Up from Slavery" Covering the Author's Experiences in Industrial Training at Tuskegee* (New York: Doubleday, Page, 1904), vi.

74. B. T. Washington to G. W. Carver, letter, February 21, 1911, in *The Papers of Booker T. Washington*, vol. 10, *1909–11,* ed. by L. R. Harlan and R. W. Smock (Urbana: University of Illinois Press, 1981), 594.

75. W.E.B. Du Bois, "Dr. Du Bois Explains," *Indianapolis (Indiana) Star*, April 8, 1912, reprinted in *Writings by W.E.B. Du Bois in Periodicals Edited by Others*, vol. 2, *1910–1934*, comp. and ed. by Herbert Aptheker (Millwood, N.Y.: Kraus-Thomson, 1982), 76–77, quote on 76.

76. W.E.B. Du Bois, *Dusk of Dawn: An Essay toward an Autobiography of a Race Concept* (New York: Harcourt, Brace, 1940).

77. W.E.B. Du Bois, "The Shape of Fear," *North American Review* 73 (June 26, 1925): 291–304, quotes on 299, 300.

78. David McBride, *Integrating the City of Medicine: Blacks in Philadelphia Health Care, 1910–1965* (Philadelphia: Temple University Press, 1989), 8–9; McBride, *From TB to AIDS*, 24–28, 69–76.

79. J. S. Butler, *Entrepreneurship and Self-Help among Black Americans: A Reconsideration of Race and Economics* (Albany: State University of New York Press, 1991), 114.

80. Mark Harrison, "'The Tender Frame of Man': Disease, Climate, and Racial Difference in India and the West Indies, 1760–1860," *Bulletin of the History of Medicine* 70 (1996): 68–93; Francois Delaporte, *The History of Yellow Fever: An Essay on the Birth of Tropical Medicine* (Cambridge, Mass.: MIT Press, 1991).

81. John Ettling, *The Germ of Laziness: Rockefeller Philanthropy and Public Health in the New South* (Cambridge, Mass.: Harvard University Press, 1981).

82. Lewis, *The Biology of the Negro*, 255–257.

83. Leon Banov, *A Sanitary Survey of the Rural Portion of Charleston County*]1921], cited in Beardsley, *History of Neglect*, 18.

84. Beardsley, *History of Neglect*, 95–96.

85. W.E.B. Du Bois, ed., *The Negro Common School: Report of a Social Study Made under the Direction of Atlanta University, Together with the Proceedings of the Sixth Conference for the Study of the Negro Problems, Held at Atlanta University, May 28, 1901* (Atlanta: Atlanta University Press, 1901), 17.

86. U.S. Census, *Negro Population: 1790–1915*, table 3, 377.

87. Rayford W. Logan, "Education in Haiti," *Journal of Negro History* 15 (1930): 430; *Report of the Survey of the Schools of the Panama Canal Zone* (Mount Hope, Canal Zone: Panama Canal Press, 1930), 39–40. The earliest statistics for Liberia (collected in 1944) suggest literacy rates of less than 3 or 4 percent.

88. W.E.B. Du Bois, ed., *The College-Bred Negro: Report of a Social Study Made under the Direction of Atlanta University, Together with the Proceedings of the Fifth Conference for the Study of the Negro Problems, Held at Atlanta University, May*

29–30, 1900 (Atlanta: Atlanta University Press, 1900), 12–13; Frazier, *The Negro in the United States*, 417–491.

89. Horace Mann Bond, *Negro Education in Alabama: A Study in Cotton and Steel* (1939; reprint, New York: Antheneum, 1969), 275–280; H. A. Bullock, *A History of Negro Education in the South from 1619 to the Present* (New York: Praeger, 1970), 83–85.

90. S. L. Smith, *Sick and Tired of Being Sick and Tired: Black Women's Health Activism in America, 1890–1950* (Philadelphia: University of Pennsylvania Press, 1995), 118–148. Traditional folk healers are most effective in communities where the use of folk healers has been a customary practice. See, for example, G. Bibeau, "From China to Africa: The Same Impossible Synthesis between Traditional and Western Medicines," *Social Science and Medicine* 21 (1985): 937–943; and M. S. Laguerre, *Afro-Caribbean Folk Medicine* (South Hadley, Mass.: Bergin & Garvey, 1987), 21, 44, 59.

91. Marie Campbell, *Folks Do Get Born* (New York: Rinehart, 1946); J. B. Litoff, *American Midwives, 1860 to the Present* (Westport: Greenwood, 1978), 30–32; Katherine Clark, introduction to *Motherwit: An Alabama Midwife's Story*, by Onnie Lee Logans as told to Katherine Clark (New York: Plume, 1989), ix–xiv; and Debra A. Susie, *In the Way of Our Grandmothers: A Cultural View of Twentieth-Century Midwifery* (Athens: University of Georgia Press, 1988). Midwives practiced widely throughout the black communities of the South until the 1940s.

CHAPTER 2 *Industry on the Isthmus*

1. "The Canal Is Built," in *The Panama Canal—Twentieth-Fifth Anniversary, August 15, 1914–August 15, 1939* (Mount Hope, Canal Zone: Governor of Panama Canal/Panama Canal Press, 1939), 17–30; D. Nuñez Polanco, "Antecendens del canal de Panamá," *Invenciones y Ensayos* 492 (June 1991): 99–104.

2. "Earlier Canal Plans," in *Panama Canal—Twenty-Fifth Anniversary*, 9–12.

3. Wilbur Zelinsky, "The Historical Geography of the Negro Population of Latin America," *Journal of Negro History* 34 (April 1941): 166; A. E. Béliz, "Los Congos: Afro-Panamanian Dance-Drama," *Americas* 11 (November 1959): 31–33; Charles Melville Pepper [1859–1930], *Panama to Patagonia: The Isthmian Canal and the West Coast Countries of South America* (Chicago: A. C. McClurg, 1906; New York: Young People's Missionary Movement, n.d.), 44–45.

4. Pepper, *Panama to Patagonia*, 50.

5. Frederick Mears, "The Reconstruction of the Panama Railroad," *Transactions of the International Engineering Congress, 1915, The Panama Canal II. Design and Erection of Structures, American Society of Civil Engineers, San Francisco, September 20–25, 1915* [hereafter abbrev. PC-ASCE], Paper no. 21 (1916): 291–292, quotes on 292; D. G. Munro, *The Five Republics of Central America: Their Political and Economic Development and Their Relations with the United States* (New York: Carnegie Endowment for International Peace, 1918), 15; "Panama Railroad Company," *Panama Canal—Twenty-Fifth Anniversary*, 40–43.

6. W. L. Partridge, "Banana County in the Wake of United Fruit: Social and Economic Linkages," *American Ethnologist* 6 (1979): 493–494; S. May and G. Plaza, *The United Fruit Company and Latin America* (National Planning Association, 1958; reprint, New York: Arno, 1976), 8–12, in reprint.

7. Charles D. Kepner, *Social Aspects of the Banana Industry* (New York: Columbia University Press, 1936), 35–42.

8. Munro, *The Five Republics*, 160.

9. Kepner, *The Banana Industry*, 110–111.

10. Ibid.,168. Munro, *The Five Republics*, 20, 160; Paula Palmer, *"What Happen": A Folk-History of Costa Rica's Talmanca Coast* (San José, Costa Rica: Ecodesarrollos, 1977), 97–114, 191–224; Kitzie McKinney, "Costa Rica's Black Body: The Politics and Poetics of Difference in Eulalia Bernard's Poetry," *Afro-Hispanic Review* 15 (1996): 11.

11. Joseph A. LePrince and A. J. Orenstein, *Mosquito Control in Panama: The Eradication of Malaria and Yellow Fever in Cuba and Panama* (New York: G. P. Putnam's, 1916), 23.

12. R. E. Wood, "The Working Force of the Panama Canal," PC-ASCE, Paper no. 7 (1916): 194.

13. Frank Freidel and Alan Brinkley, *America in the Twentieth Century* (New York: Knopf, 1976), 19–20, 71–75.

14. Ronald T. Takaki, *Iron Cages: Race and Culture in Nineteenth-Century America* (New York: Oxford University Press, 1990), 253–279.

15. Freidel and Brinkley, *America in the Twentieth Century*, 69.

16. F. Ninkovich, "Theodore Roosevelt: Civilization as Ideology," *Diplomatic History* 10 (1986): 234.

17. Emily S. Rosenberg and N. L. Rosenberg, "From Colonialism to Professionalism: The Public-Private Dynamic in United States Financial Advising, 1898–1929," *Journal of American History* 74 (1987): 63.

18. Michael L. Conniff, *Black Labor on a White Canal: Panama, 1904–1981* (Pittsburgh: University of Pittsburgh Press, 1985), 21.

19. Theodore Roosevelt, "Annual Message to the Fifty-Ninth Congress [1904]," cited in Freidel and Brinkley, *America in the Twentieth Century*, 72.

20. Theodore Roosevelt to John Byrne, September 14, 1904, cited in G. Sinkler, *The Racial Attitudes of American Presidents from Abraham Lincoln to Theodore Roosevelt* (Garden City, N.Y.: Anchor, 1972), 424.

21. Orville H. Platt, "Our Relation to the People of Cuba and Porto Rico," *Annals of the American Academy of Political and Social Science* 18 (1901): 148, 155.

22. Wolfred Nelson, "Cuba: Past, Present, and Future," *Proceedings of the American Association for the Advancement of Science* (Boston, 1898), 552, 553, cited in Alfred H. Stone, *Studies in the American Race Problem* (New York: Doubleday, Page, 1908), 394–395.

23. F. K. Mostofi, "Contributions of the Military to Tropical Medicine," *Bulletin of the New York Academy Medicine* 44 (1968): 705–708; E. R. Nye and M. E. Gibson, *Ronald Ross: Malariologist and Polymath: A Biography* (New York: St. Martin's, 1997), 86.

24. L. Wood, "The Military Government of Cuba," *Annals of the American Academy of Political and Social Science* 21 (1903): 182, emphasis in the original.

25. LePrince and Orenstein, *Mosquito Control in Panama*, 23–25; "The Valiant French Effort," *Panama Canal—Twenty-Fifth Anniversary*, 15, 32; N. E. Elton, "Yellow Fever in Panama: Historical and Contemporary," *American Journal of Tropical Medicine and Hygiene* 1 (1952): 441.

26. See, for example, Mahlon Ashford, "The Nature of Immunity to Malaria in Its Relationship to Anti-Malarial Therapy," *American Journal of Tropical Medicine* 16 (1936): 665–678.

27. Dennis G. Carlson, *African Fever: A Study of British Science, Technology, and Politics in West Africa, 1787–1864* (Canton, Mass.: Science History Publications, 1984), 27–31.

28. Todd L. Savitt, "Black Health on the Plantation: Masters, Slaves, and Physicians," in *Science and Medicine in the Old South*, ed. by Todd L. Savitt and Ronald L. Numbers (Baton Rouge: Louisiana State University Press, 1989), 327–355.

29. Conniff, *Black Labor on a White Canal*, 20–21. Many black West Indian workers "suffered violent deaths [since] most men carried arms, and frontier justice from the barrel of a gun prevailed." Ibid., 21.

30. Mears, "The Reconstruction of the Panama Railroad."

31. LePrince and Orenstein, *Mosquito Control in Panama*, 203–204, 220–221.

32. Pepper, *Panama to Patagonia*, 50.

33. "The Canal Is Built," *Panama Canal—Twenty-Fifth Anniversary*, 22–23; "Panama Railroad Company," ibid., 39–47.

34. Conniff, *Black Labor on a White Canal*, 20, includes quote; Bonham C. Richardson, *Panama Money in Barbados, 1900–1920* (Knoxville: University of Tennessee Press, 1985).

35. Harry A. Franck, *Zone Policeman 88: A Close Range Study of the Panama Canal and Its Workers* (New York: Century, 1913), 118.

36. Ibid., 118. For other examples of the negative opinion of U.S. whites to local blacks in the Canal Zone region, see Wood, "The Working Force of the Panama Canal."

37. "Labor Troubles on the Zone," *Panama Star and Herald*, August 5, 1907; "List of Complaints of Gold Employees from August 1, 1907, to May 14, 1908," in U.S. Senate, 60th Congress, *Document No. 539: Message from the President of the United States, Transmitting the Report of the Special Commission Appointed to Investigate Conditions of Labor and Housing of Government Employees on the Isthmus of Panama, December 8, 1908* (Washington, D.C.: GPO, 1908) [hereafter abbrev. *Commission to Investigate Panama*], 49–50.

38. *Commission to Investigate Panama*, 5.

39. "Report of the Assistant of the Secretary of the Isthmian Canal Commission Regarding Complaints of European Employees," *Commission to Investigate Panama*, 6–7, 19–20, 51, quote on 6–7.

40. Ibid., 51.

41. John Biesanz, "Race Relations in the Canal Zone," *Phylon* 11 (1950): 24.

42. "The Unemployed on the Isthmus," *Panama Star and Herald*, April 27, 1908.

43. Wood, "The Working Force of the Panama Canal," 196.

44. Stanley Lieberson, *A Piece of the Pie: Blacks and White Immigrants since 1880* (Berkeley: University of California Press, 1980), 311–322.

45. Remarks by William N. DeBerry cited in Stone, *Studies in the American Race Problem*, 161–162, quote on 162.

46. Elizabeth M. Petras, *Jamaican Labor Migration: White Capital and Black Labor, 1850–1930* (Boulder, Colo.: Westview, 1988), 151–155.

47. Mears, "The Reconstruction of the Panama Railroad," 23–25.

48. Wood, "The Working Force of the Panama Canal," 196.

49. W.E.B. Du Bois, letter, *New York Star*, October 21, 1948, in *Writings by W. E. B. Du Bois in Periodicals Edited by Others*, vol. 4, *1945–1961*, ed. Herbert Aptheker (Millwood, N.Y.: Kraus-Thomson Organization, 1982), 89.

50. "Health Conditions in 1914," *Canal Record* 8, no. 35 (April 21, 1915): 314.

51. Ibid.; N. L. Englehardt, *Report of the Survey of the Schools of the Panama Canal Zone Made by the Division of Educational Research, Teachers College, Columbia University* (Mount Hope, Canal Zone: 1930), 43. In their health reports, canal officials do not give specific details concerning why they viewed inadequate care by mothers as responsible for the high rate of childhood deaths. However, they listed gastrointestinal disease as the leading cause of death for children and infants. Perhaps the health officials blamed the mothers for feeding children unsafe foods and liquids.

52. Wood, "The Working Force of the Panama Canal," 197.

53. Ibid.

54. Ibid., 197.

55. P. F. Russell, "Malaria and Its Influence on World History," *Bulletin of the New York Academy of Medicine* 9 (1943): 611–623.

56. LePrince and Orenstein, *Mosquito Control in Panama*, 261–262.

57. Russell, "Malaria and Its Influence," 621–623; Marcos Cueto, "Sanitation from Above: Yellow Fever and Foreign Intervention in Peru, 1919–1922," *Hispanic American Historical Review* 72 (1992): 3–5.

58. LePrince and Orenstein, *Mosquito Control in Panama*, 24–25.

59. G. H. Bradley, "A Review of Malaria Control and Eradication in the United States," *Mosquito News* 26 (1966): 462–463, quote on 463.

60. "Sanitation and Health," *Panama Canal—Twenty-Fifth Anniversary*, 35.

61. William C. Gorgas, "Recommendation as to Sanitation Concerning Employees of the Mines on the Rand Made to the Transvaal Chamber of Mines," *Journal of the American Medical Association* 62 (June 13, 1914): 1857, includes quote; Julian H. Lewis, *The Biology of the Negro* (Chicago: University of Chicago Press, 1942), 110–111.

62. Gorgas, "Sanitation Concerning Employees," 1857, includes quote; G. E. Bushnell, *A Study in the Epidemiology of Tuberculosis with Especial Reference to Tuberculosis of the Tropics and of the Negro Race* (New York: William Wood, 1920), 119; *Commission to Investigate Panama*, 10.

63. "Health Conditions in 1914," *Canal Record* 8 (April 12, 1915): 314.

64. "Dangers of Malarial and Other Infection on Trips outside of Areas of Sanitation," *Panama Canal Record* 14 (April 6, 1921): 514.

65. William C. Gorgas, *Sanitation in Panama* (New York: D. Appleton, 1915), 209–223.

66. "Sanitation and Health," *Panama Canal—Twenty-Fifth Anniversary*, 35. By the late 1930s about 90 percent of the Canal Zone population and 75 percent of the city dwellers in Colon and Panama City had been immunized against smallpox.

67. John Biesanz and Mavis Biesanz, *The People of Panama* (New York: Columbia University Press, 1955), 52–54; Bradley, "A Review of Malaria Control and Eradication in the United States," 462–463; W. P. Chamberlain, "The Health Department of the Panama Canal," *New England Journal of Medicine* 203 (October 2, 1930): 669–680; L. E. Rozeboom, "The Role of Some Common Anopheline Mos-

quitoes of Panama in the Transmission of Malaria," *American Journal of Tropical Medicine* 18 (1938): 289; V. G. Heiser, "Reminiscences of Early Tropical Medicine," *Bulletin of the New York Academy of Medicine* 44 (1968): 655; C. C. Bass, "The Influence of Malaria on the Progress of Civilization," *Southern Medical Journal* 19 (December 1926): 856.

68. Joel Williamson, *A Rage for Order: Black/White Relations in the South since Emancipation* (New York: Oxford University Press, 1986).

69. A. A. Harper, *Tracing the Course of Growth and Development in Educational Policy for the Canal Zone Colored Schools, 1905–55*, University of Michigan Comparative Education Dissertation Series, no. 25 (Ann Arbor: University of Michigan School of Education, 1974), 32–33. Concerning school revenues, see F. S. Céspedes, "Panama," in *Educational Yearbook of the International Institute of Teachers College, Columbia University, 1942*, ed. I. L. Kandel (New York: Teachers College, Columbia University, 1942), 310.

70. Harper, *Canal Zone Colored Schools*, 33.

71. "Zone Public Schools," *Canal Record* 1 (October 1907): 58, cited in Harper, *Canal Zone Colored Schools*, 43. The *Canal Record* was the official organ of canal authority. It circulated widely throughout the Isthmus.

72. "Industrial Training and Success," *Panama Star and Herald*, June 3, 1907.

73. Harper, *Canal Zone Colored Schools*, 46 (statistics), 246–247.

74. Ibid., 46.

75. Englehardt, *Schools of the Panama Canal Zone*, 171.

76. Harper, *Canal Zone Colored Schools*, 247; E. B. Sackett, "The Negro Schools of the Canal Zone," *Journal of Negro Education* 1 (1932): 348.

77. Richardson, *Panama Money in Barbados*, 153–155; Petras, *Jamaican Labor Migration*, 259–260; Sir A. Grenfell Price, *White Settlers in the Tropics, American Geographical Society, Special Publication No. 23* (New York: The Society, 1939), 160.

78. Edward W. Said, *Culture and Imperialism* (New York: Vintage, 1993); Ann McClintock, *Imperial Leather: Race, Gender, and Sexuality in the Colonial Contest* (New York: Routledge, 1995); Ann L. Stoler, *Race and the Education of Desire: Foucault's History of Sexuality and the Colonial Order of Things* (Durham: Duke University Press, 1995); Ann L. Stoler, "Making Empire Respectable: The Politics of Race and Sexual Morality in Twentieth-Century Colonial Cultures," *American Ethnologist* 16 (August, 1989): 634–660; and Aldo Lauria-Santiago and Aviva Chomsky, eds., *Identity and Struggle at the Margins of the Nation-State: The Laboring Peoples of Central America and the Hispanic Caribbean* (Durham: Duke University Press, 1998).

79. Andres Opazo Bernales, *Panama: la iglesia y la lucha de los pobres* (San José, Costa Rica: Editorial Departamento Ecumenico de Investigaciones, 1988), 16–18. On female impoverishment in traditional economies such as Panama, see Lila E. Engberg, "Household Resources, Women, and Food Security: An Ecosystem Perspective with Case Studies from Africa," in *The World Food Crisis: Food Security in Comparative Perspective*, ed. J. I. Hans Bakker (Toronto: Canadian Scholars' Press, 1990), 229–254

80. See *The American Women in the Panama Canal, from 1804 to 1916*, compiled by (Mrs.) Ernest Urich Von Muenchow (Canal Zone: Panama Star & Herald, 1916).

81. Van Hardeveld, "Personal Experiences," in *American Women in the Panama Canal*, comp. Von Muenchow, 14.

82. Ibid., 15.

83. On this phenomena in European imperial colonies, see McClintock, *Imperial Leather*; Stoler, "Making Empire Respectable," 634–660; Stoler, *Race and the Education of Desire*.

84. Van Hardeveld, "Personal Experiences," 14.

85. Ibid., 15. Perhaps the reference is to Brunswick, Georgia.

86. Recent histories of the female club movements in the United States during the early twentieth century suggest that in the Canal Zone, wives of U.S. white males anticipated having the highest social status since their husbands were professional and skilled employees. Clubs and other women's activities allowed housewives to gain respectability among the white females in the Zone who were full-time employees (for example, teachers and clerks). See Priscilla Murolo, *The Common Ground of Womanhood: Class, Gender, and Working Girls Clubs, 1884–1928* (Urbana: University of Illinois Press, 1997); and Ann M. Krupfer, *Toward a Tenderer Humanity and Nobler Womanhood: African American Women's Clubs in Turn-of-the-Century Chicago* (New York: New York University Press, 1996).

87. H. C. Clark, "Endemic Yellow Fever in Panama and Neighboring Areas," *American Journal of Tropical Medicine and Hygiene* 1 (1952): 78–79.

88. "Organization of the Health Department," *Canal Record* 7 (August 12, 1914): 517–518; Frederick Palmer, *Central America and Its Problems: An Account of a Journey from the Rio Grande to Panama* (New York: Moffat, Yard, 1913), 46–47.

89. President William Howard Taft, "Third Annual Message, December 5, 1911," in *The State of the Union Messages of the Presidents, 1790–1966*, vol. 3, *1905–1966*, ed. F. L. Israel (New York: Chelsea House, Robert Hector, 1966), 2475–2476.

90. "Public School Year," *Canal Record* 48 (July 22, 1914): 477; "The Depopulation of the Canal Zone," *Panama Canal Record* 10 (March 21, 1917): 387–389; "Executive Order: Relating to the Exclusion of Chinese, The White House, 6 February, 1917 [signed by] President Woodrow Wilson," reprinted in *Panama Canal Record* 10 (February 28, 1917): 367–368; Price, *White Settlers in the Tropics*, 157, 166–167.

91. "New Emigration from Isthmus in Eleven Months over 15,000," *Canal Record* 7 (July 22, 1914): 477, includes quote; A. G. Price, "White Settlement in the Panama Canal Zone," *Geographical Review* 25 (1935): 10.

92. W. P. Chamberlain, "Sanitation of Canal Zone: Permanent Drainage and Mosquitoes," *Military Surgeon* (April 1927): 406.

93. J. L. Williams, "The Rise of the Banana Industry and Its Influence on Caribbean Countries" (Master's diss., Clark University, Worcester, Mass., 1925), cited by Kepner, *The Banana Industry*, 169.

94. Kepner, *The Banana Industry*, 168.

95. U. S. Bureau of the Census, "Panama Canal Zone: Introduction," *Fourteenth Census of the United States, 1920*, vol. 3, *Composition and Characteristics of the Populations by States* (Washington, D.C.: GPO, 1922), 1243.

96. "Canal Work in January, 1917," *Panama Canal Record* 10 (February 28, 1917): 357.

97. Conniff, *Black Labor on a White Canal*, 46.

98. Williams, "The Rise of the Banana Industry, cited by Kepner, *The Banana Indus-*

try, 169. On West Indian workers and the Atlantic coast banana plantations, see Elisavinda Echeverri-Gent, "Forgotten Workers: British West Indians and the Early Days of the Banana Industry in Costa Rica and Honduras," *Journal of Latin American Studies* 24 (1992): 275–308; Phillippe Bourgois, *Ethnicity at Work: Divided Labor on a Central American Banana Plantation* (Baltimore: Johns Hopkins University Press, 1989); and Aviva Chomsky, *West Indian Workers and the United Fruit Company in Costa Rica, 1870–1940* (Baton Rouge: Louisiana State University Press, 1996).

99. "Dangers of Malarial and Other Infection on Trips outside of Areas of Sanitation"; Mia Strasser de Saavedra and David Saavedra, *El libro de Oro: Estudio Completto de las posibilidades agricolas y ganaderas que Panama* (Ciudad de Panamá: R. de P. Imprenta Nacional, 1926), 173, 174.

100. Chamberlain, "Sanitation of the Canal Zone," 405.

101. "Yellow Fever in Central and South America," Rockefeller Foundation, *International Health Board Annual Report for 1919*, 159; "Yellow Fever," Rockefeller Foundation, *International Health Board Annual Report for 1925*, 104.

102. R. Boyce, "Malaria in the West Indies," in *The Prevention of Malaria*, by R. Ross (London: John Murray, 1910), 374.

103. Otis P. Starkey, *The Economic Geography of Barbados: A Study of the Relationships between Environmental Variations and Economic Development* (1939; reprint, Westport, Conn.: Negro Universities Press, 1971), 167, 190, in reprint.

104. Ross, *The Prevention of Malaria*.

105. "The Plague Scare, Disease Now in Jamaica? Alarming Report from Kingston," *Panama Star and Herald*, August 10, 1908; "Reduction in Cases of Malaria," *Panama Canal Record* 14 (March 2, 1921): 421.

106. W. T. Prout, "Malaria in Jamaica," in *The Prevention of Malaria*, ed. Ross, 377–378; quote on 377.

107. "Hookworm Campaign: Representatives of International Health Commission Organizing Work on Isthmus," *Canal Record* 8 (October 14, 1914): 94–95; "The Hookworm Campaign: Seven Months of Field Work in Panama Show Prevalence of the Disease," *Canal Record* 8 (May 3, 1915): 329.

108. W. G. Baetz, "Syphilis in Colored Canal Laborers: A Résumé of 500 Consecutive Medical Cases," *New York Medical Journal* (November 24, 1914): quote on 824; L. B. Bates, "The Wassermann Test in the Tropics," *Archives of Internal Medicine* 10 (November 1912): 470–477.

109. Baetz, "Syphilis in Colored Canal Laborers" 820.

110. Ibid., 824.

111. Ibid., 825.

112. Ibid., 820; "Treatment in Hospitals and Elsewhere," *Panama Canal Record* 10 (January 24, 1917): 288.

113. J. F. Siler, *The Prevention and Control of Venereal Diseases in the Army of the United States of America, The Army Medical Bulletin Number 67, Special Issue May 1943* (Washington, D. C.: Surgeon General, U.S. Army, 1943), 13–23; R. Minton et al., "Fifty Years of American Medicine on the Isthmus of Panama," *American Journal of Tropical Medicine and Hygiene* 3 (November 1954): 956. By 1939 the U.S. troop strength in Panama averaged 14,000 annually, with VD rates from roughly 50 to 100 percent higher than troops in the United States. Siler, *Prevention and Control*,

14, 19. On the syphilis problem during the aftermath of World War I and the U.S. military generally, see Toni P. Miles and David McBride, "World War I Origins of the Syphilis Epidemic among 20th-Century Black Americans: A Biohistorical Analysis," *Social Science and Medicine* 45 (1997): 61–69.

114. "Health Conditions on the Isthmus," *Panama Star and Herald*, May 18, 1908. By 1917 the health department of Panama City had begun an all-out campaign to encourage residents to properly ventilate their homes. "Public Health Lectures," *Canal Record* (Ancon) 6 (March 5, 1913): 230; "The Cause and Cure of Consumption," *Panama Canal Record* 11 (December 12, 1917): 185–186.

115. McCarthy, untitled, *Boston Medical and Surgical Journal* 166 (1912): 207, cited in Bushnell, *Epidemiology of Tuberculosis*, 59, 78 (statement is by Bushnell on 59).

116. [H. C.] Clark, *American Journal of Tropical Disease and Preventive Medicine* 3 (1915–1916): 331, cited in Bushnell, *Epidemiology of Tuberculosis*, 82. During 1920 the influenza pandemic also tore through Panama. Minton et al., "Fifty Years of American Medicine on the Isthmus of Panama," 957.

117. "Panama Health Office Emphasizes Importance of Fresh Air," *Panama Canal Record* 11 (November 28, 1917): 166; "Effects of Antituberculosis Campaign," *Panama Canal Record* 11 (February 27, 1918): [1], includes quote.

118. "Vital Statistics," *Panama Canal Record* 17 (September 26, 1923): 111. Other "Vital Statistics" reports that I sampled in the *Panama Canal Record* revealed roughly similar monthly mortality and malaria patterns for 1921 through 1923. The issues sampled were vol. 14 (July 27, 1921): 789; vol. 16 (August 30, 1922): 58–59, and (September 27): 102; and vol. 17 (November 28, 1923): 241.

119. Rockefeller Foundation, *International Health Division [formerly International Health Board], Annual Report, 1934*, 89, 93. The Canal Zone project involved a tuberculin survey of the school children of the Zone. It is credited with helping the TB death rate in Panama drop from 250 per 100,000 in 1933 to 192 per 100,000 in 1934. (Ibid., 93).

120. I. S. Falk, *Health in Panama: A Survey and A Program—Prepared for the Government of the Republic of Panama* (Stonington, Conn., and Washington, D. C.: n.p., January 1957), 40, 64.

121. "Traffic in Native Produce over Gatun Lake," *Panama Canal Record* 10 (March 7, 1917): 371; Richardson, *Panama Money in Barbados*, 143–144.

122. W. J. Mayo, "Observations on South America: I, Jamaica and Canal Zone," *Journal of the American Medical Association* 75 (July 31, 1920): 311–315; "Observations on South America: II, Peru," *Journal of the American Medical Association* 75 (August 7, 1920): 377–378.

123. Mayo, "Observations on South America: I," 311– 313.

124. Ibid., 314, includes quote; LePrince and Orenstein, *Mosquito Control in Panama*, 283–284, 286–288.

125. Ibid., 314, 315.

126. Rockefeller Foundation, *International Health Board, Annual Report, 1926* (New York: [The Foundation, 1926]), 83–85.

127. Clark, "Endemic Yellow Fever in Panama and Neighboring Areas," 78.

128. "Sanitation and Health," *Panama Canal—Twenty-Fifth Anniversary*, 34.

129. Rene Dubos, *Man Adapting* (1965; reprint, enlarged ed., New Haven: Yale University Press, 1980), 374, in reprint.

130. *Report of the Health Department of the Panama Canal for the Calendar Year 1944* [H. C. Dooling, Chief Health Officer], review by R. F. Tredre in "Reports, Surveys, and Miscellaneous Papers," *Tropical Diseases Bulletin* 47 (1950): 99; Falk, *Health in Panama*, 75–76, quote on 75.

131. "Sanitation and Health," *Panama Canal—Twenty-Fifth Anniversary*, 34–35, quote on 35.

132. Price, "White Settlement in the Panama Canal Zone," 1, 10, quote on 1.

133. Marshall A. Barber, *A Malariologist in Many Lands* (Lawrence: University of Kansas Press, 1946), 9, 63, 26, includes quotes. Barber worked on projects for both the U.S. Public Health Service and the International Health Division.

134. Franklin D. Roosevelt, "Informal, Extemporaneous Remarks before the Roosevelt Home Club, Hyde Park, New York, August 27, 1938," in *The Public Papers and Addresses of Franklin D. Roosevelt with a Special Introduction and Explanatory Notes by President Roosevelt*, 1938 volume, *The Continuing Struggle for Liberalism*, compiled by S. I. Rosenman (New York: Macmillan, 1941), 504.

135. C. F. Reid, "Federal Support and Control of Education in the Territories and Outlying Possessions," *Journal of Negro Education, Yearbook Issue* 7 (1938): 409.

136. L. G. Sutherland, "Panama Gold: The Story of Negroes on the Panama Canal," *Opportunity* 12 (1934): 336–339; Sharon P. Collasos, *Labor and Politics in Panama: The Torrijos Years* (Boulder, Colo.: Westview, 1991); Steve C. Ropp, *Panamanian Politics: From Guarded Nation to National Guard* (New York: Praeger, 1982).

137. John Biesanz, "Race Relations in the Canal Zone," *Phylon* 11 (1950): 30; John Biesanz and Mavis Biesanz, *The People of Panama* (New York: Columbia University Press, 1955), 223–226; D. J. Davis, "Panama," in *No Longer Invisible: Afro-Latin Americans Today*, ed. Minority Rights Group (London: Minority Rights Publication, 1995), 204–206.

138. L. E. Blauch, "Education in the Territories and Outlying Possessions of the United States," *Journal of Negro Education* 15 (1946): 479; Biesanz, "Race Relations in the Canal Zone," 25, 30, includes quotes.

139. Ibid.; G. W. Westerman, "School Segregation on the Panama Canal Zone," *Phylon* 15 (1954): 276–287.

140. Canal Zone schools were finally integrated during the public school desegregation struggles surrounding the *Brown v. Board of Education* decision on the U.S. mainland.

CHAPTER 3   *Curing the Caribbean*

1. F. Douglass, "Dispatches, Haiti, 1889–1890, No. 13, Nov. 18, 1889," cited in L. M. Sears, "Frederick Douglass and the Mission to Haiti, 1889–1891," *Hispanic American Historical Review* 21 (1941): 226.

2. Jacques M. Léger, *Haiti: Her History and Her Detractors* (1907; reprint, New York: Negro Universities Press, 1970). Léger was a Haitian ambassador to the United States. This work offers a cogent chronicle of the distortions about Haitian history and society widespread in the United States and Europe. See also Rayford W. Logan, "Education in Haiti," *Journal of Negro History* 15 (1930): 401.

3. English-language studies that wrestled with these issues early in this century were, for example, W. A. Pusey, "The Beginning of Syphilis," *Journal of the American*

Medical Association 64 (June 12, 1915): 1961; C. S. Butler, "The Medical Needs of the Republic of Haiti at the Present Time," *United States Naval Medical Bulletin* 24 (1929): 270–273; Melville J. Herskovits, *Life in a Haitian Valley* (New York: Knopf, 1937), and M. J. Herskovits, *Myth of the Negro Past* (1941; reprint, Boston: Beacon Press, 1990). Among current sources, see, for example, Anne Greene, *The Catholic Church in Haiti: Political and Social Change* (East Lansing: Michigan State University Press, 1993), 104–105.

4. Rodolphe Charmont (M.D.), *La vie incroyable d'Alcius* (Port-au-Prince: Société d'éditions et de librairie, [1946?]), 129–314; General [F. Déus] Légitime, "Some General Considerations on the People and the Government of Haiti," in *Papers on Inter-Racial Problems Communicated to the First Universal Races Congress, University of London,* July 26–29, 1911, ed. G. Spiller (London: P. S. King & Son, 1911), 182, includes quote. Légitime was Haiti's president from 1888 to 1889.

5. For accounts of the events leading to U.S. intervention in Haiti, see Arthur C. Millspaugh, *Haiti under American Control, 1915–1930* (Boston: World Peace Foundation, 1931); Rayford W. Logan, *Haiti and the Dominican Republic* (London: Oxford University Press/Royal Institute of International Affairs, 1968); Hans Schmidt, *The United States Occupation of Haiti, 1915–1934* (New Brunswick, N.J.: Rutgers University Press, 1971), 42–63; Patrick Bellegarde-Smith, *Haiti: The Breached Citadel* (Boulder, Colo.: Westview Press, 1990); Brenda G. Plummer, *Haiti and the Great Powers* (Baton Rouge: Louisiana State University Press, 1988).

6. David Nicholls, *From Dessalines to Duvalier: Race, Colour, and National Independence in Haiti* (Cambridge, U.K.: Cambridge University Press, 1979), 142–145.

7. Walter A. McDougall, *Promised Land, Crusader State: The American Encounter with the World since 1776* (Boston: Houghton Mifflin, 1997).

8. Leonard Wood, "The Military Government of Cuba," *Annals of the American Academy of Political and Social Science* 21 (1903): 153–182; R. C. Ileto, "Cholera and the Origins of the American Sanitary Order in the Philippines," in *Imperial Medicine and Indigenous Societies*, ed. David Arnold (Manchester, U.K.: Manchester University Press, 1988), 125–148.

9. Arthur Kleinman, "Concepts and a Model for the Comparison of Medical Systems as Cultural Systems," *Concepts of Health, Illness, and Disease: A Comparative Perspective*, ed. Caroline Currer and Meg Stacey (Oxford: Berg, 1993), 29–47.

10. C. E. Chapman, "The Development of the Intervention in Haiti," *Hispanic American Historical Review* 7 (1927): 299.

11. Peter Berger, B. B. Berger, and H. Kellner, *The Homeless Mind: Modernization and Consciousness* (New York: Vintage, 1973), 8–9.

12. Robert I. Rotberg with Christopher K. Clague, *Haiti: The Politics of Squalor* (Boston: Houghton Mifflin, 1971), 107–146; quote on 128; Logan, *Haiti and the Dominican Republic,* 136–137; Ludwell L. Montague, *Haiti and the United States, 1714–1938* (1940; reprint, New York: Russell & Russell, 1966), 209–292.

13. Michel S. Laguerre, *The Military and Society in Haiti* (Knoxville: University of Tennessee Press, 1993), 63, 66.

14. U.S. Department of State, *Seventh Annual Report of the American High Commissioner at Port Au Prince, Haiti: 1928,* 9–10; U.S. Department of State, *Eighth Annual Report of the American High Commissioner: 1929,* 14–17.

15. Maurice de Young, "Class Parameters in Haitian Society," *Journal of Inter-Ameri-*

can Studies 1 (October 1959): 455; Hans Schmidt, *Maverick Marine: General Smedley D. Butler and the Contradictions of American Military History* (Lexington: University Press of Kentucky, 1987), 84; Laguerre, *Military and Society in Haiti*, 70–71.

16. Qtd. in P. H. Douglas, "The Political History of the Occupation," in *Occupied Haiti: Being the Report of a Committee of Six Disinterested Americans Representing Organizations Exclusively American [Who] Personally Studied Conditions in Haiti in 1926,* ed. Emily Greene Balch (1927; reprint, New York: Negro Universities Press, 1969), 24, in reprint; S. D. Butler to John A. McIlhenny, June 23, 1917, in *General Smedley Darlington Butler: The Letters of a Leatherneck, 1898–1931,* ed. A. C. Venzon (New York: Praeger, 1992), 194, includes quote.

17. Hans Schmidt, *Maverick Marine: General Smedley D. Butler and the Contradictions of American Military History* (Lexington: University Press of Kentucky, 1987), 85. U.S. Marines held rank simultaneously in both the Marine Corps and the Gendarmerie. In the Gendarmerie, Marines were given higher ranks than they held in the Marine Corps as well as generous pay and supervisory inducements.

18. S. D. Butler to James R. Mann, April 4, 1916, in *General Smedley Darlington Butler*, ed. Venzon, 197.

19. U.S. Department of State, *Eighth Annual Report of the American High Commissioner: 1929,* 25–32; Rotberg, *Haiti: The Politics of Squalor*, 131; G. E. Simpson, "Haitian Politics," *Social Forces* 20 (May 1942): 488. The *Eighth Annual Report* details the exact location and numbers associated with these public works installations. An exception to the road network was the stretch from Jacmel to Port-au-Prince, which had no road built as of 1929.

20. The emphasis by U.S. officials on an engineering and technical approach to public health campaigns, as opposed to social, political, and economic challenges underlying epidemics and poor health care, became the central characteristic of U.S. public health campaigns in the non-West through the 1950s. Nancy E. Gallagher, *Egypt's Other War: Epidemics and the Politics of Public Health* (Syracuse, N.Y.: Syracuse University Press, 1990), 178.

21. Butler, "The Medical Needs of the Republic of Haiti," 277. Butler's estimate for yaws and syphilis rates in this document must be interpreted most cautiously since these rates evidently varied greatly from specific urban and rural population groups and locales. For example, a 1946 estimate of syphilis rates for rural Haitians was about 5 percent. James H. Dwindelle et al., "Preliminary Report on the Evaluation of Penicillin in the Treatment of Yaws," *American Journal of Tropical Medicine* 26 (May 1946): 317.

22. Butler, "The Medical Needs of the Republic of Haiti," 274–275. Butler was the commander of the U.S. Navy medical corps.

23. K. C. Melhorn, "Public Health in Haiti: A Resume of 10 Years' Work," *U.S. Naval Medical Bulletin* 27 (1929): 568–573; U.S. Department of State, *Seventh Annual Report of the American High Commissioner: 1928*, 28; U.S. Department of State, *Eighth Annual Report of the American High Commissioner: 1929*, 34.

24. U.S. Department of State, *Eighth Annual Report of the American High Commissioner: 1929*, 34.

25. The strike among primary and secondary school youth occurred in Cayes in 1929.

It resulted in at least ten deaths. U.S. Department of State, *Eighth Annual Report of the American High Commissioner: 1929*, 6–13.

26. Frederic M. Wise and M. O. Frost, *A Marine Tells It to You* (New York: J. H. Sears, 1929), 307.

27. James W. Johnson, *Along This Way: The Autobiography of James Weldon Johnson* (New York: Viking, 1933), 346.

28. Ibid., 347, 348.

29. Ibid., 347–348; Dantes Bellegarde, *La Résistance haïtienne (L'Occupation américaine d'Haïti)* (Montréal: Editions Beauchemin, 1937), 70–71; L. D. Pamphile, "The NAACP and the American Occupation of Haiti," *Phylon* 47 (1986): 96–97; Nicholls, *From Dessalines to Duvalier*, 149. Union Patriotique had ceased operation as one of the terms of the Treaty of 1916.

30. Bellegarde, *La Résistance haïtienne*, 142–147.

31. Charmont, *La vie incroyable d'Alcius*, 297, 302, 306–308.

32. Union Patriotique d'Haïti, *Memoir on the Political, Economic, and Financial Conditions Existing in the Republic of Haiti under the American Occupation by the Delegates to the United States of the Union Patriotique d'Haïti* (hereafter abbrev. *MUPH*), reprinted in *Nation*, International Relations Section 62 (May 25, 1921): 775.

33. Union Patriotique d'Haïti, *MUPH*, 768.

34. Tuberculosis had been largely ignored by U.S. and European public health programs operating in the Caribbean through the 1920s. Finally, the Rockefeller Foundation sponsored an investigation of tuberculosis in Jamaica in the 1930s. It was not until the beginning of World War II that the high rates of tuberculosis in the Caribbean nations became noticed by the public health circles in the United States and Europe. Ivy Portland et al., introduction to *Tuberculosis in the West Indies: Report of the Sociological and Clinical Survey,* by W. Stanton Gilmour (London: National Association for the Prevention of Tuberculosis, [c. 1944]),12; R.A.S. Cory, "Review of *Tuberculosis in the West Indies: Report of the Sociological and Clinical Survey* by W. S. Gilmour," *West Indian Review* 3, no. 4, New Series (Last Quarter, 1946), 29.

35. Union Patriotique d'Haïti, *MUPH*, 755, 757.

36. U.S. Department of State, *Eighth Annual Report of the American High Commissioner: 1929*, 14, 15.

37. Ibid., 15, 17.

38. In this regard, U.S. authorities in Haiti were repeating what had occurred during the initial occupation of the Philippines. According to R. C. Ileto, the Philippine people came to dread military surgeons after the U.S. Army had entered their country. A cholera epidemic "introduced the stern figure of the American army surgeon, less open than the regular military officers to compromise with the local elite. Detailed to most local health boards, army surgeons had little regard for their Filipino colleagues and generally ignored local knowledge gained from previous epidemics. Ileto, "Cholera and the Origins of the American Sanitary Order in the Philippines," 139.

39. Nancy E. Gallagher, *Medicine and Power in Tunisia: 1780–1900* (Cambridge, U.K.: Cambridge University Press, 1984); Amira el Azhary Sonbol, "The Creation of a Medical Profession in Egypt during the Nineteenth Century: A Study in Modernization" (Ph.D. diss., Georgetown University, 1981).

40. R. P. Parsons and E. R. Stitt, *History of Haitian Medicine* (New York: Paul B. Hoeber, 1930), 69–71, quote on 69.
41. E. R. Stitt, Foreword to Parsons and Stitt, *History of Haitian Medicine*, xxii; Butler, "The Medical Needs of the Republic of Haiti," 274, 275; Melhorn, "Public Health in Haiti," 572.
42. C. S. Butler, "Coordination of Medical Problems, Medical Education, Public Health, and Hospitals in the Republic of Haiti," *Bulletin of the Association of American Medical Colleges* 3 (1928): 48.
43. Bellegarde, *La Résistance haïtienne*, 98–101; Rotberg, *Haiti: The Politics of Squalor*, 132; Selden Rodman, *Haiti: The Black Republic* (New York: Devin-Adair, 1954), 55; Logan, *Haiti and the Dominican Republic*, 141; Rayford Logan, "Education in Haiti," *Journal of Negro History* 15 (1930): 450–451; Parsons and Stitt, *History of Haitian Medicine*, 77, includes quote.
44. J. R. Hawke, "Three Years of Vocational Industrial Education in Haiti" (M.S. thesis, Pennsylvania State University, 1930), 52–55.
45. Ibid., 24–25.
46. Marc Aurele Holly, *Agriculture in Haiti with Special Reference to Rural Economy and Agricultural Education*, Sesquicentennial Collection (New York: Vantage, 1955), 200, 211–212; U.S. Department of State, *Eighth Annual Report of the American High Commissioner: 1929*, 72–74; Hawke, "Vocational Industrial Education in Haiti," 25.
47. C. K. Streit, "Haiti: Intervention in Operation," *Foreign Affairs* 6 (1928): 624; Hawke, "Vocational Industrial Education in Haiti," 112–120; Holly, *Agriculture in Haiti*, 199–200; Zonia Baber and E. G. Balch, "Problems of Education," in *Occupied Haiti*, ed. Balch, 95.
48. Baber and Balch, "Problems of Education," 104, includes quotes.
49. Logan, *Haiti and the Dominican Republic*, 137.
50. Dantes Bellegarde, "Haiti under the Rule of the United States," *Opportunity* 5 (1927): 354–357.
51. Ibid., 357.
52. Streit, "Haiti: Intervention in Operation," 618, 619.
53. U.S. Department of State, *Eighth Annual Report of the American High Commissioner: 1929*, 32.
54. Johnson, *Along This Way*, 349 includes quote, 350. Similar observations were made by the Women's International League for Peace and Freedom in its report, *Occupied Haiti: Being the Report of a Committee of Six Disinterested Americans*, ed. E. G. Balch.
55. Geographers, anthropologists, and nutritionists have catalogued this information for decades. See, for example, J. M. May and D. L. McLellan, *The Ecology of Malnutrition in the Caribbean: The Bahamas, Cuba, Jamaica, Hispaniola (Haiti and the Dominican Republic), Puerto Rico, The Lesser Antilles, and Trinidad and Tobago* (New York: Hafner Press, 1973), 155–156, quote on 156; M. de Young, *Man and Land in the Haitian Economy*, Latin American Monograph No. 3 (Gainesville: University of Florida Press, 1958); and I. D. Beghin, W. Fougère, K. W. King, *L'Alimentation et la nutrition en Haïti* (Paris: Presses Universitaires de France, 1970). The indigenous Haitian economy also was largely run by women.
56. E. G. Balch, "Land and Living," in *Occupied Haiti*, ed. E. G. Balch, 61.
57. Ibid.

58. Holly, *Agriculture in Haiti,* 57–63.
59. U.S. Department of State, *Seventh Annual Report of the American High Commissioner at Port Au Prince, Haiti: 1928*, 9.
60. Holly, *Agriculture in Haiti*, 75–82.
61. Wise and Frost, *A Marine Tells It to You*, 136–137.
62. Ibid., 314–315. Through an elaborate spy scheme involving Gendarmes posing as Caco fighters, Charlemagne was assassinated. Ibid., 323.
63. P. W. Wilson and M. S. Mathis, "Epidemiology and Pathology of Yaws: A Report Based on a Study of 1,423 Consecutive Cases in Haiti," *Journal of the American Medical Association* 94 (1930): 1291–1292.
64. A. A. Buck, "Yaws (Nonvenereal Treponematosis)," in *Maxcy-Rosenau Public Health and Preventive Medicine,* ed. J. M. Last (Norwalk, Conn.: Appleton-Century-Crofts, 1986), 246–247; M. Dawson, "Socio-Economic and Epidemiological Change in Kenya: 1880–1925" (Ph.D. diss., University of Wisconsin, 1983), 191–192; Megan Vaughan, "Syphilis in Colonial East and Central Africa: The Social Construction of an Epidemic," in *Epidemics and Ideas: Essays on the Historical Perception of Pestilence* (Cambridge, U.K.: Cambridge University Press, 1992), 279–281, 286; Claude Quetel, *History of Syphilis*, trans. Judith Braddock and Brian Pike (Cambridge, U.K.: Polity, 1990), 258. The causative agent for yaws, Treponema pertenue, cannot be distinguished morphologically or by blood tests from Treponema pallidum, which causes syphilis. Endemic syphilis is also known as nonvenereal syphilis or treponaridosis.
65. Butler, "The Medical Needs of the Republic of Haiti," 277–278, includes quote; Parsons and Stitt, *History of Haitian Medicine*, 143.
66. "Notes and Commentaries: Diagnosis and Treatment of Yaws," *United States Naval Medical Bulletin* 28 (1930): 796–797. Two of Butler's studies appeared in medical journals published in France: C. S. Butler and E. Peterson, "La Treponematose et Hygiéne Publique," *Presse Médicale* 60 (July 27, 1927); and C. S. Butler, "De la Treponematose," *Annales de dermatologie et de syphiligriphie,* 7 Serie, tome 11 (November 1931). See also Camille Lhérisson. "Traduction," *Annales de dermatologie et de syphiligriphie,* 7 Serie, tome 11 (November 1931).
67. C. J. Hackett, "Yaws Eradication," *Transactions of the Royal Society of Tropical Medicine and Hygiene* 61 (1967): 148–152.
68. A.G.B., Review of "Pathology in the Tropics: A Study Based on the Review of 700 Consecutive Autopsies in Haiti," by R. M. Choisser, *United States Naval Medicine Bulletin* 27 (1929): 551–568, cited in *Tropical Disease Bulletin* 27 ([London], 1931): 587–589, quote on 587. Haitian General was the nation's major hospital. Its patients came from all over Haiti.
69. P. W. Wilson, "The Frontal Attack on Yaws—A Plea for a Change in Strategy," *United States Naval Medical Bulletin* 18 (1930): 2.
70. Rotberg, *Haiti: The Politics of Squalor*, 132; Rodman, *Haiti: The Black Republic*, 55; Logan, *Haiti and the Dominican Republic*, 141.
71. Wilson, "The Frontal Attack on Yaws," 2–4; Wilson and Mathis, "Epidemiology and Pathology of Yaws," 1291.
72. Wilson, "The Frontal Attack on Yaws," 4. These volunteer doctors probably were from religious missions or academic and philanthropic organizations.
73. Wilson and Mathis, "Epidemiology and Pathology of Yaws," 1231; K. R. Hill, "Non-Specific Factors in the Epidemiology of Yaws," in *First International Symposium*

on Yaws Control, WHO Monograph Series no. 15 (Geneva: WHO, 1953), 41; H. F. Dowling, *Fighting Infection: Conquests of the Twentieth Century* (Cambridge, Mass.: Harvard University Press, 1977), 95. Comprehensive surveys of yaws were conducted in Jamaica in early 1930s by the government and the International Health Division of the Rockefeller Foundation. In one clinic 2,101 of 2,853 patients were treated with a total of 8,556 injections (about three per patient on average). These treatments with neoarsphenamine and bismuth were associated with negative reactions about once per one thousand treatments. These side effects included jaundice, hemorrhagic encephalitis, diarrhea, stomatitis, and local reactions at the site of the injection. *Report of the Jamaica Yaws Commission for 1933* (Kingston, Jamaica: Government Printing Office, 1933), 15.

74. Rotberg, *Haiti: The Politics of Squalor*, 132; Rodman, *Haiti: The Black Republic*, 55; Logan, *Haiti and the Dominican Republic*, 141.
75. Butler, "The Medical Needs of the Republic of Haiti," 275–277.
76. Camille Lhérisson, "Diseases of the Peasants of Haiti," *American Journal of Public Health* 25 (1935): 924, 927, 929; Melhorn, "Public Health in Haiti," 570; S. S. Cook, "Mosquito Control in Haiti, with Especial Reference to the Use of Paris Green," *Southern Medical Journal* 24 (May 1931): 431–433; U.S. Army, Medical Department, *Preventive Medicine in World War II*, vol. 6, *Communicable Diseases, Malaria* (Washington, D.C.: Office of the Surgeon General, Department of the Army, 1963), 240.
77. Ibid. Quote is from Melhorn, "Public Health in Haiti," 570.
78. Balch, ed., *Occupied Haiti*, 89–91.
79. Stitt, Foreword to Parsons and Stitt, *History of Haitian Medicine*, xx.
80. Lhérisson, "Diseases of the Peasants of Haiti," 928.
81. Lhérisson, "Diseases of the Peasants of Haiti," 924–925. In 1935, Lhérisson was the president of the Haitian Medical Society.
82. Lhérisson, "Diseases of the Peasants of Haiti," 928.
83. L. Jones, C. Jones, and R. Greenhill, "Public Utilities," in *Business Imperialism: 1840–1930*, ed. D.C.M. Platt (Oxford: Clarendon Press, 1977), 81–82.
84. Ibid., 77–118.
85. F. D. Roosevelt, "The President Accepts Invitation of Haiti to Help Mediate Dispute between Haiti and the Dominican Republic, November 14, 1937," in *Public Papers and Addresses of Franklin D. Roosevelt*, vol. 6, compiled by Samuel I. Rosenman (New York: Macmillan, 1941), Item 151, 488–489; F. D. Roosevelt, "The President Hails the Acceptance by the Dominican Republic of the Mediation Proposals of Haiti, December 20, 1937," ibid., Item 165, 548–549; Thomas Fiehrer, "Political Violence in the Periphery: The Haitian Massacre of 1937," *Race and Class* 32 (1990): 1–20.

CHAPTER 4   *Out of the Shadows*

1. Harold R. Isaacs, *The New World of Negro Americans* (New York: Viking, 1963), 129–132; J. R. Hooker, "The Negro American Press and Africa in the 1930s," *Canadian Journal of African Studies* 1 (March 1967): 43–50; Sylvia M. Jacobs, *The African Nexus: Black American Perspectives on the European Partitioning of Africa, 1880–1920* (Westport, Conn.: Greenwood, 1981), 207–233.

2. Crummell lived in Liberia for twenty years, returning to America in 1873. On Crummell's exceptional intellect and his dedication to the Christianization of Africa, see W. L. Williams, *Black Americans and the Evangelization of Africa, 1877–1900* (Madison: University of Wisconsin Press, 1982); Wilson J. Moses, *Alexander Crummell: A Study in Civilization and Discontents* (New York: Oxford University Press, 1989). On other African American emigration projects involving Liberia, see Edwin S. Redkey, *Black Exodus: Black Nationalist and Back-to-Africa Movements, 1890–1910* (New Haven: Yale University Press, 1969); Wilson J. Moses, ed., *Liberian Dreams: Back-to-Africa Narratives from the 1850s* (University Park, Pa.: Penn State University Press, 1998).

3. Rayford W. Logan, "Liberia in the Family of Nations," *Phylon* 7 (1946): 5–11; W. R. Stanley, "Changing Export Patterns of Liberia's Seaports," in *Seaport Systems and Spatial Change*, ed. Brian S. Hoyle and K. Hilling (Chichester, N.Y.: John Wiley & Sons, 1984), 435.

4. W. R. Stanley, "Transport Expansion in Liberia," *Geographical Review* 60 (1970): 529–531.

5. Sumner is paraphrased by African educator and activist Nnamdi Azikiwe, *Liberia in World Politics* (London: Stockwell, 1934; reprint, Westport, Conn.: Negro Universities Press, 1970), 143.

6. S. M. Jacobs, "Afro-American Women Missionaries Confront the African Way of Life," in *Women in Africa and the African Diaspora*, ed. Rosalyn Terborg-Penn, Sharon Haley, A. B. Rushing (Washington, D.C.: Howard University Press, 1987), 121–132; Sandy D. Martin, *Black Baptists and African Missions: The Origins of a Movement, 1880–1915* (Macon, Ga.: Mercer University Press, 1989).

7. H. G. Brittain, *Scenes and Incidents of Every-Day Life in Africa* (1860; reprint, New York: Negro Universities Press, 1969); "Projected Hospital at Monrovia," *African Repository* 53 (July 1887): 93; R. A. Corby, "Cuttington College, Liberia: Years at Cape Palmas, 1889–1901," *Liberian Studies Journal* 17 (1992): 1–24; S. C. Saha, *A History of Agriculture in Liberia, 1822–1970: Transference of American Values* (Lewiston, N.Y.: Mellen, 1990), 60–62; quotes on 61.

8. William H. Heard, *The Bright Side of African Life Illustrated* (Philadelphia: AME Publishing House, 1898), 99.

9. Ibid., includes quote; William H. Heard, *From Slavery to the Bishopric in the A.M.E. Church* (Philadelphia: A.M.E. Book Concern, 1924), 78–79.

10. Clinton C. Boone, *Liberia as I Know It* (Richmond, Va.: 1929; reprint, Westport, Conn.: Negro Universities Press, 1970); Martin, *Black Baptists and African Missions*, 173, 177–178, 180–181, 195–196, 198–199, 201–202, 212. He also interrupted his African service to return to the United States to complete dental studies at the Bodee Dental School in New York City. Boone's first wife, Eva Roberta Coles, had died in the Congo. Boone, *Liberia as I Know It*, 112–113, 119; Martin, *Black Baptists and African Missions*, 180–181, 195–196, 198–199, 201–202, 212.

11. Saha, *History of Agriculture in Liberia*, 79.

12. W. R. Stanley, "Changing Export Patterns of Liberia's Seaports," in *Seaport Systems and Spatial Change*, ed. Hoyle and Hilling, 435; Y. Gershoni, "An Historical Examination of Liberia's Economic Policies, 1900–1944," *Liberian Studies Journal* 11 (1986): 20–21.

13. B. Hoyle, "Ports, Cities, and Coastal Zones: Competition and Change in a

Multimodal Environment," in *Cityports, Coastal Zones, and Regional Change*, ed. B. Hoyle (Chichester, N.Y.: John Wiley & Sons, 1996), 1–6.

14. Stanley, "Changing Export Patterns of Liberia's Seaports," 435; Saha, *History of Agriculture in Liberia*, 80.

15. U.S. Senate, 61st Congress, *Affairs in Liberia*, Senate Document No. 457, vol. 60 (Washington, D.C.: GPO, 1910).

16. Ibid., 31.

17. Ibid., 26.

18. Ibid.

19. Ibid., 35, 36, includes quotes; Azikiwe, *Liberia in World Politics*; R. A. Smith, *The American Foreign Policy in Liberia, 1822–1971* (Monrovia: Providence Publications, 1972), 73–74.

20. W.E.B. Du Bois, "Liberia, the League, and the United States," *Foreign Affairs* 9 (July 1933): 682–695, cited in *Writings by W.E.B. Du Bois in Periodicals Edited by Others*, vol. 2, *1910–1934*, comp. and ed. Herbert Aptheker (Millwood, N.Y.: Kraus-Thomson, 1982), 323, includes quote; R. U. McLaughlin, *Foreign Investment and Development in Liberia* (New York: Praeger, 1966), 144; Smith, *The American Foreign Policy in Liberia*, 73–74; D. P. Kilroy, "Extending the American Sphere to West Africa: Dollar Diplomacy in Liberia, 1908–1926" (Ph.D. diss., University of Iowa, 1995).

21. Robert R. Moton to Woodrow Wilson, February 27, 1918, in *Papers of Woodrow Wilson*, vol. 46, ed. A. S. Link (Princeton, N.J.: Princeton University Press, 1966), 480

22. R. R. Moton, introduction to *Liberia as I Know It*, by Boone, viii.

23. The expedition's findings were published in *The African Republic of Liberia and the Belgian Congo Based on the Observations Made and the Material Collected during the Harvard African Expedition, 1926–1927*, 2 vols., ed. Richard P. Strong (Cambridge, Mass.: Harvard University Press, 1930). On the survey's epidemiological limitations, see H. A. Poindexter, "A Laboratory Epidemiological Study of Certain Infectious Diseases in Liberia," *American Journal of Tropical Medicine* 29 (1949): 435. On Harvey Firestone's contributions, see "Statement Submitted on June 23, 1933, to the Council Committee on Liberia by Mr. Harvey S. Firestone, Jr.," in *Liberia: Documents Relating to the Plan of Assistance Proposed by the League of Nations*, U.S. Department of State (Washington, D.C.: GPO, 1933), 38–39.

24. Donald Spivey, *The Politics of Miseducation: The Booker Washington Institute of Liberia, 1929–1984* (Lexington: University Press of Kentucky, 1986), 19–21. For a critique of Spivey's political perspective by a Liberian historian, see Mary A. B. Sherman, "Review: *The Politics of Miseducation* by Donald Spivey," *Liberian Studies Journal* 12 (1987): 88–93.

25. J. H. Dillard et al., *Twenty Year Report of the Phelps-Stokes Fund, 1911–1931, with a Series of Studies of Negro Progress and of Developments of Race Relations in the United States and Africa* (New York: Phelps-Stokes Fund, 1932), 28–29; Spivey, *The Politics of Miseducation*, 38–43; Sherman, "Review."

26. Kilroy, "Extending the American Sphere," 272–273; J. L. Sibley and D. Westermann, *Liberia—Old and New: A Study of Its Social and Economic Background and Possibilities of Development* (Garden City, N.Y.: Doubleday, Doran, 1928), 106–107.

27. Azikiwe, *Liberia in World Politics*, 123–168; R. L. Buell, *The Native Problem in*

Africa, vol. 2 (New York: Macmillan, 1928), 833–834; George Padmore, *Pan-Africanism or Communism: The Coming Struggle for Africa* (London: Dennis Dobson, 1956), 63–68; Saha, *History of Agriculture in Liberia*, 97–99.

28. "Liberia's Future in Rubber," *Opportunity* 3 (November 1925): 324.

29. Kilroy, "Extending the American Sphere," 191–195.

30. Cedric Robinson, "Du Bois and Black Sovereignty: The Case of Liberia," *Race and Class* 32 (1990): 39–50; D. McBride, "Solomon Porter Hood, 1853–1943: Black Missionary, Educator, and Minister to Liberia," *Quarterly Publication of the Lancaster Historical Society* 84 (1984): 2–9; Kilroy, "Extending the American Sphere," 204; Du Bois, "Liberia, the League, and the United States," in *Writings by W.E.B. Du Bois in Periodicals Edited by Others*, vol. 2, ed. Aptheker, 323, includes quote.

31. J. H. Randolph, Foreword to *Liberia as I Know It*, by Boone, ix.

32. Moton, introduction to *Liberia as I Know It*, by Boone, viii.

33. "Current Events in Negro Education: The Plight of Liberia," *Journal of Negro Education* 2 (October 1933): 518, includes quote.

34. Ibid.

35. Sir Andrew McFadyean, ed., *The History of Rubber Regulation, 1934–1943* (London: International Rubber Regulation Committee/George Allen & Unwin, 1944), 14–15, 21.

36. Saha, *History of Agriculture in Liberia*, 63–67; E. T. Thompson, "Mines and Plantations and Movement of People," *American Journal of Sociology* 37 (January 1932): 609.

37. "Statement submitted on June 23, 1933, to the Council Committee on Liberia by Mr. Harvey S. Firestone, Jr.," in *Liberia: Documents Relating to the Plan of Assistance Proposed by the League of Nations*, 38–39; G. Proctor Cooper and Samuel J. Record, *The Evergreen Forests of Liberia: A Report on Investigations Made in the West African Republic of Liberia by the Yale University School of Forestry in Cooperation with the Firestone Plantations Company*, Yale University School of Forestry, Bulletin no. 31 (New Haven, Conn.: Yale University, 1931), 6–7.

38. Cooper and Record, *The Evergreen Forests*, 5–6; quote on 6.

39. Pete Daniel, *Breaking the Land: The Transformation of Cotton, Tobacco, and Rice Cultures since 1880* (Urbana: University of Illinois Press, 1985).

40. Rupert P. Vance, *Human Factors in Cotton Culture: A Study in the Social Geography of the American South* (Chapel Hill: University of North Carolina Press, 1929), 34.

41. D.P.S. Ahluwalia, *Plantation and the Politics of Sugar in Uganda* (Kampala, Uganda: Fountain, 1995), 7–9; George E. Beckford, *Persistent Poverty: Underdevelopment in Plantation Economies of the Third World* (New York: Oxford University Press, 1972); Edgar Graham with Ingrid Floering, *The Modern Plantation in the Third World* (New York: St. Martin's Press, 1984), 33–36, 59.

42. W. C. Taylor, *The Firestone Operations in Liberia* (Washington, D.C.: National Planning Association, 1956) 15; Stanley, "Transport Expansion in Liberia," 533.

43. Robert W. Clower et al., *Growth without Development: An Economic Survey of Liberia* (Evanston: Northwestern University Press, 1966), 143.

44. Beckford, *Persistent Poverty*, 6.

45. E. T. Thompson, "Comparative Education in Colonial Areas, with Special Reference to Plantation and Mission Frontiers," *American Journal of Sociology* 48 (May 1943): 710–721.

46. Gordon Gaskill, "Anybody Can Make Money in Liberia," *Negro Digest* 7 (November 1948): 76; Saha, *History of Agriculture in Liberia*, 107–110.

47. Sibley and Westermann, *Liberia—Old and New*, 106–107, includes quote; Graham with Floering, *The Modern Plantation*.

48. A. J. Knoll, "Firestone's Labor Policy, 1924–1939," *Liberian Studies Journal* 16 (1991): 49–75.

49. W. L Partridge, "Banana County in the Wake of United Fruit: Social and Economic Linkages," *American Ethnologist* 6 (August 1979): 495; M. Edelman, *The Logic of the Latifundio: The Large Estates of Northwestern Costa Rica since the Late Nineteenth Century* (Stanford, Calif.: Stanford University Press, 1992).

50. Merran Fraenkel, *Tribe and Class in Monrovia* (London: International African Institute/Oxford University Press, 1964), 24.

51. Y. Hayami, "Peasant and Plantation in Asia," in *From Classical Economics to Development Economics*, ed. G. M. Meier (New York: St. Martin's Press, 1994), 121–134; M. Wilkins, "The Role of Private Business in the International Diffusion of Technology," *Journal of Economic History* 34 (1974): 177; Gaskill, "Anybody Can Make Money," 76.

52. D. Felix, "Technological Dualism in Late Industrializers: On Theory, History, and Policy," *Journal of Economic History* 34 (1974): 232.

53. Daniel R. Headrick, *The Tentacles of Progress: Technological Transfer in the Age of Imperialism, 1850–1940* (New York: Oxford, 1988), 14.

54. Arthur W. Lewis, "Economic Development with Unlimited Supplies of Labour," in *The Economics of Underdevelopment*, ed. A. N. Agarwala and S. P. Singh (London: Oxford University Press, 1970), 400–449; A. W. Lewis, "Unlimited Labour: Further Notes," in *Selected Economic Writings of W. Arthur Lewis*, ed. Mark Gersovitz (New York: New York University Press, 1985), 365–390.

55. E. F. Frazier, *Race and Cultural Contacts in the Modern World* (Boston: Beacon, 1957), 117–126, quote on 120.

56. Ibid.

57. I. K. Sundiata, *Black Scandal: America and the Liberian Labor Crisis, 1929–1936* (Philadelphia: ISHI, 1980), chaps. 1–2.

58. Fraenkel, *Tribe and Class in Monrovia*, 27, 29; A. Konneh, "Citizenship at the Margins: Status, Ambiguity, and the Mandingo of Liberia," *African Studies Review* 39 (September 1996): 141–154.

59. "Statement Submitted by the American Representative on the Liberian Committee of the Council, October 9, 1933," in *Liberia: Documents Relating to the Plan of Assistance Proposed by the League of Nations*, 31–32.

60. Ibid., 35.

61. Henry B. Cole, ed., *The Liberian Year Book 1956* (London: Diplomatic Press, [1956?]), 207; Fraenkel, *Tribe and Class in Monrovia*, 22.

62. See the U.S. State Department's correspondence, Robert Woods Bliss to Joseph L. Johnson, U.S. Minister Resident and Consul General, Liberia, May 25, 1921, in *The Marcus Garvey and Universal Negro Improvement Association Papers*, vol. 2, ed. Robert Hill (Berkeley: University of California at Berkeley, 1983), 425.

63. Statement by President Charles D. B. King, cited in Boone, *Liberia as I Know It*, 136.

64. Frank Chalk, "Du Bois and Garvey Confront Liberia: Two Incidents in the Coolidge Years," *Canadian Journal of African Studies* 1 (November 1967): 135–142; Amos

Sawyer, *The Emergence of Autocracy in Liberia: Tragedy and Challenge* (San Francisco: Institute for Contemporary Studies, 1992), 182–184.

65. Fraenkel, *Tribe and Class in Monrovia*, 23–24.
66. M. A. Barber, J. B. Rice, and J. Y. Brown, "Malaria Studies on the Firestone Plantation in Liberia, West Africa," *American Journal of Hygiene* 15 (May 1932): 601–633; Cooper and Record, *The Evergreen Forests*.
67. "Statement by Mr. Harvey S. Firestone, Jr.," 38–39.
68. Strong, *The African Republic of Liberia and the Harvard African Expedition*, 1: 199, 210; Barber et al., "Malaria Studies," 606–607.
69. Barber et al., "Malaria Studies," 606–607, includes quote.
70. Strong, *The African Republic of Liberia and the Harvard African Expedition*, 1: 272–307.
71. J. Knott, "Yaws in Liberia," *Porto Rico Journal of Public Health and Tropical Medicine* 6 (June 1931): 385–390, quote on 385.
72. See *Report of the Jamaica Yaws Commission for 1932* (Kingston, Jamaica: GPO, 1934); *Report of the Jamaica Yaws Commission for 1934* (Kingston, Jamaica: GPO, 1935); *Report of the Jamaica Yaws Commission for 1936* (Kingston, Jamaica: GPO, 1936).
73. G. W. Harley, "The Symptomatology of Yaws in Liberia, Part I—The Seven Cardinal Symptoms," *Transactions of the Royal Society of Tropical Medical Hygiene* 27 (1933): 99.
74. Ibid.; G. W. Harley, "Ganta Dispensary Patients: A Statistical Study of 6,291 Consecutive Out-Patients in North-Eastern Liberia," *American Journal of Tropical Medicine* 13 (1933): 67; C. J. Hackett, "Extent and Nature of the Yaws Problem in Africa," in *First International Symposium on Yaws Control: Bangkok, 1952,* ed. World Health Organization (Geneva: World Health Organization, 1953), 146.
75. Ludwik Anigstein, *First Report on the Medical Survey of Liberia (Central Province of the Hinterland)* (Monrovia, Liberia: Department of State, GPO, January 1936), mimeo-typescript, 2–3; L. Anigstein, *Second Report on the Medical Survey of Liberia (Central Province of the Hinterland)* ([Monrovia, Liberia: Department of State, GPO], January 1936). mimeo-typescript, 3.
76. Anigstein, *First Report on the Medical Survey of Liberia*, 5.
77. Ibid.
78. Ibid.; Anigstein, *Second Report on the Medical Survey of Liberia;* Anigstein, *Third Report on the Medical Survey of Liberia (Western and Central Provinces)* ([Monrovia, Liberia: Department of State, GPO], April 23, 1936), mimeo-typescript.
79. Anigstein, *First Report on the Medical Survey of Liberia*, 8.
80. Anigstein's research on yaws remained the international medical community's only formal study of this disease in Liberia through the 1950s. See Hackett, "Extent and Nature of the Yaws Problem in Africa," 140.
81. For a groundbreaking comparative study of racial politics, labor repression, and industry in South Africa and Alabama, see S. B. Greenberg, *Race and State in Capitalist Development: Comparative Perspectives* (New Haven: Yale University Press, 1980).
82. Chinweizu, *The West and the Rest of Us: White Predators, Black Slavers, and the African Elite* (New York: Vintage, 1975), 91, 92.
83. Basil Davidson, *The Black Man's Burden: Africa and the Curse of the Nation-State* (New York: Times Books, 1992), 43–44, 200, 243–249.

84. P. F. Russell, *Man's Mastery of Malaria [University of London, Heath Clark Lectures, 1953]* (London: Oxford University Press, 1955), 136, 184–186. Dr. Russell (M.D. and M.P.H.) was an officer at the Rockefeller Foundation's medicine and public health division.

CHAPTER 5 *Malaria and Modernization*

1. Steel production, for example, surged following the war. Under the control of U.S. Steel, the nation's largest corporation, the steel supply grew at 15 to 20 percent annually. Frank Freidel and Alan Brinkley, *America in the Twentieth Century* (New York: Knopf, 1982), 159–163; Heather Dean, "Scarce Resources: The Dynamic of American Imperialism," in *Imperialism, Intervention, and Development* (London: Croom Helm, 1979), ed. Andrew Mack et al., 143–178.

2. W.E.B. Du Bois, "A Field for Socialist," *New Review* (New York), January 11, 1913, 54–57, reprinted in–*Writings by W.E.B. Du Bois in Periodicals Edited by Others*, vol. 2, *1910–1934*, ed. H. Aptheker (Millwood, N.Y.: Kraus-Thomson, 1982), quote on 80.

3. "Arrival of Southern Businessmen," *Panama Star and Herald*, April 29, 1907; E. R. Johnson, "Influence of the Panama Canal on Southern Agriculture, Industry, and Commerce," in *The South in the Building of the Nation* (Richmond, Va.: Southern Historical Publication Society, 1909), 642–647.

4. Charles H. Wood, *The Demography of Inequality in Brazil* (Cambridge: Cambridge University Press, 1988), 207–211.

5. Allen Isaacman and Richard Roberts, eds., *Cotton, Colonialism, and Social History in Sub-Saharan Africa* (Portsmouth, N.H.: Heinemann, 1995); R. D. Wolff, *The Economics of Colonialism: Britain and Kenya, 1870–1930* (New Haven, Conn.: Yale University Press, 1974).

6. For an excellent historical analysis of these factors, see Daniel R. Headrick, *The Tentacles of Progress: Technology Transfer in the Age of Imperialism, 1850–1940* (New York: Oxford University Press, 1988).

7. The scholarship on the strength of the plantation in twentieth-century Atlantic nations varies widely in quality and time of origin. Key studies on the subject relating to the United States are R. W. Shugg, "Survival of the Plantation System in Louisiana," *Journal of Southern History* 3 (August 1937): 311–325; and G. D. Jaynes, *Branches with Roots: Genesis for the Black Working Class in the American South, 1862–1882* (New York: Oxford University Press, 1986). On the enduring plantation and racial inequality, despite the abolition of slavery or legal racial inequality in Latin America, see Magnus Morner, "Historical Research on Race Relations in Latin America during the National Period," in *Race and Class in Latin America*, ed. M. Morner (New York: Columbia University Press, 1970), 199–230; and M. K. Huggins, *From Slavery to Vagrancy in Brazil: Crime and Social Control in the Third World* (New Brunswick: Rutgers University Press, 1985).

8. Rupert B. Vance, *Human Factors in Cotton Culture: A Study in the Social Geography of the American South* (Chapel Hill: University of North Carolina Press, 1929), 63–65.

9. Vance, *Human Factors in Cotton Culture*, 79, 174, quote on 79.

10. M. R. Davie, *Negroes in American Society* (New York: McGraw-Hill, 1949), 77.

11. Rexford G. Tugwell, *Industry's Coming of Age* (New York: Harcourt, Brace and Co., 1927), 4, includes quote. Tugwell became a member of Roosevelt's New Deal braintrust. David E. Conrad, *The Forgotten Farmers: The Story of Sharecroppers in the New Deal* (Urbana: University of Illinois Press, 1965), 111–112.

12. Vance, *Human Factors in Cotton Culture*, 63–65; quote on 65.

13. U.S. Public Health Service, Mississippi Coastal District, Gulfport, Mississippi, *Final Report on the Sanitary Operations in the Mississippi Coastal District by the Mississippi Board of Health and the United States Public Health Service, 1918–1919,* by L. C. Frank [1919].

14. Richard Sterner, *The Negro's Share: A Study of Income, Consumption, Housing, and Public Assistance* (New York: Harper, 1943), 25–27. This was a Carnegie study.

15. Basil Miller, *George Washington Carver: God's Ebony Scientist* (Grand Rapids: Zondervan, 1943), 80.

16. Leon F. Litwack, *Trouble in Mind: Black Southerners in the Age of Jim Crow* (New York: Alfred A. Knopf, 1998), 374, 477, includes quotes.

17. Quote is in an essay, by an unidentified author, that appeared in *The Outlook* 74 (n.d.): 687–688, cited in Carter G. Woodson, *The Rural Negro* (Washington, D.C.: Association for the Study of Negro Life and History, 1930), 79.

18. Woodson, *The Rural Negro*, 67–88; quote on 72.

19. B. T. Washington, "Extracts from an Address at the Black Belt Fair, Demopolis, Alabama," September 27, 1912, in *The Booker T. Washington Papers,* vol. 12, *1912–1914,* ed. L. R. Harlan and R. W. Smock (Urbana: University of Illinois Press, 1982), 19.

20. Gunnar Myrdal, *An American Dilemma: The Negro Problem and Modern Democracy,* Twentieth Anniversary ed. (1944; reprint, New York: Harper & Row, 1962), 228, in reprint.

21. Myrdal, *An American Dilemma*, 554–555; Frank Tannenbaum, *Darker Phases of the South* (1924; reprint, New York: Negro Universities Press, 1969) 74–115.

22. Tannenbaum, *Darker Phases of the South*, 43–44, 55.

23. Paul F. Russell, *Man's Mastery of Malaria* (London: Oxford University Press, 1955), 171.

24. Woodson, *The Rural Negro*, 16–19.

25. H. M. Dart, *Maternity and Child Care in Selected Rural Areas of Mississippi,* Rural Child Welfare Series no. 5, Children's Bureau Publication no. 88 (Washington, D.C.: GPO, 1921), 24, 27; Arthur F. Raper, *Preface to Peasantry: A Tale of Two Black Belt Counties* (1936; reprint, New York: Atheneum, 1968), 48, in reprint; Elizabeth C. Tandy, "Infant and Maternal Mortality," *Journal of Negro Education* 6 (1937): 328. Tennessee was the exception to the rule. In 1935 midwives delivered 54 percent of the black infants (Tandy, ibid.).

26. Charles S. Johnson et al., *Statistical Atlas of Southern Counties: Listing and Analysis of Socio-Economic Indices of 1,104 Southern Counties* (Chapel Hill: University of North Carolina Press, 1941), 23–32, quote on 26. The council was funded by the Julius Rosenwald Fund.

27. Ibid., 26.

28. Works Progress Administration, Division of Social Research, *Landlord and Tenant on the Cotton Plantation,* Research Monograph 5, by T. J. Woofter, Jr. (Washington, D.C.: GPO, 1936), 125–144, quote on 131; Henry A. Bullock, *A History of*

Negro Education in the South from 1619 to the Present (New York: Praeger, 1967), 169–185.

29. Allison Davis, B. B. Gardner, and M. R. Gardner, *Deep South: A Social Anthropological Study of Caste and Class* (1941; reprint, abridged ed., Chicago: University of Chicago Press, 1969), 249, includes quote, in reprint.

30. Ibid.; Charles H. Wood and J. A. Magno de Carvalho, *The Demography of Inequality in Brazil* (Cambridge, U.K.: Cambridge University Press, 1988), 89–90; Charles S. Johnson, *The Shadow of the Plantation* (Chicago: University of Chicago Press, 1934), 139.

31. Woodson, *The Rural Negro*, 204; Davis et al., *Deep South*, 249.

32. Horace M. Bond, *Negro Education in Alabama: A Study in Cotton and Steel* (1939; reprint, New York: Atheneum, 1969), 246, in reprint.

33. Raper, *Preface to Peasantry*, 186; Stanley B. Greenberg, *Race and State in Capitalist Development: Comparative Perspectives* (New Haven, Conn.: Yale University Press, 1980), 218–219.

34. See, for example, Leonard Dinnerstein, Roger L. Nichols, and David M. Reimers, *Natives and Strangers: Blacks, Indians, and Immigrants in America,* 2nd ed. (New York: Oxford University Press, 1990), 144–147. For a valuable case study that refutes the generally accepted perspective, see Earl Lewis, "Expectations, Economic Opportunities, and Life in the Industrial Age: Black Migration to Norfolk, Virginia, 1910–1945," in *The Great Migration in Historical Perspective*, ed. J. W. Trotter (Bloomington: Indiana University Press, 1991), 23–45, quote on 23.

35. Bond, *Negro Education in Alabama*, 234–237.

36. T. R. Snavely, "The Exodus of Negroes from the Southern States: Alabama and North Carolina," in *Negro Migration in 1916–17, Reports by R. H. Leavell et al.,* U.S. Department of Labor, Division of Negro Economics (1919; reprint, New York: Negro Universities Press, 1969), 53, in reprint.

37. Bond, *Negro Education in Alabama*, 234–237, quote on 234.

38. Lawrence Gordon, "Document: A Brief Look at Blacks in Depression Mississippi, 1929–1934: Eyewitness Accounts," *Journal of Negro History* 64 (1979): 385.

39. Ibid.

40. Raper, *Preface to Peasantry*, 195, includes quote; Snavely, "The Exodus of Negroes," 55–57; G. W. Williams and J. D. Applewhite, "Tuberculosis in the Negroes of Georgia: Economic, Racial, and Constitutional Aspects," *American Journal of Hygiene* 29 (March 1939): 96–97.

41. W. T. Harn, "Farm Tenancy," in *Farmers in a Changing World, Yearbook of Agriculture, 1940,* U.S. Department of Agriculture (Washington, D.C.: GPO, 1940), 890; Woodson, *The Rural Negro*, 30; Myrdal, *An American Dilemma*, 242–250.

42. R. A. Rusca and C. A. Bennett, "Problems of Machine Picking," in *Crops in Peace and War: The 1950–1951 Yearbook of Agriculture*, U.S. Department of Agriculture (Washington, D.C.: GPO, 1951), 441–444; G. C. Fite, "Recent Progress in the Mechanization of Cotton Production in the United States," *Agricultural History* 24 (1950): 19–28; G. C. Fite, "Mechanization of Crop Production since World War II," *Agricultural History* 54 (1980): 190–207; J. T. Kirby, "The Transformation of Southern Plantations c. 1920–1960," *Agricultural History* 57 (1983): 257–276.

43. Greenberg, *Race and State in Capitalist Development*, 428, table II A, 430, table II C.2.

44. Works Progress Administration, *Landlord and Tenant on the Cotton Plantation*, xx; H. W. Odum, *Southern Regions of the United States* (Chapel Hill: University of North Carolina Press, Southern Regional Committee of the National Research Council, 1936), 61. The exhaustive study of the South by the Southern Regional Committee barely mentions the word "plantation." During the Great Depression plantations had been vigorously denounced outside the South by liberal and leftist political activists, academics, and labor activists. The committee avoided the term altogether probably to avoid playing into the hands of these critics during this volatile political era. For examples of the critical or leftist perspectives, see Vance, *Human Factors in Cotton Culture*; Johnson, *The Shadow of the Plantation* (1934); Charles S. Johnson, Edwin E. Embree, and Will Alexander, *The Collapse of Cotton Tenancy* (Chapel Hill: University of North Carolina Press, 1935); and J. S. Allen, *The Negro Question in the United States* (New York: International Publishers, 1936). The idea that the South was a colony was even advanced by southern social scientists, writers, and business owners of the interwar years. But their view was linked to sectional nationalism and conspiracy thinking traced back to pre–Civil War southern radicalism. G. B. Tindall, *The Emergence of the New South* (Baton Rouge: Louisiana State University Press, 1967), 594–606.

45. E. Franklin Frazier, *The Negro in the United States,* rev. ed. (New York: Macmillan, 1959), 191.

46. Matthew Tayback, "Demographic Trends in the South—Implications for Public Health Administration," *American Journal of Public Health* 46 (1956): 1299.

47. T. A. Hill, "Negroes in Southern Industry," *Annals of the American Society of Political and Social Science* 153 (January 1931): 171, 175–177; quote on 171.

48. Ibid., 174; H. R. Cayton and G. S. Mitchell, *Black Workers and the New Unions* (College Park, Md.: McGrath, 1939), 315; Ray Marshall, *The Negro and Organized Labor* (New York: John Wiley & Sons, 1965), 186; David Rosner and Gerald Markowitz, *Deadly Dust: Silicosis and the Politics of Occupational Disease in Twentieth-Century America* (Princeton, N.J.: Princeton University Press, 1991), 70–74.

49. Hill, "Negroes in Southern Industry," 171.

50. Elizabeth Fee, *Disease and Discovery: A History of the Johns Hopkins School of Hygiene and Public Health, 1916–1939* (Baltimore: Johns Hopkins University Press, 1987).

51. H. Bradley, "A Review of Malaria Control and Eradication in the United States," *Mosquito News* 26 (December 1966): 462; Paul F. Russell, "The United States and Malaria: Debits and Credits," *Bulletin of the New York Academy of Medicine* 44 (1968): 633.

52. T. F. Abercrombie, "Malaria and Its Relation to the Economic Development of Georgia," *Emory University Quarterly* 10 (October 1954): 168–171.

53. M. R. Davie, *Negroes in American Society* (New York: McGraw-Hill, 1949), 77.

54. A. W. Sellards, "Bonds of Union between Tropical Medicine and General Medicine," *Science* 66 (July 29, 1927): 93–100; Nathan Reingold and I. H. Reingold, eds., *Science in America: A Documentary History, 1900–1939* (Chicago: University of Chicago Press, 1981), 287–288 (comment by editors).

55. G. W. Corner, *A History of the Rockefeller Institute, 1901–1953* (New York: Rockefeller Institute Press, 1964), 64, cited in W. W. Lowrance, *Modern Science and Human Values* (New York: Oxford University Press, 1985), 9.

56. R. H. von Ezdorf, "Malaria in the United States: Its Prevalence and Geographic Distribution," *Public Health Reports* 30 (May 28, 1915): 1603.

57. Russell, "The United States and Malaria," 628.

58. Nathan Reingold and I. H. Reingold, "Biology since 1915: Defining the True Way and Locating Its Home," in *Science in America*, ed. Reingold and Reingold, 284; Fee, *Disease and Discovery*; J. R. Matthews, *Quantification and the Quest for Medical Certainty* (Princeton, N.J.: Princeton University Press, 1995); Katheryn Ott, *Fevered Lives: Tuberculosis in American Culture since 1870* (Cambridge: Harvard University Press, 1996), 53–68, 93–94.

59. Rosner and Markowitz, *Deadly Dust,* 77.

60. Russell, *Man's Mastery of Malaria*, 171.

61. Mary Gover, "Trend of Mortality among Southern Negroes since 1920," *Journal of Negro Education* 6 (1937): 281. These rates were adjusted for age.

62. Fredrich Rapp, "Technology and Natural Science—A Methodological Investigation," in *Contributions to a Philosophy of Technology*, ed. F. Rapp (Dordrecht, 1974), cited in Nicolas Rescher, *Scientific Progress: A Philosophical Essay on the Economics of Research in Natural Science* (Pittsburgh: University of Pittsburgh Press, 1978), 135, includes quote. Rescher lucidly describes this dimension of advancing technology. He calls it the "technology of inquiry itself" and "the technology of data-generation and of information acquisition and processing" (Ibid., 134). Also see, J. D. Howell, *Technology in the Hospital: Transforming Patient Care in the Early Twentieth Century* (Baltimore: Johns Hopkins University Press, 1995).

63. Sellards, "Bonds of Union," 94.

64. Joseph A. LePrince and A. J. Orenstein, *Mosquito Control in Panama: The Eradication of Malaria and Yellow Fever in Cuba and Panama* (New York: Putnam's, 1916), 218. At about the same time, William Gorgas published a thorough account of the campaigns against yellow fever in Cuba and against mosquitoes in the Canal Zone. William C. Gorgas, *Sanitation in Panama* (New York: D. Appleton, 1915).

65. Comments by Malcolm Watson, cited in LePrince and Orenstein, *Mosquito Control in Panama*, 225.

66. M. A. Barber, W.H.W. Komp, and T. B. Hayne, "Prevalence of Malaria (1925) in Parts of Delta of Mississippi and Arkansas: Economic Conditions," *Southern Medical Journal* 19 (May 1926): 373–377; quote on 373.

67. Margaret Humphreys, *Yellow Fever and the South* (New Brunswick, N.J.: Rutgers University Press, 1992), 150–156.

68. K. F. Maxcy, "Epidemiological Principles Affecting the Distribution of Malaria in the Southern United States," *Public Health Reports* 39 (May 16, 1924): 1118.

69. Ibid., 1119.

70. Ibid., 1118, 1119.

71. Ibid., 1120.

72. Ibid., 1120–1121, 1126, quote on 1126.

73. See note 58.

74. League of Nations, *Second General Report of the Malaria Commission* (1927), cited in F. Abercrombie, *History of Public Health in Georgia, 1733–1950* ([Atlanta?]: n.p., 1951), 70–71; League of Nations, *The Treatment of Malaria: Fourth General Report of the Malaria Commission*, no. 6 (December 1937): 1014–1015; I. J. Wolman, "The Global Attack on Malaria," *Clinical Pediatrics* 2 (November 1963): 653–655.

75. "The Treatment of Malaria: Study of Synthetic Drugs, as Compared with Quinine, in the Therapeutics and Prophylaxis of Malaria," *Fourth General Report of the Malaria Commission, League of Nations, Bulletin of the Health Organisation* 6, no. 6 (December 1937): 1014–1015, quote on 1015.

76. M. E. Winchester, "Atabrine Prohylaxis in Malaria: Report of Third Year's Investigation," *American Journal of Tropical Medicine* 18 (1938): 625, includes quote; L. J. Bruce-Chwatt, *Essential Malariology* (New York: John Wiley, 1985), 3–4.

77. Katherine Clark, introduction to *Motherwit: An Alabama Midwife's Story by Onnie Lee Logan as Told to Katherine Clark* (New York: Plume, 1989), xii–xiii; G. J. Fraser, *African American Midwifery in the South: Dialogues of Birth, Race, and Memory* (Cambridge, Mass.: Harvard University Press, 1998), 59–77; R. D. Devries and Rebecca Barroso, "Midwives among the Machines: Re-creating Midwifery in the Late Twentieth Century," in *Midwives, Society, and Childbirth: Debates and Controversies in the Modern Period*, ed. Hilary Marland and A. M. Rafferty (New York: Routledge, 1997), 248–472.

78. H. A. Poindexter, "Special Health Problems of Negroes in Rural Areas," *Journal of Negro Education* 6 (1939): 399–412, quote on 400.

79. Ibid., 401.

80. Ibid.

81. "Malaria in the Last Two Years," *Southern Medical Journal* 22 (April 1929): 413–414.

82. See, for example, H. P. Guzda, "Social Experiment of the Labor Department: The Division of Negro Economics," *Public Historian* 4 (fall 1982): 7–37; N. E. Woodruff, "Debates over Mechanization, Labor, and Civil Rights in the 1940s," *Journal of Southern History* 60 (May 1994): 263–284.

83. M. A. Barber, "The History of Malaria in the United States," *Public Health Reports* 44 (October 25, 1929): 2586.

84. J. A. Ferrell, "Malaria Activities," *Southern Medical Journal* 19 (May 1926): 396–397; J. A. LePrince, "Historical Review of Development of Control of Disease-Bearing Mosquitos," *Transactions of the American Society of Civil Engineers* 92 (1928): 1262; H. E. Meleney et al., "Observations on the Malaria Problem of West Tennessee," *Southern Medical Journal* 22 (April 1929): 384.

85. D. L. Seckinger, "Atabrine and Plasmochin in the Treatment and Control of Malaria," *American Journal of Tropical Medicine* 15 (1935): 631–645; M. E. Winchester, "Individual Chemoprophylaxis against Malaria; Preliminary Report" *Southern Medical Journal* 29 (1936): 1029; Winchester, "Atabrine Prophylaxis in Malaria, 625–640.

86. Douglas L. Smith, *The New Deal in the Urban South* (Baton Rouge: Louisiana State University Press, 1988), 70.

87. J. M. Andrews, "What's Happening to Malaria in the U.S.A.?" *American Journal of Public Health* 38 (1948): 936–937.

88. Russell, "The United States and Malaria," 634.

89. Quoted in Arthur M. Schlesinger Jr., *The Age of Roosevelt: The Crisis of the Old Order, 1919–1933* (Boston: Houghton Mifflin, 1957), 89.

90. John A. Hobson, *The Evolution of Modern Capitalism: A Study in Machine Production,* 4th ed. (1926; reprint, London: George Allen & Unwin, 1954), 469–471, in reprint.

91. William E. Leuchtenberg, *Franklin D. Roosevelt and the New Deal, 1932–1940* (New York: Harper Colophon, 1963), 54–55; Freidel and Brinkley, *America in the Twentieth Century*, 232–234.

92. Jacqueline Jones, *The Dispossessed: America's Underclasses from the Civil War to the Present* (New York: Basic Books, 1992), 218.

93. J. P. Davis, "A Black Inventory of the New Deal," *The Crisis* 42 (May 1935): 141–142, 154–155, quote on 142.

94. Nancy L. Grant, *TVA and Black Americans: Planning for the Status Quo* (Philadelphia: Temple University Press, 1990), 152–153.

95. Bradley, "Review of Malaria Control," 464.

96. Ibid., includes quote; Russell, "The United States and Malaria," 634; Andrews, "What's Happening to Malaria?" 938–939.

97. E. C. Faust, "Malaria Mortality in the Southern United States for the Year 1937," *American Journal of Tropical Medical* 19 (1939): 447–455; Russell, "The United States and Malaria," 628; editorial, "Malaria in the Last Two Years," *Southern Medical Journal* 22 (April 1929): 413–414; Russell, "The United States and Malaria," 648.

98. Ibid.

99. Faust, "Malaria Mortality in the Southern United States," 455.

100. Andrews, "What's Happening to Malaria?" 934, includes quote; Peter Newman, "Malaria Control and Population Growth," *Journal of Development Studies* 6 (January 1970): 133–134.

101. F. Y. Wiselogle, ed., *A Survey of Antimalarial Drugs, 1941–1945,* subsidized by the Office of Scientific Research and Development on Recommendation by the Committee on Medical Research (Ann Arbor, Mich.: J. W. Edwards, 1946), 7, includes quote; T. T. Mackie, "Tropical Disease Problems among Veterans of World War II," *American Journal of Tropical Medicine* 29 (1949): 443.

102. Russell, "The United States and Malaria," 636.

103. Ibid.

104. William H. McNeill, *Plagues and Peoples* (New York: Anchor, 1976), 249.

105. Gover, "Trend of Mortality among Southern Negroes since 1920," 281. Also see Julian H. Lewis, *The Biology of the Negro* (Chicago: University of Chicago Press, 1942), 192–196.

106. Quentin M. Geiman, "A Half Century of Tropical Medicine," *American Journal of Tropical Medicine and Hygiene* 3 (1954): 406–407.

107. For a summary of these conferences that occurred from 1941 to 1945, see Wiselogle, *A Survey of Antimalarial Drugs*.

108. L. L. Williams, "The Anti-Malaria Program in North America," in *A Symposium on Human Malaria*, American Association for the Advancement of Science, ed. Forest Ray Moulton (Washington, D.C.: American Association for the Advancement of Science, 1941), 369.

109. Ibid. The exoerythrocytic stage is that phase in the development of the malaria parasite when cells other than red blood cells are infected. This phase occurs right after the mosquito bite.

110. Stanley C. Oaks Jr., Violaine S. Mitchell, Greg W. Pearson, and Charles C. J. Carpenter, eds., *Malaria: Obstacles and Opportunities* (Washington, D.C.: National Academy Press, 1991), 211–212.

111. Geiman, "A Half Century of Tropical Medicine," 407.

112. James G. Maddox et al., *The Advancing South: Manpower Prospects and Problems* (New York: Twentieth Century Fund, 1967), 69–74.

113. Maddox et al., *The Advancing South*, 66.

114. W. B. Brierly, "Malaria and Socio-Economic Conditions in Mississippi," *Social Forces* 23 (1944–1945): 457.

115. Maddox et al., *The Advancing South*, 20–21; Ladd Haystead and G. C. Fite, *The Agricultural Regions of the United States* (Norman: University of Oklahoma Press, 1955), 113–114.

116. Sam Jones and James Aswell, "The South Wants the Negro," *Negro Digest* 4 (February 1946): 5. Sam Jones had served as governor of Louisiana before he co-wrote this article.

117. U.S. Office of Scientific Research and Development, *Science, the Endless Frontier: A Report to the President on a Program for Postwar Scientific Research and Development, Vannevar Bush, Director* (Washington, D.C.: GPO, 1945), quote on 159.

118. Enoc P. Waters, "Atom City Black Belt," *Negro Digest* 4 (February 1946): 81.

119. Ibid. For an analysis of the problem of segregation at Oak Ridge from the 1940s through the 1960s, see Charles W. Johnson and Charles O. Jackson, *City behind a Fence: Oak Ridge, Tennessee, 1942–1946* (Knoxville: University of Tennessee Press, 1981), 111–115, 117–118, 173–175.

120. Fannie Cook, "An Atomic Approach to Racism," *Negro Digest* 4 (August 1946): 23.

121. Walter White, "Race in the Atom World," *Negro Digest* 6 (October 1948): 67. Also see Richard B. Gehman, "How Negroes Live in Atom City," *Negro Digest* 6 (October 1948): 4–8.

122. Maddox et al., *The Advancing South*, 124–125; P. M. Hauser, "Demographic Factors in the Integration of the Negro," in *The Negro American*, ed. Talcott Parsons and K. B. Clark (Boston: Beacon, 1970), 76.

CHAPTER 6   *Alliances and Utopias*

1. For details of the U.S.-Liberia military treaties, see notes 3 and 4 below.

2. Franklin D. Roosevelt, "The Seven Hundred and Eighty-First Press Conference (Excerpts), November 3, 1941," in *Public Papers and Addresses of Franklin D. Roosevelt, 1941*, vol. 10, compiled by S. I. Rosenman (New York: Harpers, 1943), 463.

3. Dwight D. Eisenhower to Lesley James McNair, "Memo for General Headquarters, Subject: Directive for War Plan Black," February 25, 1942, in *The Papers of David Dwight Eisenhower*, vol. 1, ed. A. D. Chandler (Baltimore: Johns Hopkins University Press, 1970), 137–140; J. K. Penfield, "The Role of the United States in Africa: Our Interests and Operations," *Department of State Bulletin* 40, no. 1041 (June 8, 1959): 843, 845; R. W. Howe, *Along the Afric Shore: An Historic Review of Two Centuries of U.S.-African Relations* (New York: Barnes & Noble, 1975), 94; Weischhoff, "Liberia in the World Today," 498–499; R. A. Smith, *The American Foreign Policy in Liberia* (Monrovia: Providence Publications, 1972), 66–69. The accord was "U.S. Executive Agreement Series 275."

4. Ibid. The name of the agreement was "Principles Applying to Mutual Aid for Defense, U.S. Executive Agreement Series 324."

5. Merran Fraenkel, *Tribe and Class in Monrovia* (London: International African Institute/Oxford University Press, 1964) 28, 29; Sir Andrew McFadyean, ed., *The History of Rubber Regulation, 1934–1943* (London: International Rubber Regulation Committee/George Allen & Unwin, 1944), 12; H. B. Cole, ed., *The Liberian Year Book 1956* (London: Diplomatic Press, [1956]), 92.

6. T. T. Mackie and B. Sonnenberg, "Tropical Disease Problems among Veterans of World War II," *American Journal of Tropical Medicine* 29 (1949): 443.

7. R. M. Prothero, *Migrants and Malaria* (London: Longmans, Green, 1965), 8, 19, 23, 102, 106, 109; Gordon Harrison, *Mosquitoes, Malaria, and Man: A History of the Hostilities since 1880* (London: John Murray, 1978), 239–249.

8. U.S. Army, Medical Department, *Preventive Medicine in World War II,* vol. 6, *Communicable Diseases: Malaria* (Washington, D.C.: Office of the Surgeon General, 1963), 303, 308–309.

9. Ibid., 21–22; U.S. Army, Medical Department, *Preventive Medicine in World War II,* vol. 8, *Civil Affairs/Military Government Public Health Activities* (Washington, D.C.: Office of the Surgeon General, 1976), 220; Fraenkel, *Tribe and Class in Monrovia,* 29.

10. U.S. Army, *Preventive Medicine in World War II,* vol. 8, 226, 236.

11. "The First Inaugural—1944: Tubman's Promise of Progress," in "President William V. S. Tubman: Pace-Setter for a Democratic and Stable Government in Africa," *New York Times,* November 27, 1966, Special Section 11, p. 9.

12. Quote from President Tubman's graduation address to the College of West Africa, Monrovia, 1950, cited by Robert A. Smith, in his *William V. S. Tubman: The Life and Work of an African Statesman* (Monrovia, 1966; second ed., Amsterdam: Van Ditmar, 1967), 23–24, in 2nd ed.

13. Smith, *William V. S. Tubman,* Tubman quote on 103, Smith quote on 126.

14. C. C. Adams, "Christianity and Education Bringing Light to Liberia," *Pittsburgh Courier,* August 11, 1945.

15. Ibid.

16. R. E. Anderson, *Liberia: America's African Friend* (Chapel Hill: University of North Carolina Press, 1952), 197–210; A. F. Cane, "Education: Cornerstone of National Reconstruction," in "President William V. S. Tubman: Pace-Setter for a Democratic and Stable Government in Africa," *New York Times,* November 27, 1966, Special Section 11, pp. 22, 29. The National Baptist Convention was comprised of nearly 4 million black Baptists.

17. R. O'Hara Lanier, "The Problem of Mass Education in Liberia," *Journal of Negro Education* 30 (1961): 254–255, 258. Tubman's comment was at his first inaugural address in 1944 (ibid., 254–255).

18. George Rosen, *From Medical Police to Social Medicine: Essays on the History of Health Care* (New York: Science History Publications, 1974), 142–158; Bruno Latour, *The Pasteurization of France* (Cambridge, Mass.: Harvard University Press, 1988).

19. L. S. Senghor, "William Tubman: The Building of a Nation," introduction to *The New Liberia: A History and Political Survey,* by L. A. Marinelli (New York: Praeger, 1964), 4.

20. Smith, *The American Foreign Policy in Liberia,* 74.

21. W. V. S. Tubman, "Speech Delivered at the Dedication of Phoebe Hospital, Suacoco, Liberia, 16 May 1965," in *The Official Papers of William V. S. Tubman President*

of the Republic of Liberia, Covering Address, Messages, Speeches, and Statements, 1960–1967 (London: Longmans Green, 1968), 440.

22. W. V. S. Tubman, "Address Delivered by President Tubman during the Dedication of the St. Joseph Hospital and Medical College, Monrovia, 19 March 1967," in *The Official Papers of William V. S. Tubman, 1960–1967*, 446.

23. Ibid., 446–447.

24. In contrast to Tubman, leaders of other non-Western nations, leaders such as Jawaharlal Nehru and Kwame Nkrumah, focused on the need for their domestic technological resources to "catch up" with those of the modern West, so that foreign control or dependence could be minimized. Although this idea was excoriated in U.S. and Western foreign policy circles during the Cold War era, it is an emphasis now advanced by many in the Western foreign-affairs intellectual and policy-maker community. See, for example, Martin Fransman, *Technology and Economic Development* (Boulder, Colo.: Westview, 1986), 67–69; and N. Rosenberg, *Inside the Black Box: Technology and Economics* (Cambridge, U.K.: Cambridge University Press, 1982).

25. J. B. West, "United States Health Missions in Liberia," *Public Health Reports* 63 (October 15, 1948): 1352–1353.

26. Ibid., 1352.

27. For an overview of this practice throughout Liberia's history, see Amos Sawyer, *The Emergence of Autocracy in Liberia: Tragedy and Challenge* (San Francisco: ICS Press, 1992).

28. "The African as Wage-Earner," *Negro Year Book: A Review of Events Affecting Negro Life, 1941–1946*, ed. J. P. Guzman (Tuskegee: Tuskegee Institute, Alabama, 1947), 561–562. On the Tubman administration's dislike for unions, see R. W. Clower et al., *Growth without Development: An Economic Survey of Liberia* (Evanston: Northwestern University Press,1966), 282–284; Dew Tuan-Wich Mayson and Amos Sawyer, "Labour in Liberia," *Review of African Political Economy* 14 (January–April 1979): 3–14.

29. "The African as Wage-Earner," 561. Among the strikers, the Liberian worker hired as the public relations officer was Morris Massaquoi.

30. Mayson and Sawyer, "Labour in Liberia," 8; R. U. McLaughlin, *Foreign Investment and Development in Liberia* (New York: Praeger, 1966), 105–108.

31. A full analysis of Tubman's administration and domestic development is outside the scope of this book. For a balanced assessment see Sawyer, *The Emergence of Autocracy in Liberia*. Sawyer is a prominent Liberian scholar and former political activist. For a scathing, short, and short-sighted critique of Tubman's rule, see R. J. Jackson and C. G. Rosberg, "William Tubman and William Tolbert," in their *Personal Rule in Black Africa: Prince, Autocrat, Prophet, Tyrant* (Berkeley: University of California Press, 1982), 112–115.

32. D. E. Dunn, *The Foreign Policy of Liberia during the Tubman Era, 1944–1971* (London: Hutchinson Benham, 1979), 55.

33. R. E. Blount, "Malaria: An Unsolved Military Medical Problem," *American Journal of the Medical Sciences* 247 (1964): 407; West, "United States Health Missions in Liberia," 1352; Anderson, *Liberia: America's African Friend*, 270–275.

34. J. L. Brand, "The United States Public Health Service and International Health, 1945–1950," *Bulletin of the History of Medicine* 63 (1989): 596; Anderson, *Liberia: America's African Friend*, 268–269.

35. "Information Re Proposed Trip of U.S. Public Health Service Employees to Liberia, April 30, 1945," U.S. State Department, National Archives, RG 59, box 7138, typescript, 1 page, includes first quote; Loy W. Henderson [Director, Office of Near Eastern and African Affairs] to Dean Acheson [AA], June 25, 1945, ibid., includes second quote.

36. John B. West [Director, U.S. Public Health Service Mission to Liberia], "U.S. Public Health Service Monrovia Sub-Station, Monthly Report, March 1945," mimeo dated April 17, 1945, 9 pages, U.S. State Department, National Archives, RG 59, box 7138.

37. West, "United States Health Missions in Liberia," quote on 1352.

38. West, "U.S. Public Health Service Monrovia Sub-Station."

39. West, "United States Health Missions in Liberia," 1354–1355; "U.S. Public Health Service, Malaria Control Office—Monrovia Station, Entomological Survey of Bushrod Island, May 6, 1947," 4 pages, mimeo, U.S. State Department, National Archives, RG 59, box 7138. U.S. Federal Security Agency, *Annual Report of the Federal Security Agency, 1950* (Washington, D.C.: GPO, n.d.), 166.

40. Adell Patton Jr., "Howard University and Meharry Medical Schools in the Training of African Physicians, 1868–1978," in *Global Dimensions of the African Diaspora,* ed. J. E. Harris (Washington, D.C.: Howard University Press, 1982), 142, 159 n. 1; Brand, "The United States Public Health Service and International Health," 596.

41. Mackie and Sonnenberg, "Tropical Disease Problems among Veterans of World War II," 444. Hookworm, filariasis, sandfly fever, scrub typhus, amebic dysentery, and schistosomiasis also were significant problems for U.S. armed forces.

42. Louis Weinsten and M. J. Barza, "Bacterial Infections," in *The Health Horizons,* ed. H. Weschler, J. Gurin, and G. F. Cahill Jr. (Cambridge: Harvard University Press, 1978), 65–78; S. S. Hughes, *The Virus: A History of the Concept* (New York: Science History Publications, 1977), 84–97.

43. U.S. Department of the Army, Foreign Areas Studies Division, *U.S. Army Area Handbook for Liberia* (Washington, D.C.: GPO, 1964), 102; National Academy of Sciences, National Research Council, Division of Medical Sciences, *Tropical Health: A Report on a Study of Needs and Resources,* Publication no. 996 (Washington, D.C.: National Academy of Sciences, National Research Council, 1962), 272; Cole, *Liberian Year Book 1956,* 65; R. M. Fox, "*Anopheles Gambiae* in Relation to Malaria and Filariasis in Coastal Liberia," *American Journal of Tropical Medicine and Hygiene* 6 (1957): 601.

44. "Liberia, Scene of New International Institute for Tropical Medicine," *American Journal of Tropical Medicine* 27 (1947): 517, includes quote; Cole, *Liberian Year Book 1956,* 65–66.

45. Mackie cited in "Liberia, Scene of New International Institute," *American Journal of Tropical Medicine,* 517, includes quote.

46. U.S. Department of the Army, *U.S. Army Area Handbook for Liberia,* 102; National Academy of Sciences, National Research Council, Division of Medical Sciences, *Tropical Health: A Report on a Study of Needs and Resources,* Publication no. 996 (Washington, D.C.: National Academy of Sciences, National Research Council, 1962), 272; Fox, "*Anopheles Gambiae* in Relation to Malaria and Filariasis in Coastal Liberia," 601. Other important researchers at the institute included Jean Walker Fox, an entomologist and wife of Richard K. Fox; and Max J. Miller, a

parasitologist (Cole, *Liberian Year Book 1956*, 65–66). In addition to the Fox study, published research by institute staff included R. S. Bray, "*Plasmodium Ovale* in Liberia," *American Journal of Tropical Medicine and Hygiene* 6 (1957): 961–970; and R. C. Muirhead-Thomson, "The Malarial Infectivity of an African Village Population to Mosquitoes (*Anopheles Gambiae*)," ibid., 971–979.

47. Cole, *Liberian Year Book 1956*, 65.

48. Gordon Gaskill, "Anybody Can Make Money in Liberia," *Negro Digest* 7 (November 1948): 77. This article originally appeared in *American Magazine*.

49. H. A. Poindexter, "A Laboratory Epidemiological Study of Certain Infectious Diseases in Liberia," *American Journal of Tropical Medicine* 29 (1949): 435–442.

50. Ibid., quote on 441.

51. Smith, *The American Foreign Policy in Liberia*, 73–77.

52. West, "United States Health Missions in Liberia," 1364. By 1950, West had left the U.S. Public Health Service and settled permanently in Monrovia, running a pharmacy business. Poindexter took charge of the PHML. Anderson, *Liberia: America's African Friend*, 269.

53. Marinelli, *The New Liberia*, 79–81.

54. U.S. Department of the Army, *U.S. Army Area Handbook for Liberia*, 106.

55. Franz Schurmann, *The Logic of World Power: An Inquiry into the Origins, Currents, and Contradictions of World Politics* (New York: Pantheon, 1974), 39–113; Kurt Gottfried and B. G. Blair, *Crisis Stability and Nuclear War* (New York: Oxford University Press, 1988).

56. For an overview of these developments, and the historical studies on these particular aspects of U.S. international relations, see E. B. Skolnikoff, *The Elusive Transformation: Science, Technology, and the Evolution of International Politics* (Princeton, N.J.: Princeton University Press, 1993); Catherine Gwen, *U.S. Role in the World Bank, 1945–1992* (Washington, D.C.: Brookings Institution, 1994); Javed Siddiqi, *World Health and World Politics: The World Health Organization and the UN System* (Columbia: University of South Carolina Press, 1995).

57. Keith Pavitt and Michael Worboys, *Science, Technology, and the Modern Industrial State* (London: Butterworths, 1977), 54.

58. Seymour M. Lipset and Reinhard Bendix, *Social Mobility in Industrial Society* (Berkeley: University of California Press, 1964), 58; S. P. Huntington, "Power, Expertise, and the Military Profession," in *The Professions in America*, ed. K. Lynn and the editors of *Daedalus* (Boston: Beacon, 1965), 131–153.

59. J. M. Henderson, "The Present Status of Sanitary Engineering in the Tropics," *American Journal of Tropical Medicine and Hygiene* 6 (January 1957): 1–3; U.S. Department of State, Department of Defense, International Cooperation Administration, "Background for Mutual Security," in *The Mutual Security Program, Fiscal Year 1959: A Summary Presentation, February 1958* (Washington, D.C.: GPO, 1958), 3–28.

60. Jonathan B. Bingham, *Shirt-Sleeve Diplomacy: Point 4 in Action* (New York: John Day, 1954), includes Truman quote on 10

61. Ibid.; Henderson, "Sanitary Engineering in the Tropics," 8–9; Smith, *The American Foreign Policy in Liberia*, 78–79.

62. Clower et al., *Growth without Development*, 360–375; U.S. Federal Security Agency, *Annual Report for the Federal Security Agency*, 166. A detailed breakdown of this U.S. assistance is provided by Clower et al., *Growth without Development*.

63. For a description of a WHO malaria control project in a Liberia site, Kpain, see Prothero, *Migrants and Malaria*, 109–110. On other aid from WHO as well as UNESCO and FAO in Liberia, see Henry B. Cole, ed., *The Liberian Yearbook for 1962* (Liberian Review Publication, 1962 [1963?]), 154, 158, 166–168; R. O. Lanier, "The Problem of Mass Education in Liberia," *Journal of Negro Education* 30 (1961): 255, 257; Republic of Liberia, National Planning Agency, Office of National Planning, Bureau of Statistics, *Revenues of the Government of Liberia, 1939–1962* (Monrovia, July 15, 1963), 9, 18; American University, Foreign Area Studies, *Area Handbook for Liberia* (Washington, D.C.: GPO, 1972), 99–101; Clower et al., *Growth without Development*, 363, 364–367.

64. Clower et al., *Growth without Development*, 363.

65. Ibid., 360, 361; "Problems of Malaria Eradication in Africa," *WHO Chronicle* 17 (October 1963): 370; I. J. Wolman, "The Global Attack on Malaria," *Clinical Pediatrics* 2 (November 1963): 653–655.

66. For a list of these specific incentives, see Jerker Carlsson, *The Limits to Structural Change: A Comparative Study of Foreign Direct Investments in Liberia and Ghana, 1950–1971* (Uppsala: Scandinavian Institute of African Studies, 1981), 66–68. On the Open Door Policy in general, see, Martin Lowenkopf, *Politics in Liberia: The Conservative Road to Development* (Stanford: Hoover Institution Press, 1976), 43–58; M. B. Apkan, "Black Imperialism: Americo-Liberian Rule over the African Peoples of Liberia, 1841–1964," *Canadian Journal of African Studies* 7 (1973): 234–236; J. Mills Jones, "Development Planning, Politics, and the Bureaucracy: The Liberian Experience," *Liberian Studies Journal* 11 (1986): 3.

67. For a detailed history and description of these mining companies in Liberia, see Clower et al., *Growth without Development*, 117–125, 197–225; Carlsonn, *Limits of Structural Change*, 285–299; James T. Tarpeh, "The Liberian-LAMCO Joint Venture Partnerships: The Future of Less Developed Country and Multinational Corporation Collaboration as a National Strategy for Host Country Development," 2 vols. (Ph.D. diss., University of Pittsburgh, 1978).

68. Arthur J. Knoll, "Harvey Firestone's Liberian Investment (1923–1933)," *Liberian Studies Journal* 14 (1989): 13; Cole, *Liberian Year Book 1956*, 92. The Liberian government was able to negotiate the building of harbor facilities in Greenville. An agreement with the German-owned African Fruit Company, which operated a banana plantation near Sinoe, resulted in the opening of these harbor facilities in 1963. S. C. Saha, *A History of Agriculture in Liberia 1822–1970* (Lewiston, N.Y.: Edwin Mellen, 1990), 82.

69. Cole, *Liberian Yearbook for 1962*, 170–171; U.S. Department of the Army, *U.S. Army Area Handbook for Liberia* [1964], 304–305.

70. Smith, *William V. S. Tubman*, 142.

71. Carlsson, *Limits to Structural Change*, 99; Clower et al., *Growth without Development*, 132; Tarpeh, "Liberian-LAMCO Joint Venture Partnerships," 2.

72. Cole, *Liberian Yearbook for 1962*, 160; Tarpeh, "Liberian-LAMCO Joint Venture Partnerships," 205, 231–233, 325–328, 527; Clower et al., *Growth without Development*, 125. According to Tarpeh, LAMCO was charged some fees relating to internal transport of its iron ore.

73. Ibid.

74. Tarpeh, "Liberian-LAMCO Joint Venture Partnerships," 533.

75. Tarpeh, "Liberian-LAMCO Joint Venture Partnerships," 205; Clower et al., *Growth without Development*, 125; Carlsonn, *Limits of Structural Change*, 123–124.

76. Mary H. Moran, *Civilized Women: Gender and Prestige in Southeastern Liberia* (Ithaca, N.Y.: Cornell University Press, 1990), 167; Fraenkel, *Tribe and Class in Monrovia*, 43, 131–150, 178–184; Clower et al., *Growth without Development*, 312; Saha, *History of Agriculture in Liberia*, 107–110.

77. U.S. Agency for International Development, *The AID Story* (Washington, D.C.: The Agency, 1966), 36, 54.

78. Sawyer, *The Emergence of Autocracy in Liberia*, 257; Cole, *Liberian Yearbook for 1962*, 145; Clower et al., *Growth without Development*, 136. Prairie View is a historically black college.

79. Dunn, *The Foreign Policy of Liberia during the Tubman Era*, 41–43, quote on 43; Clower et al., *Growth without Development*, 363.

80. Cole, *Liberian Yearbook for 1962*, 161.

81. Ibid., 161–163; Republic of Liberia, Bureau of Statistics, Office of National Planning, *1962 Population Census of Liberia/Population Characteristics of Major Areas* [c. 1964]. This census lists 148 physicians in Montserrado County and 28 in Grand Bassa County (ibid., sec. 2, p. 14, and sec. 5, p. 13).

82. L. L. Alexander, "Hospitals of West Africa: Liberia, Ghana, and Nigeria," *Journal of the National Medical Association* 65 (1973): 373–374.

83. Ibid.; American University, *Area Handbook for Liberia*, 102; Clower et al., *Growth without Development*, 160.

84. McLaughlin, *Foreign Investment and Development*, 104.

85. Tubman, "Speech Delivered at the Dedication of Phoebe Hospital," 440.

86. Cole, *Liberian Yearbook for 1962*, 154. See also American University, *Area Handbook for Liberia* [1972], 101.

87. Clower et al., *Growth without Development*, 337–357, quote on 346.

88. J. M. May, *Ecology of Malnutrition in Eastern Africa and Four Countries of West Africa, Studies in Medical Geography*, vol. 9 (New York: Hefner, 1970), 69–70, includes quote; U.S. Department of the Army, *U.S. Army Area Handbook for Liberia*, 113–116.

89. Clower et al., *Growth without Development*, 337–357.

90. Cole, *Liberian Yearbook for 1962*, 148–150, 161–163. One Liberian medical student in 1962 was studying in Egypt.

91. Sawyer, *The Emergence of Autocracy in Liberia*, 261.

92. Dunn, *The Foreign Policy of Liberia during the Tubman Era*, 43.

93. V. W. Henderson, "Region, Race, and Jobs," in *Employment, Race, and Poverty*, ed. A. M. Ross and Herbert Hill (New York: Harcourt, Brace & World, 1967), 83.

94. NATO, the North Atlantic Treaty Organization, had fifteen members in its early decades of operation: the United States, Canada, Iceland, Norway, the United Kingdom, Denmark, Belgium, Luxembourg, Portugal, France, Italy, Spain, Greece, Turkey, and West Germany.

CHAPTER 7   *Technology of Reaction*

1. Pierre Noel, "Recent Research in Public Health in Haiti," in *The Haitian Potential—Research and Resources of Haiti*, ed. Vera Rubin and R. P. Schnaedel (New

York: Teachers College Press, 1975), 162; Remy Bastien, "The Negro in Haiti," in *Negro Year Book: A Review of Events Affecting Negro Life*, ed. J. P Guznam (Atlanta: Tuskegee Institute, 1947), 620 (Bastien was the secretary of the Haitian Bureau of Ethnology); J. H. Paul and A. Bellerive, "A Malaria Reconnaissance of the Republic of Haiti," *Journal of the National Malaria Society* 6 (March 1947): 65.

2. Ibid.

3. Bastien, "The Negro in Haiti," 618, 620–621; Camille Lhérisson, "Diseases of the Peasants of Haiti," *American Journal of Public Health* 25, no. 8 (1935); Selden Rodman, *Haiti: The Black Republic* (New York: Devin-Adair, 1954), 56; François Duvalier, "Contribution à l'étude du pian en Haïti: L'Aspect médico-social et l'oeuvre de la Mission Sanitaire Américaine," *Union Médical du Canada* 74 (June 1945): 782.

4. I. L. Kandel, ed., *Educational Yearbook of the International Institute of Teachers College, Columbia University, 1942: Education in the Latin American Countries* (New York: Teachers College, 1943), 238–239, quote on 238; T. E. Weil et al., *Area Handbook for Haiti* (Washington, D.C.: Foreign Area Studies, American University, 1973), 86.

5. John Biesanz, "The Economy of Panama," *Inter-American Economic Affairs* 6 (summer 1952): 3–28, quotes of leader and 1946 embassy report are on 3 and 4.

6. U.S. Army, Medical Department, *Preventive Medicine in World War II*, vol. 6, *Communicable Diseases: Malaria* (Washington, D.C.: Office of the Surgeon General, 1963), 173, 236; R. D. Heinl Jr. and N. G. Heinl, *Written in Blood: The Story of the Haitian People, 1492–1971* (Boston: Houghton Mifflin, 1978), 539–543.

7. The American Foundation, *Medical Research: A Midcentury Survey,* vol. 1, *American Medical Research in Principle and Practice* (Boston: Little, Brown, 1955), 168.

8. The American Foundation, *Medical Research: A Midcentury Survey,* vol. 2, *Unsolved Clinical Problems: In Biological Perspective* (Boston: Little, Brown, 1955), 6.

9. L. E. Cluff, "America's Romance with Medicine and Medical Science," *Daedalus* 115 (spring 1986): 137–159, quote on 139.

10. G. E. Simpson, "Haitian Politics," *Social Forces* 20 (May 1942): 489, includes quote. The Haitian intellectuals' disdain for U.S. heavy-handedness is illustrated by J. A. Porter in his piece, "Picturesque Haiti," *Opportunity* 24, no. 4 (October–December 1946), 178–180, 210.

11. Ludwell L. Montague, *Haiti and the United States, 1714–1938* (1940; reprint, New York: Russell & Russell, 1966), 282, in reprint; Bastien, "The Negro in Haiti," 626; Rodman, *Haiti: The Black Republic*, 27–28; M. H. Dorsinville, "Haiti and Its Institutions: From Colonial Times to 1957," in *The Haitian Potential*, ed. Rubin and Schnaedel, 211; Anne Greene, *The Catholic Church in Haiti: Political and Social Change* (East Lansing: Michigan State University Press, 1993), 65 n. 112, 106–107, 202.

12. *Haiti, Pioneer of Freedom*, pamphlet, 9 pages, (Washington: Coordinator of Inter-American Affairs, July 1944); Dantes Bellegarde, "Haiti's Voice at the Peace Table," *Opportunity* 21 (October 1943): 154; Bastien, "The Negro in Haiti," 621, 628.

13. C. I. Bevans, comp., *Health and Sanitation Program—June 29 and July 12, 1944; September 25 and 27, 1947, Treaties and Other International Agreements of the U.S.A., 1776–1949* (Washington, D.C.: GPO/Department of State Publication,

1971), 774–775, 817–823; United Nations, Mission to Haiti, *Report of the United Nations Mission of Technical Assistance to the Republic of Haiti* ([Lake Success, N.Y.: United Nations,] 1949), 67; F. Duvalier, "La Valeur de la pénicilline dans le traitement du pian, en Haïti," *Union Médicale du Canada* 77 (January 17, 1948): 17; J. H. Dwindelle et al., "Preliminary Report on the Evaluation of Penicillin in the Treatment of Yaws," *American Journal of Tropical Medicine* 26 (May 1946): 311–318; Rodman, *Haiti: The Black Republic*, 55; Bastien, "The Negro in Haiti," 621.

14. P. L. Perine et al., *Handbook of Endemic Treponematoses: Yaws, Endemic Syphilis, and Pinta* (Geneva: World Health Organization, 1984), 3, 6; H. F. Dowling, *Fighting Infection: Conquests of the Twentieth Century* (Cambridge, Mass.: Harvard University Press, 1977).

15. Duvalier has been the subject of several biographies, political studies, and journalists' exposés. Best among the English-language literature are James Ferguson, *Papa Doc, Baby Doc: Haiti and the Duvaliers* (Oxford, U.K.: B. Blackwell, 1987); Elizabeth Abbot, *Haiti, the Duvaliers, and Their Legacy* (New York: McGraw-Hill, 1988); and Bernard Diederich and Al Burt, *Papa Doc: Haiti and Its Dictator* (Maplewood, N.J.: Waterfront Press, 1991). Some details of Duvalier's medical career can be gleaned from a volume of his collected works, *OEuvres essentielles* (3 ed.; revu, augmente et corrige. Port-au-Prince: Presses nationales d'Haiti, 1968). See especially "Curriculum Vitae du Docteur François Duvalier Président à vie de la République d'Haïti," ibid., 8–9; and the materials in the section, "Etudes Médicales," ibid., 473–601.

16. Duvalier, "Contribution à l'étude du pian en Haïti," 781–798; Duvalier, "La Valeur de la pénicilline dans le traitement du pian," 17.

17. Duvalier, "La Valeur de la pénicilline dans le traitement du pian," 17, includes quote. On some of Duvalier's work in this campaign, see Robert I. Rotberg with Christopher K. Clague, *Haiti: The Politics of Squalor* (Boston: Houghton Mifflin, 1971), 178–179.

18. C. R. Rein, "Diagnostic Aids in Mass-Treatment Campaigns against Yaws," in *First International Symposium on Yaws Control, Bangkok, 1952*, World Health Organization, Monograph Series (Geneva: WHO, 1953), 245–254. Rein was a professor at New York University Postgraduate Medical School.

19. Rene Dubos, *The Mirage of Health: Utopias, Progress, and Biological Change* (Garden City, N.Y.: Anchor, 1959), 104–109. Dubos was an internationally prominent biologist at Rockefeller University. His research focused on tropical diseases, comparative pathology, and public health.

20. For example, no information on indicators referring to work injuries or violent deaths appear in the Agency for International Development's annual reports on Haiti. These reports appeared in its annual publication begun in 1967: Agency for International Development, *A. I. D. Economic Data Book: Latin America* (Washington: The Agency, 1967).

21. Edouard Petrus et al., "La campagne antipianique en Haiti," in *First International Symposium*, World Health Organization, 261–271; Scaha Levitan et al., "The Treatment of Infectious Yaws with One Injection of Penicillin," ibid., 55–75. Petrus was a physician with the Haitian Public Health Service; Levitan was a physician with WHO.

22. G. E. Samame, "Treponematosis Eradication with Special Reference to Yaws Eradication in Haiti," *Bulletin of the World Health Organization* 15 (1956): 906, includes quote. Samame was a physician and representative of the Pan American Sanitary Bureau in the WHO Regional Office of the Americas. The office was located in Mexico City.

23. J. M. Hunter and W. H. Knowles, "Ten Problems of Point Four," *Inter-American Economic Affairs* 7 (summer 1953): 65

24. Arturo Escobar, "Power and Visibility: Development and the Invention and Management of the Third World," *Cultural Anthropology* 3 (1988): 428–443; Catherine Owen, *U.S. Relations with the World Bank, 1945–1992* (Washington, D.C.: Brookings Institution, 1994); D. A. Rondinelli, *Development Administration and U.S. Foreign Aid Policy* (Boulder, Colo.: Lynne Rienner, 1987), 17–28.

25. "Survey Shows How Point Four Program Tackles Basic Problems in Under-developed East," *New York Times*, January 12, 1953, 10–11, quote on 10.

26. J. K. Gailbraith, *The New Industrial State* (New York: New American Library, 1967).

27. The phrase "chaos or Castroism" is coined from a remark by Senator Stephen M. Young (Ohio) criticizing U.S. military aid to President Duvalier. He was quoted in an editorial, "Travesties of Foreign Aid," *The Nation* 194 (June 23, 1964): 546.

28. Leslie F. Manigat, *Haiti of the Sixties, Object of International Concern* (Washington, D.C.: Washington Center of Foreign Policy Research, 36–39; Rodman, *Haiti: The Black Republic*, 28; Rotberg, *Haiti: The Politics of Squalor*, 172. Manigat was a Haiti-based political scientist.

29. Cary Hector, Claude Moise, and Emile Ollivier, eds., *1946–1976: Trente Ans de Pouvoir Noir en Haïti: Tome Premier/L'Explosion de 1946—Bilan et Perspectives* (Lassalle, Canada: Collectifs Paroles, 1976).

30. Maurice Dartigue, *An Economic Program for Haiti: A Special Report [for] the Institute for Inter-American Affairs, Food Supply Division* ([Washington, D.C.:] The Institute, [1946]), 20; H.L.T. Koren to I. S. Falk, October 10, 1950, Falk Papers, Sterling Library Archives, Yale University, New Haven, Conn. (hereafter abbrev. Falk Papers); Bastien, "The Negro in Haiti," 630.

31. *Who Was Who in America with World Notables*, vol. 8, *1982–1985* (Chicago: Marquis Who's Who, Inc., 1985), 132.

32. Koren to Falk, October 10, 1950, Falk Papers; I. S. Falk, *Report on the Social Insurance Program in Haiti* (n.p., April 1951), 1. This report was translated and published in Haiti as *Rapport sur le régime d'assurance sociale d'Haïti* [on cover:] (Institut d'assurances sociales d'Haïti [S. I.: s.n. , 1951]).

33. I. S. Falk, "Memorandum of conference with employers," October 26, 1950, Falk Papers; I. S. Falk to Ambassador De Courcy, October 31, 1950, Falk Papers.

34. I. S. Falk to A. J. Altmeyer/W. L. Mitchell, October 21, 1950, Falk Papers, pp. 1, 2, emphasis by Falk; Falk, *Report on the Social Insurance*, 14.

35. Auguste Denize [President of the Haitian Medical Association], *La Crise médicale: A la recherche d'une Solution* (Port-au-Prince: Bibliothèque de l'Association Médicale Haïtienne, [1948]), 12–13.

36. Falk, *Report on Social Insurance*, 10.

37. United Nations, *Report of the United Nations Mission of Technical Assistance to the Republic of Haiti*, 31–32. The agricultural sector contained about 84 percent of Haiti's total workforce.

38. Ibid., 15, 130; U.S. Department of Labor, Bureau of International Labor Affairs/ Bureau of Labor Statistics, "Haiti," in *Directory of Labor Organizations in the Western Hemisphere*, vol. 2 (U.S. Bureau of International Labor Affairs, 1964), looseleaf, ii.

39. Falk, *Report on Social Insurance*, 130–133.

40. Inter-American Development Bank, *Socio-Economic Progress in Latin America: Social Progress Trust Fund, Tenth Annual Report, 1970* (Washington, D.C.: Inter-American Development Bank, 1971), 247; O. Ernest Moore, *Haiti: Its Stagnant Society and Shackled Economy: A Survey* (New York: Exposition, [1972]), 191. In 1965 the National Office of Old Age Insurance was established and incorporated into IDASH. At the end of 1970, about 28,000 persons from 2,500 employment sites were receiving accident insurance under the IDASH program. About 1,950 employees were receiving allowances and 2,000 were receiving medical benefits. By late 1970, some 33,000 participants were getting old age insurance. (Ibid.)

41. UNESCO, *The Haiti Pilot Project: Phase One, 1947–1949—Monographs on Fundamental Education* (Paris: UNESCO, 1951), 58, includes quote; Rotberg, *Haiti: The Politics of Squalor*, 172–173.

42. G. R. Smucker, "Peasant Councils and the Politics of Community," in *Politics, Projects, and People: Institutional Development in Haiti*, ed. D. W. Brinkerhoff and Jean-Claude Garcia-Zamor (New York: Praeger, 1986), 98; Rotberg, *Haiti: The Politics of Squalor*, 172–173.

43. Smucker, "Peasant Councils and the Politics of Community," 98.

44. R. W. Logan, "The United States Mission in Haiti, 1915–1952," *Inter-American Economic Affairs* 6 (spring 1953): 26–28.

45. United Nations, *Report of the United Nations Mission of Technical Assistance to the Republic of Haiti*, 64–66, 73–75; Marian Neal, "United Nations Technical Assistance Programs in Haiti," *International Conciliation*, no. 468 (February 1951), 105; Rotberg, *Haiti: The Politics of Squalor*, 178. For international organizations' health assistance in the 1960s and 1970s, see Ary Bordes and Andrea Couture, *For the People, For a Change: Bringing Health to the Families of Haiti* (Boston: Beacon Press, 1978). Dr. Bordes is a Haitian physician active in public health projects and services for Haiti's poor communities for the past several decades.

46. "In Haiti: Reverence for Life" [editorial correspondence by H.E.F.?], *Christian Century* 74 (March 20, 1957): 351–352; E. Shocket, "The Mellon Mission in Haiti," *Florida Medicine* 55 (December 1968): 1098–1100; W. L. Bergen, "Administration and Evaluation of Rural Health Services," *American Journal of Tropical Medicine and Hygiene* 23 (1974): 936–937. Mennonites served as lab technicians.

47. Ibid.

48. C. E. Wallace, F. N. Marshall, and C. Robinson, "A Retrospective Review of the Maternal and Child Health Services at the Albert Schweitzer Hospital in Rural Haiti," *Bulletin of the Pan American Health Organization* 16 (1982): 7–16; Barry Paris, "Song of Haiti: The State of Haiti's Health Spurred a Pittsburgh Mellon to Action," *Pittsburgh Post-Gazette*, March 7, 1993, includes quotes.

49. "Haiti," *Hispanic American Report: An Analysis of Developments in Spain, Portugal, and Latin America* 10 (March 1957): 129.

50. See, for example, Mats Lundahl, *Peasants and Poverty: A Study of Haiti* (New York: St. Martin's, 1979); M. S. Laguerre, *Voodoo and Politics in Haiti* (New York: St.

Martin's, 1989); R. I. Rotberg, ed., *Haiti Renewed: Political and Economic Prospects* (Washington, D.C.: Brookings Institution Press/World Peace Foundation, 1997). Also, see the works cited in note 15 above.

51. A complete official list of Duvalier's scholarly publications on African ethnography in Haiti, as well as his medical articles, is found in his collection '*OEuvres essentielles*. For a systematic study of Duvalier's political ties to Voodoo practitioners, see M. Laguerre, *Voodoo and Politics* (New York: St. Martin's, 1989), 101–120.

52. This pattern has been cogently identified and documented in the works of the political historian Barrington Moore Jr. See, for example, his early, still outstanding studies, *Political Power and Social Theory: Six Studies* (Cambridge: Harvard University Press, 1958); and *Social Origins of Dictatorship and Democracy: Lord and Peasant in the Making of the Modern World* (Boston: Beacon, 1966).

53. M. S. Laguerre, *The Military and Society in Haiti* (Knoxville: University of Tennessee Press, 1993), 110–123, 136, 154–155, 174; Laguerre, *Voodoo and Politics*.

54. Lundahl, *Peasants and Poverty*, 343.

55. Robert Proctor, *Racial Hygiene: Medicine under the Nazis* (Cambridge, Mass.: Harvard University Press, 1988); R. J. Lifton, *The Nazi Doctors: Medical Killing and the Psychology of Genocide* (New York: Basic Books, 1986); R. J. Lifton and Eric Makusen, *The Genocidal Mentality: Nazi Holocaust and Nuclear Threat* (New York: Basic Books, 1990).

56. Jeffrey Herf, *Reactionary Modernism: Technology, Culture, and Politics in Weimar and the Third Reich* (New York: Cambridge University Press, 1984), quote is on 224.

57. Dagmar Engels and Shula Marks, "Introduction: Hegemony in a Colonial Context," in *Contesting Colonial Hegemony: State and Society in Africa and India* (London: British Academic Press/German Historical Institute, 1994), ed. Engels and Marks, 1–15.

58. F. Duvalier, "Considerations sur cent cinquate ans d'évolution du régime alimentaire dans le proleterariat urbain et rural en Haiti," *Extrait du Bulletin de l'Association des Médicins de Langue Française du Canada,* L'Union Médicale du Canada, Tome 82 (653), June 1953, reprinted in Duvalier, *OEuvres essentielles*, 584.

59. Duvalier, "Contribution à l'étude du pian en Haïti [1945]," version reprinted in his *OEuvres essentielles*, 523.

60. *Hispanic American Report* was published monthly by the Hispanic American Studies program at Stanford University and the Hispanic American Society. Each issue carried a nation-by-nation account of domestic political affairs and crises.

61. "Haiti," *Hispanic American Report* 13 (September 1960): 611; "Haiti," *Hispanic American Report* 14 (May 1961): 410–411. See also S. Romualdi, "Haiti: Forgotten Dictatorship: [Francois] Duvalier Rules His Little Country with a Ruthless Hand," *New Leader* 43 (July 18–25, 1960): 15–16; and A. Kalbe, "Iron-Fisted 'Papa': Haitian Dictator's 'Re-election' Strengthens His Corrupt Rule over Impoverished Island," *New Leader* 44 (August 14–21, 1961): 15–16.

62. Statement by Dean Rusk was made in 1973 and is presented in Heinl and Heinl, *Written in Blood*, 622.

63. L. Ronald Scheman, "The Alliance for Progress: Concept and Creativity," in *The Alliance for Progress: A Retrospective,* ed. Scheman (New York: Praeger, 1988); 3–62, quote on 3.

64. Richard Levins and Richard Lewontin, *The Dialectical Biologist* (Cambridge, Mass.: Harvard University Press, 1985), 229, 234. G. Gutíerrez quoted by Levins and Lewontin, ibid., 228.

65. U.S. Agency for International Development, *The AID Story* (Washington, D.C.: The Agency, 1966), includes quote on 12.

66. U.S. Department of State, Mutual Security Agency, *Report to Congress on the Mutual Security Program for the Fiscal Year 1961* (Washington, D.C.: GPO, 1962), 6.

67. "Haiti," *Hispanic American Report* 14 (October, 1961): 894; "Haiti," *Hispanic American Report*, 15 (June 1962): 512–513.

68. Sidney W. Mintz, "The Employment of Capital by Market Women of Haiti," in *Capital, Saving, and Credit in Peasant Societies: Studies from Asia, Oceania, the Caribbean, and Middle America: Essays Edited with Two General Essays*, ed. Raymond W. Firth and B. S. Yamey (Chicago: Aldine, 1964), 256–257.

69. Ibid., 282.

70. Ibid., 283. See also Russ Symontowne, "Haiti Learns the 3 R's," *Negro Digest* 5 (October 1947): 25.

71. For the social science studies of coumbites published through the early 1970s, see J. M. May and D. L. McLellan, *The Ecology of Malnutrition in the Caribbean: The Bahamas, Cuba, Jamaica, Hispaniola (Haiti and the Dominican Republic), Puerto Rico, the Lesser Antilles, and Trinidad and Tobago*, vol. 12 of *Studies in Medical Geography* (New York: Hafner Press, 1973), 156–157.

72. UNESCO, *The Haiti Pilot Project: Phase One, 1947–1949—Monographs on Fundamental Education* (Paris: UNESCO, 1951), 17–18.

73. O. E. Moore, *Haiti*, 151–153; Lundahl, *Peasants and Poverty*, 557–574.

74. Inter-American Development Bank, *Social Progress Trust Fund, First Annual Report, 1961* (Washington, D.C.: Inter-American Development Bank, 1962), 133; Inter-American Development Bank, *Social Progress Trust Fund, Third Annual Report, 1963* (Washington, D.C.: Inter-American Development Bank, 1964), 275.

75. Inter-American Development Bank, *Social Progress Trust Fund, Fourth Annual Report, 1964* (Washington, D.C.: Inter-American Development Bank, 1965), 343; Inter-American Development Bank, *Social Progress Trust Fund, Fifth Annual Report, 1965* (Washington, D.C.: Inter-American Development Bank, 1966), 377, 383–385; "Haiti," *Hispanic American Report* 14 (December 1961): 1093. The Tripartite Mission was administered by the Organization of American States (OAS), the Inter-American Development Bank (IDB), and the UN Economic Commission for Latin America (ECLA).

76. Inter-American Development Bank, *Social Progress Trust Fund, Third Annual Report, 1963*, 275.

77. "Haiti," *Hispanic American Report* 13 (August 1960): 527–528; Lynn Grossberg, "Haiti: The Explosive Buffer," *The Nation* 191 (August 20, 1960): 83–85; "Most Aid to Haiti Suspended by the U.S.," *New York Times*, August 1, 1962; "Haiti," *Hispanic American Report* 15 (August 1962): 712–713; "Haiti," *Hispanic American Report* 15 (October 1962): 916–917: "Haiti," *Hispanic American Report* 15 (December 1962): 1110; "Haiti," *Hispanic American Report* 16 (August 1963): 771–722; O. E. Moore, *Haiti*, 202–203; Inter-American Development Bank, *Socio-Economic Progress in Latin America: Social Progress Trust Fund, Ninth Annual Report, 1969* (Washington, D.C.: Inter-American Development Bank, 1970), 355;

Inter-American Development Bank, *Socio-Economic Progress in Latin America: Social Progress Trust Fund, Tenth Annual Report, 1970*, 241; Inter-American Development Bank, *Economic and Social Progress in Latin America, Annual Report, 1972* (Washington, D.C.: Inter-American Development Bank, [1973?]), 219.

78. "Haiti," *Hispanic American Report* 16 (June 1963): 568; Inter-American Development Bank, *Socio-Economic Progress in Latin America, Tenth Annual Report, 1970*, 241; Weil et al., *Area Handbook for Haiti*, 63.

79. Weil et al., *Area Handbook for Haiti*, 63, includes quote.

80. Simon M. Fass, *Political Economy in Haiti: The Drama of Survival* (New Brunswick, N.J.: Transaction Books, 1988), 14.

81. "Haiti," *Hispanic American Report* 15 (October 1962): 916; "Haiti," *Hispanic American Report* 16 (February 1963): 134; "Haiti," *Hispanic American Report* 17 (March 1964): 228; Inter-American Development Bank, *Social Progress Trust Fund, Fifth Annual Report, 1965*, 378–379; Inter-American Development Bank, *Economic and Social Progress in Latin America, Annual Report, 1972*, 220.

82. For these criticisms of the mass-impact eradication programs, see L. J. Casazza and C. D. Williams, "Family Health versus Family Planning," *The Lancet*, no. 7805 (March 31, 1973): 712–713; Javed Siddiqi, *World Health and World Politics: The World Health Organization and the UN System* (Columbia: University of South Carolina, 1995), 152–160, 179–191. According to a national nutrition survey conducted in 1978, 73 percent of Haitian children under the age of five suffered from some form of malnutrition. George Bicego et al., "Trends, Age Patterns, and Differentials in Childhood Mortality in Haiti (1960–1987)," *Population Studies* 45 (1991): 235.

83. Inter-American Development Bank, *Social Progress Trust Fund, Second Annual Report, 1962* (Washington, D.C.: Inter-American Development Bank, 1963), 301–302.

84. Leslie F. Manigat, *Haiti of the Sixties, Object of International Concern: A Tentative Global Analysis of the Potentially Explosive Situation of a Crisis Country in the Caribbean* (Washington, D.C.: Washington Center of Foreign Policy Research, [1964]), 78.

85. Manigat, *Haiti of the Sixties*, 78, includes quote.

86. W. G. O'Neill, "No Longer a Pipe Dream," *Haiti Renewed*, ed. Rotberg, 199, includes quote; Heinl and Heinl, *Written in Blood*, 648–649, includes same quote.

87. Flore Zephir, *Haitian Immigrants in Black America: A Sociological and Sociolinguistic Portrait* (Westport, Conn.: Bergin & Garvey, 1996), 5; Heinl and Heinl, *Written in Blood*, 648. For detailed analysis of the Haitian immigrants in the United States, see the valuable work by Zephir (ibid.); as well as A. Stepick, "The Haitian Exodus: Flight from Terror and Poverty," in *The Caribbean Exodus*, ed. B. B. Levine (New York: Praeger, 1987), 131–151; Jake C. Miller, *The Plight of Haitian Refugees* (New York: Praeger, 1984); Michel S. Laguerre, *Diasporic Citizenship: Haitian Americans in Transnational America* (New York: St. Martin's Press, 1998). There has been no writing of similar substance on the modern Haitian immigrants who have settled in other parts of the Atlantic world.

88. "Liberian Agony Continues," *Africa News* 34 (September 17, 1990): 1–3, quote on 2; G. A. Cornia and Germano Mwabu, "Health Status and Health Policy in Sub-Saharan Africa: A Long-Term Perspective," working paper no. 141, United Nations

University World Institute for Development Economics Research, September 1997, 22. Internally displaced people are refugees who must flee their homes to other locations inside their country. They are referred to technically as "IDPS" populations by UN policy experts.

CONCLUSION   *Science and Hope*

1. Executive Office of the President, Office of Science and Technology, *Scientific and Educational Basis for Improving Health, Report of the Panel on Biological and Medical Science of the President's Science Advisory Committee* (Washington, D.C.: GPO, 1972), 18; R. Nixon, "The State of the Union, January 30, 1974," in *Public Papers of the Presidents of the United States: Richard Nixon, January 1 to August 9, 1974* (Washington, D.C.: GPO, 1975), 91; *Public Papers of the Presidents of the United States: Jimmy Carter, 1980–81, Book III—September 29, 1980 to January 20, 1981* (Washington, D.C.: GPO, 1981), 2961.

2. H. R. Isaacs, *The New World of Negro Americans: A Study for the Center for International Studies, Massachusetts Institute of Technology* (New York: Viking, 1963), 288.

3. T. R. Dunlap, *Nature and the English Diaspora: Environment and History in the United States, Canada, Australia, and New Zealand* (New York: Cambridge University Press, 1999).

4. See Daniel J. Boorstin, *Hidden History* (New York: Harper & Row, 1987).

5. For analysis of the different schools of thought regarding U.S. international dominance, see C. Newbury, "The Semantics of International Influence: Informal Empires Reconsidered," in *Imperialism, the State, and the Third World*, ed. Michael Twaddle (London: British Academic Press, 1992), 27–37; P. J. Schrader, *United States Foreign Policy toward Africa: Incrementalism, Crisis, and Change* (Cambridge, U.K.: Cambridge University Press, 1994).

6. It was maligned because it projected images of blacks possessing economic power. This upset southern whites more comfortable with the stereotype of the illiterate black sharecropper. The new black middle class was also criticized by black intellectuals and leftists because of its powerlessness inside the U.S. corporate economy and major political parties. This maelstrom of criticism is the basis of sociologist E. F. Frazier's still controversial study, *Black Bourgeoisie: The Rise of a New American Middle-Class in the United States* (New York: Collier, 1962).

7. M. L. King, Jr., "Facing the Challenge of a New Age," *Phylon* 18 (1957): 25–34; quote on 28.

8. Francis Fukuyama, *The End of History and the Last Man* (New York: Free Press, 1992), xiv.

Selected References

Abbot, Elizabeth. *Haiti, the Duvaliers and Their Legacy.* New York: McGraw-Hill, 1988.

Abercrombie, Thomas F. *History of Public Health in Georgia, 1733–1950.* Atlanta: n.p., 1951.

Abercrombie, Thomas F. "Malaria and Its Relation to the Economic Development of Georgia." *Emory University Quarterly* 10 (October 1954): 168–171.

Adas, Michael. *Machines and the Measure of Men: Science, Technology, and Ideologies of Western Domination.* Ithaca, N.Y.: Cornell University Press, 1989.

Ahluwalia, D.P.S. *Plantation and the Politics of Sugar in Uganda.* Kampala, Uganda: Fountain, 1995.

Allen, James S. *The Negro Question in the United States.* New York: International Publishers, 1936.

American Association for the Advancement of Science. *A Symposium on Human Malaria: With Special Reference to North America and the Caribbean Region.* Ed. F. R. Moulton. Washington, D.C.: The Association, 1941.

Anderson, Robert E. *Liberia, America's African Friend.* Chapel Hill: University of North Carolina Press, 1952.

Anderson, Robert. *Innovation Systems in a Global Context: The North American Experience.* Montreal: McGill-Queen's University, 1998.

Apkan, M. B. "Black Imperialism: Americo-Liberian Rule over the African Peoples of Liberia, 1841–1964." *Canadian Journal of African Studies* 7 (1973): 232–239.

Aptheker, Herbert, comp. and ed. *Writings by W.E.B. Du Bois in Periodicals Edited by Others.* Vol. 2, *1910–1934.* Millwood, N.Y.: Kraus-Thomson Organization, 1982.

———, ed. *Writings by W.E.B. Du Bois in Periodicals Edited by Others.* Vol. 4, *1945–1961.* Millwood, N.Y.: Kraus-Thomson Organization, 1982.

Arnold, David, ed. *Imperial Medicine and Indigenous Societies.* Manchester: Manchester University Press, 1988.

Ashford, Mahlon. "The Nature of Immunity to Malaria in Its Relationship to Anti-Malarial Therapy." *American Journal of Tropical Medicine* 16 (1936): 665–678.

Azikiwe, Nnamdi. *Liberia in World Politics.* London: Stockwell, 1934; Westport, Conn.: Negro Universities Press, 1970.

Bakker, J. I. Hans, ed. *The World Food Crisis: Food Security in Comparative Perspective.* Toronto: Canadian Scholars' Press, 1990.

Balch, Emily Greene, ed. *Occupied Haiti: Being the Report of a Committee of Six Disinterested Americans Representing Organizations Exclusively American [Who]*

Personally Studied Conditions in Haiti in 1926. 1927; New York: Negro Universities Press, 1969.

Barber, Bernard. *Science and the Social Order.* New York: Collier, 1970.

Barber, M. A., J. B. Rice, and J. Y. Brown. "Malaria Studies on the Firestone Plantation in Liberia, West Africa." *American Journal of Hygiene* 15 (May 1932): 601–633.

Barber, Marshall A. *A Malariologist in Many Lands.* Lawrence: University of Kansas Press, 1946.

Barrett, William. *The Illusion of Technique: A Search for Meaning in a Technological Civilization.* Garden City: Anchor Press/Doubleday, 1978.

Beardsley, Edward H. *A History of Neglect: Health Care for Blacks and Mill Workers in the Twentieth-Century South.* Knoxville: University of Tennessee Press, 1987.

Beckford, George E. *Persistent Poverty: Underdevelopment in Plantation Economies of the Third World.* New York: Oxford University Press, 1972.

Béliz, A. E. "Los Congos: Afro-Panamanian Dance-Drama." *Americas* 11 (1959): 31–33.

Bellegarde, Dantes. "Haiti's Voice at the Peace Table." *Opportunity* 21 (October 1943): 154–159.

Bellegarde-Smith, Patrick. *Haiti: The Breached Citadel.* Boulder, Colo.: Westview, 1990.

Berger, Peter, B. B. Berger, and H. Kellner. *The Homeless Mind: Modernization and Consciousness.* New York: Vintage, 1973.

Bernales, Andres Opazo. *Panama: la iglesia y la lucha de los pobres.* San José, Costa Rica: Editorial Departamento Ecumenico de Investigaciones, 1988.

Berry, Mary F., and John W. Blassingame. *Long Memory: The Black Experience in America.* New York: Oxford University Press, 1982.

Bibeau, G. "From China to Africa: The Same Impossible Synthesis between Traditional and Western Medicines." *Social Science and Medicine* 21 (1985): 937–943.

Bicego, George, A. Chahnazarian, K. Hill, and M. Cayemittes. "Trends, Age Patterns, and Differentials in Childhood Mortality in Haiti (1960–1987)." *Population Studies* 45 (1991): 235–252.

Biesanz, John. "Race Relations in the Canal Zone." *Phylon* 11 (1950): 22–32.

———. "The Economy of Panama." *Inter-American Economic Affairs* 6 (1952): 3–28.

Biesanz, John, and Mavis Biesanz. *The People of Panama.* New York: Columbia University Press, 1955.

Bijker, W. E., Thomas P. Hughes, and T. J. Pinch, eds. *The Social Construction of Technological Systems: New Directions in the Sociology and History of Technology.* Cambridge, Mass.: MIT Press, 1987.

Bingham, Jonathan B. *Shirt-Sleeve Diplomacy: Point 4 in Action.* New York: John Day, 1954.

Blauch, L. E. "Education in the Territories and Outlying Possessions of the United States." *Journal of Negro Education* 15 (1946): 479–489.

Bond, Horace M. *Negro Education in Alabama: A Study in Cotton and Steel.* 1939; New York: Atheneum, 1969.

Boone, Clinton C. *Liberia as I Know It.* Richmond, Va.: n.p., 1929; Westport, Conn.: Negro Universities Press, 1970.

Bordes, Ary, and Andrea Couture. *For the People, for a Change: Bringing Health to the Families of Haiti.* Boston: Beacon Press, 1978.

Bourgois, Phillippe. *Ethnicity at Work: Divided Labor on a Central American Banana Plantation.* Baltimore: Johns Hopkins University Press, 1989.

Bowman, Shearer D. *Masters and Lords: Mid-Nineteenth-Century U.S. Planters and Prussian Junkers*. New York: Oxford University. Press, 1993.

Brand, J. L. "The United States Public Health Service and International Health, 1945–1950." *Bulletin of the History of Medicine* 63 (1989): 579–578.

Brandfon, Robert L. *Cotton Kingdom of the New South: A History of the Yazoo Mississippi Delta from Reconstruction to the Twentieth Century*. Cambridge, Mass.: Harvard University Press, 1967.

Brittain, H. G. *Scenes and Incidents of Every-Day Life in Africa*. 1860; New York: Negro Universities Press, 1969.

Brock, Thomas D. *Robert Koch: A Life in Medicine and Bacteriology*. Madison, Wis.: Science Tech Publishers, 1988.

Bruce-Chwatt, Leonard J. *Essential Malariology*. New York: John Wiley, 1985.

Bullock, Henry A. *A History of Negro Education in the South from 1619 to the Present*. New York: Praeger, 1970.

Burlingham, Roger. *Engines of Democracy: Inventions and Society in Mature America*. New York: Charles Scribner's Sons, 1940.

Bushnell, G. E. *A Study in the Epidemiology of Tuberculosis with Especial Reference to Tuberculosis of the Tropics and of the Negro Race*. New York: William Wood, 1920.

Butler, John S. *Entrepreneurship and Self-Help among Black Americans: A Reconsideration of Race and Economics*. Albany: State University of New York Press, 1991.

Campbell, James T. *Songs of Zion: The African Methodist Episcopal Church in the United States and South Africa*. New York: Oxford University Press, 1995.

Campbell, Marie. *Folks Do Get Born*. New York: Rinehart, 1946.

Carleton, Mark T. *Politics and Punishment: The History of the Louisiana State Penal System*. Baton Rouge: Louisiana State University Press, 1971.

Carlson, Dennis G. *African Fever: A Study of British Science, Technology, and Politics in West Africa, 1787–1864*. Canton, Mass.: Science History Publications, 1984.

Carlsson, Jerker. *The Limits to Structural Change: A Comparative Study of Foreign Direct Investments in Liberia and Ghana, 1950–1971*. Uppsala, Sweden: Scandinavian Institute of African Studies, 1981.

Cayton, Horace R., and George S. Mitchell. *Black Workers and the New Unions*. 1939; College Park, Md.: McGrath, 1969.

Chalk, Frank. "Du Bois and Garvey Confront Liberia: Two Incidents in the Coolidge Years." *Canadian Journal of African Studies* 1 (1967): 135–142.

Chamberlain, W. P. "The Health Department of the Panama Canal." *New England Journal of Medicine* 203 (October 2, 1930): 669–680.

Chinweizu. *The West and the Rest of Us: White Predators, Black Slavers, and the African Elite*. New York: Vintage, 1975.

Chomsky, Aviva. *West Indian Workers and the United Fruit Company in Costa Rica, 1870–1940*. Baton Rouge: Louisiana State University Press, 1996.

Clark, Katherine. *Motherwit: An Alabama Midwife's Story by Onnie Lee Logan as Told to Katherine Clark*. New York: Plume, 1989.

Clower, Robert W., et al., *Growth without Development: An Economic Survey of Liberia*. Evanston, Ill.: Northwestern University Press, 1966.

Cluff, L. E. "America's Romance with Medicine and Medical Science." *Daedalus* 115 (spring 1986): 137–159.

Collasos, Sharon P. *Labor and Politics in Panama: The Torrijos Years*. Boulder, Colo.: Westview, 1991.

Conniff, Michael L. *Black Labor on a White Canal: Panama, 1904–1981.* Pittsburgh: University of Pittsburgh Press, 1985.

Conniff, Michael L., and Thomas J. Davis, eds. *Africans in the Americas: A History of the Black Diaspora.* New York: St. Martin's Press, 1994.

Conrad, David E. *The Forgotten Farmers: The Story of Share-Croppers in the New Deal.* Urbana: University of Illinois Press, 1965.

Cooper, G. Proctor, and Samuel J. Record. *The Evergreen Forests of Liberia: A Report on Investigations Made in the West African Republic of Liberia by the Yale University School of Forestry in Cooperation with the Firestone Plantations Company.* Yale University School of Forestry, Bulletin No. 31. New Haven, Conn.: Yale University, 1931.

Corby, R. A. "Cuttington College, Liberia: Years at Cape Palmas, 1889–1901." *Liberian Studies Journal* 17 (1992): 1–24.

Cosby, Alfred W. *The Measure of Reality: Quantification and Western Society, 1250–1600.* New York: Cambridge University Press, 1997.

Cueto, Marcos. "Sanitation from Above: Yellow Fever and Foreign Intervention in Peru, 1919–1922." *Hispanic American Historical Review* 72 (1992): 1–22.

Currer, Caroline, and Meg Stacey, eds. *Concepts of Health, Illness, and Disease: A Comparative Perspective.* Oxford: Berg, 1993.

Daniel, Pete. *Breaking the Land: The Transformation of Cotton, Tobacco, and Rice Cultures since 1880.* Urbana: University of Illinois Press, 1985.

Dart, H. M. *Maternity and Child Care in Selected Rural Areas of Mississippi.* Rural Child Welfare Series No. 5. Children's Bureau Publication No. 88. Washington, D.C.: GPO, 1921.

David, C. S. *The Cotton Kingdom in Alabama.* Montgomery: Alabama State Department of Archives and History, 1939.

Davidson, Basil. *The Black Man's Burden: Africa and the Curse of the Nation-State.* New York: Times Books, 1992.

Davie, Maurice R. *Negroes in American Society.* New York: McGraw-Hill, 1949.

Davis, Allison, B. B. Gardner, and M. R. Gardner. *Deep South: A Social Anthropological Study of Caste and Class.* 1941; abridged ed., Chicago: University of Chicago Press, 1969.

Davis, J. P. "A Black Inventory of the New Deal." *Crisis* 42 (May 1935): 141–142, 154–155.

Delaporte, Francois. *The History of Yellow Fever: An Essay on the Birth of Tropical Medicine.* Cambridge, Mass.: MIT Press, 1991.

Delgado, Richard, and D. R. Millen. "God, Galileo, and Government: Toward Constitutional Protection for Scientific Inquiry." *Washington Law Review* 53 (1978): 349–361.

Diederich, Bernard, and Al Burt. *Papa Doc: Haiti and Its Dictator.* Maplewood, N.J.: Waterfront Press, 1991.

Dillard, J. H., et al. *Twenty-Year Report of the Phelps-Stokes Fund, 1911–1931, with a Series of Studies of Negro Progress and of Developments of Race Relations in the United States and Africa.* New York: Phelps-Stokes Fund, 1932.

Dinnerstein, Leonard, Roger L. Nichols, and David M. Reimars. *Natives and Strangers: Blacks, Indians, and Immigrants in America.* 2nd ed. New York: Oxford University Press, 1990.

Dowling, Harry F. *Fighting Infection: Conquests of the Twentieth Century.* Cambridge, Mass.: Harvard University Press, 1977.

Du Bois, W.E.B. *Black Reconstruction: An Essay toward a History of the Part Which Black Folk Played in an Attempt to Reconstruct Democracy in America, 1860–1880.* New York: Harcourt, Brace, 1935.

———. *Dusk of Dawn: An Essay toward an Autobiography of a Race Concept.* New York: Harcourt, Brace, 1940.

———, ed. *The College-Bred Negro: Report of a Social Study Made under the Direction of Atlanta University, Together with the Proceedings of the Fifth Conference for the Study of the Negro Problems, Held at Atlanta University, May 29–30, 1900.* Atlanta: Atlanta University Press, 1900.

———, ed. T*he Negro Common School: Report of a Social Study Made under the Direction of Atlanta University, Together with the Proceedings of the Sixth Conference for the Study of the Negro Problems, Held at Atlanta University, May 28, 1901.* Atlanta: Atlanta University Press, 1901.

———, ed. *The Negro Artisan: Report of a Social Study Made under the Direction of Atlanta University.* Atlanta: Atlanta University Press, 1902.

Dubos, Rene. *The Mirage of Health: Utopias, Progress, and Biological Change.* Garden City, N.Y.: Anchor, 1959.

———. *Man Adapting.* 1965; New Haven: Yale University Press, 1980.

Dunn, D. Elwood *The Foreign Policy of Liberia during the Tubman Era, 1944–1971.* London: Hutchinson Benham, 1979.

Duvalier, Francois. "Contribution à l'étude du pian en Haïti: L'Aspect médico-social et l'oeuvre de la Mission Sanitaire Amèricaine." *L'Union Médicale du Canada—Bulletin AMLFAN* 74 (Montréal, June 1945): 781–798.

———. *OEuvres essentielles.* 3rd ed., revues, augmentées, et corrigées. Port-au-Prince, Haiti: Presses nationales d'Haïti, 1968.

Echeverri-Gent, Elisavinda. "Forgotten Workers: British West Indians and the Early Days of the Banana Industry in Costa Rica and Honduras." *Journal of Latin American Studies* 24 (1992): 275–308.

Edelman, Marc. *The Logic of the Latifundio: The Large Estates of Northwestern Costa Rica since the Late Nineteenth Century.* Stanford, Calif.: Stanford University Press, 1992.

Eggert, Gerald G. *Harrisburg Industrializes: The Coming of Factories to an American Community.* University Park: Penn State Press, 1993.

Ellis, John. *The Social History of the Machine Gun.* Baltimore: Johns Hopkins University Press, 1975.

Embree, Edwin R. *Indians of the Americas.* 1939; New York: Macmillan, 1970.

Engels, Dagmar, and Shula Marks. "Introduction: Hegemony in a Colonial Context." In *Contesting Colonial Hegemony: State and Society in Africa and India,* ed. D. Engels and S. Marks. London: British Academic Press/German Historical Institute, 1994.

Englehardt, N. L. *Report of the Survey of the Schools of the Panama Canal Zone Made by the Division of Educational Research, Teachers College, Columbia University.* Mount Hope, Canal Zone: n.p., 1930.

Escobar, Arturo. "Power and Visibility: Development and the Invention and Management of the Third World." *Cultural Anthropology* 3 (1988): 428–443.

Ettling, John. *The Germ of Laziness: Rockefeller Philanthropy and Public Health in the New South.* Cambridge, Mass.: Harvard University Press, 1981.

Farley, James J. *Making Arms in the Machine Age: Philadelphia's Frankford Arsenal, 1816–1870.* University Park: Penn State University Press, 1994.

Fass, S. M. *Political Economy in Haiti: The Drama of Survival*. New Brunswick, N.J.: Transaction Books, 1988.

Fee, Elizabeth. *Disease and Discovery: A History of the Johns Hopkins School of Hygiene and Public Health, 1916–1939*. Baltimore: Johns Hopkins University Press, 1987.

Felix, D. "Technological Dualism in Late Industrializers: On Theory, History, and Policy." *Journal of Economic History* 34 (1974): 232–240.

Ferguson, James. *Papa Doc, Baby Doc, Haiti and the Duvaliers*. Oxford: B. Blackwell, 1987.

Fiehrer, Thomas. "Political Violence in the Periphery: The Haitian Massacre of 1937." *Race and Class* 32 (1990): 1–20.

Fierce, Milfred C. *Slavery Revisited: Blacks and the Southern Convict Lease System, 1865–1933*. [New York]: Africana Studies Research Center, Brooklyn College, City University of New York, 1994.

Fisher, Lawrence E. *Colonial Madness: Mental Health in the Barbadian Social Order*. New Brunswick, N.J.: Rutgers University Press, 1985.

Fite, G. C. "Recent Progress in the Mechanization of Cotton Production in the United States." *Agricultural History* 24 (1950): 19–28.

———. "Mechanization of Crop Production since World War II." *Agricultural History* 54 (1980): 190–207.

Fogel, Robert W. *Without Consent or Contract: The Rise and Fall of American Slavery*. New York: Norton, 1989.

Fogel, Robert W., and Stanley Engerman. *Time on the Cross: The Economics of American Negro Slavery*. Boston: Little, Brown, 1974.

Fogel, Robert W., Ralph A. Galantine, and Richard L. Manning, eds. *Without Consent or Contract: The Rise and Fall of American Slavery: Evidence and Methods*. New York: Norton, 1992.

Fraenkel, Merran. *Tribe and Class in Monrovia*. London: International African Institute/ Oxford University Press, 1964.

Franck, Harry A. *Zone Policeman 88: A Close Range Study of the Panama Canal and Its Workers*. New York: Century, 1913.

Frangsmyr, Tore, J. L. Heilbron, and R. E. Rider, eds. *The Quantifying Spirit in the Eighteenth Century*. Berkeley: University of California Press, 1990.

Frank, L. C. *Final Report on the Sanitary Operations in the Mississippi Coastal District by the Mississippi Board of Health and the United States Public Health Service, 1918–1919*. Gulfport, Miss.: United States Public Health Service, Mississippi Coastal District, [1919].

Fraser, Gertrude J. *African American Midwifery in the South: Dialogues of Birth, Race, and Memory*. Cambridge, Mass.: Harvard University Press, 1998.

Frazier, E. Franklin. *The Negro in the United States*. 1949; rev. ed., New York: Macmillan, 1957.

———. *Race and Cultural Contacts in the Modern World*. Boston: Beacon, 1957.

Fredrickson, George M. *White Supremacy: A Comparative Study in American and South African History*. New York: Oxford University Press, 1981.

———. *Black Liberation: A Comparative History of Black Ideologies in the United States and South Africa*. New York: Oxford University Press, 1995.

Freidel, Frank, and Alan Brinkley. *America in the Twentieth Century*. New York: Knopf, 1982.

Galbraith, John K. *The New Industrial State*. New York: New American Library, 1967.

Gaillard, Jacque, V. V. Krishna, and Roland Waast, eds. *Scientific Communities in the Developing World*. New Delhi, India: Sage, 1997.

Gallagher, Nancy E. *Medicine and Power in Tunisia: 1780–1900*. Cambridge: Cambridge University Press, 1984.

———. *Egypt's Other War: Epidemics and the Politics of Public Health*. Syracuse: Syracuse University Press, 1990.

Geiman, Quentin M. "A Half Century of Tropical Medicine." *American Journal of Tropical Medicine and Hygiene* 3 (May 1954): 397–411.

Genovese, Eugene D. *The Political Economy of Slavery: Studies in the Economy and Society of the Slave South*. New York: Vintage, 1965.

Gershoni, Yekutiel. "An Historical Examination of Liberia's Economic Policies, 1900–1944." *Liberian Studies Journal* 11 (1986): 20–34.

Gilje, P. A. "The Rise of Capitalism in the Early Republic." *Journal of the Early Republic* 16 (1996): 159–181.

Gontar, D. "A Version of the South, or Deconstructing Reconstruction: V. S. Naipaul's *A Turn in the South*." *Plantation Society in the Americas* 3 (1993): 93–113.

Gorgas, William C. *Sanitation in Panama*. New York: D. Appleton, 1915.

Gottfried, Kurt, and B. G. Blair. *Crisis Stability and Nuclear War*. New York: Oxford University Press, 1988.

Gover, Mary. "Trend of Mortality among Southern Negroes since 1920." *Journal of Negro Education* 6 (1937): 276–288.

Graham, Edgar, with Ingrid Floering. *The Modern Plantation in the Third World*. New York: St. Martin's Press, 1984.

Grant, Nancy L. *TVA and Black Americans: Planning for the Status Quo*. Philadelphia: Temple University Press, 1990.

Greenberg, Stanley B. *Race and State in Capitalist Development: Comparative Perspectives*. New Haven, Conn.: Yale University Press, 1980.

Greene, Anne. *The Catholic Church in Haiti: Political and Social Change*. East Lansing: Michigan State University Press, 1993.

Guzda, H. P. "Social Experiment of the Labor Department: The Division of Negro Economics." *Public Historian* 4 (1982): 7–37.

Gwen, Catherine. *U.S. Role in the World Bank, 1945–1992*. Washington, D.C.: Brookings Institution, 1994.

Harlan, Louis R., and R. W. Smock, eds. *The Booker T. Washington Papers*. Vol. 10, *1909–1911*. Urbana: University of Illinois Press, 1981.

———. *The Booker T. Washington Papers*. Vol. 12, *1912–1914*. Urbana: University of Illinois Press, 1982.

Harris, Joseph E., ed. *Global Dimensions of the African Diaspora*. Washington, D.C.: Howard University Press, 1993.

Harrison, Gordon. *Mosquitoes, Malaria, and Man: A History of the Hostilities since 1880*. London: John Murray, 1978.

Harrison, Mark. "'The Tender Frame of Man': Disease, Climate, and Racial Difference in India and the West Indies, 1760–1860." *Bulletin of the History of Medicine* 70 (1996): 68–93.

Hayami, Y. "Peasant and Plantation in Asia." In *From Classical Economics to Development Economics*, ed. Gerald M. Meier, 121–134. New York: St. Martin's Press, 1994.

Haystead, Ladd, and G. C. Fite. *The Agricultural Regions of the United States*. Norman: University of Oklahoma Press, 1955.

Headrick, Daniel R. *The Tentacles of Progress: Technology Transfer in the Age of Imperialism, 1850–1940*. New York: Oxford University Press, 1988.

Heard, William H. *The Bright Side of African Life Illustrated*. Philadelphia: AME Publishing House, 1898.

———. *From Slavery to the Bishopric in the A. M. E. Church*. Philadelphia: A. M. E. Book Concern, 1924.

Heinl, R. D., Jr., and N. G. Heinl. *Written in Blood: The Story of the Haitian People, 1492–1971*. Boston: Houghton Mifflin, 1978.

Herf, Jeffrey. *Reactionary Modernism: Technology, Culture, and Politics in Weimar and the Third Reich*. New York: Cambridge University Press, 1984.

Herskovits, Melville J. *Life in a Haitian Valley*. New York: Knopf, 1937.

———. *Myth of the Negro Past*. 1941; Boston: Beacon Press, 1990.

Hill, Robert, ed. *The Marcus Garvey and Universal Negro Improvement Association Papers*. Vol. 2. Berkeley: University of California at Berkeley, 1983.

Hobson, John A. *The Evolution of Modern Capitalism: A Study in Machine Production*. 1926; 4th ed., London: George Allen & Unwin, 1954.

Holly, Marc Aurele. *Agriculture in Haiti with Special Reference to Rural Economy and Agricultural Education*. Sesquicentennial Collection. New York: Vantage, 1955.

Hooker, J. R. "The Negro American Press and Africa in the Nineteen Thirties." *Canadian Journal of African Studies* 1 (1967): 43–50.

Howard, William T. *Public Health Administration and the Natural History of Disease in Baltimore, Maryland, 1797–1920*. Washington, D.C.: Carnegie Institution, 1924.

Howe, R. W. *Along the Afric Shore: An Historic Review of Two Centuries of U.S.-African Relations*. New York: Harper & Row, 1975.

Howell, Joel D. *Technology in the Hospital: Transforming Patient Care in the Early Twentieth Century*. Baltimore: Johns Hopkins University Press, 1995.

Hoyle, Brian S., ed. *Cityports, Coastal Zones, and Regional Change*. Chichester, N.Y.: John Wiley & Sons, 1996.

Huges, Sally S. *The Virus: A History of the Concept*. New York: Science History Publications, 1977.

Huggins, Martha K. *From Slavery to Vagrancy in Brazil: Crime and Social Control in the Third World*. New Brunswick, N.J.: Rutgers University Press, 1985.

Hughes, Thomas P. *Networks of Power: Electrification in the West, 1880–1930*. Baltimore: Johns Hopkins University Press, 1993.

Humphreys, Margaret. *Yellow Fever and the South*. New Brunswick, N.J.: Rutgers University Press, 1992.

Hunter, Louis C. *A History of Industrial Power in the United States, 1780–1930*. Vol. 2, *Steam Power*. Charlottesville: University Press of Virginia, 1986.

Isaacman, Allen, and Richard Roberts, eds. *Cotton, Colonialism, and Social History in Sub-Saharan Africa*. Portsmouth, N.H.: Heinemann, 1995.

Isaacs, Harold R. *The New World of Negro Americans*. New York: Viking, 1963.

Jackson, Luther P. *Free Negro Labor and Property Holding in Virginia, 1830–1860*. New York: Russell and Russell, 1942.

Jackson, Robert J., and Carl G. Rosberg. *Personal Rule in Black Africa: Prince, Autocrat, Prophet, Tyrant*. Berkeley: University of California Press, 1982.

Jacobs, Sylvia M. *The African Nexus: Black American Perspectives on the European Partitioning of Africa, 1880–1920*. Westport, Conn.: Greenwood, 1981.

Jaynes, G. D. *Branches without Roots: Genesis of the Black Working Class in the American South, 1862–1882*. New York: Oxford University Press, 1986.

Johnson, Charles S. *Shadow of the Plantation*. Chicago: University of Chicago Press, 1934.

———. *Statistical Atlas of Southern Counties: Listing and Analysis of Socio-Economic Indices of 1,104 Southern Counties*. Chapel Hill: University of North Carolina Press, 1941.

Johnson, Charles S., Edwin E. Embree, and Will W. Alexander. *The Collapse of Cotton Tenancy*. Chapel Hill: University of North Carolina Press, 1935.

Johnson, Charles W., and Charles O. Jackson. *City behind a Fence: Oak Ridge, Tennessee, 1942–1946*. Knoxville: University of Tennessee Press, 1981.

Johnson, James W. *Along This Way: The Autobiography of James Weldon Johnson*. New York: Viking, 1933.

Johnson, Robert C. "Science, Technology, and Black Community Development." *Black Scholar* 15 (1984): 32–44.

Jones, J. Mills. "Development Planning, Politics, and the Bureaucracy: The Liberian Experience." *Liberian Studies Journal* 11 (1986): 1–19.

Jones, Jacqueline. *Labor of Love, Labor of Sorrow: Black Women, Work, and the Family from Slavery to the Present*. New York: Vintage, 1985.

———. "Work Now, Get Paid Much Later: 'Free Labor' in the Postbellum South." *Reviews in American History* 15 (1987): 265–271.

———. *The Dispossessed: America's Underclasses from the Civil War to the Present*. New York: Basic Books, 1992.

Kalbe, A. "Iron-Fisted 'Papa': Haitian Dictator's 'Re-election' Strengthens His Corrupt Rule over Impoverished Island." *New Leader* 44 (August 14–21, 1961): 15–16.

Kandel, I. L., ed. *Educational Yearbook of the International Institute of Teachers College, Columbia University, 1942: Education in the Latin American Countries*. New York: Teachers College, 1943.

Kepner, Charles D. *Social Aspects of the Banana Industry*. New York: Columbia University Press, 1936.

Kilroy, D. P. "Extending the American Sphere to West Africa: Dollar Diplomacy in Liberia, 1908–1926." Ph.D. diss., University of Iowa, 1995.

Kiple, Dalila de Sousa. "Darwin and Medical Perceptions of the Black: A Comparative Study of the United States and Brazil, 1871–1918." Ph.D. diss., Bowling Green State University, 1987.

Kiple, Kenneth F. *The African Exchange: Toward a Biological History of Black People*. Durham, N.C.: Duke University Press, 1988.

Kirby, Jack T. "The Transformation of Southern Plantations c. 1920–1960." *Agricultural History* 57 (1983): 257–276.

Klein, Herbert S. *Slavery in the Americas: A Comparative Study of Virginia and Cuba*. Chicago: University of Chicago Press, 1967.

Knoll, Arthur J. "Harvey Firestone's Liberian Investment (1923–1933)." *Liberian Studies Journal* 14 (1989): 13–33.

Kolchin, Peter. *Unfree Labor: American Slavery and Russian Serfdom*. Cambridge, Mass.: Harvard University Press, 1987.

Konneh, A. "Citizenship at the Margins: Status, Ambiguity, and the Mandingo of Liberia." *African Studies Review* 39 (1996): 141–154.

Kraut, Alan M. *Silent Travelers: Germs, Genes, and the "Immigrant Menace."* Baltimore: Johns Hopkins University Press, 1995.

Krupfer, Ann M. *Toward a Tenderer Humanity and Nobler Womanhood: African American Women's Clubs in Turn-of-the-Century Chicago.* New York: New York University Press, 1996.

Laguerre, Michel S. *Afro-Caribbean Folk Medicine.* South Hadley, Mass.: Bergin & Garvey, 1987.

———. *Voodoo and Politics in Haiti.* New York: St. Martin's Press, 1989.

———. *The Military and Society in Haiti.* Knoxville: University of Tennessee Press, 1993.

———. *Diasporic Citizenship: Haitian Americans in Transnational America.* New York: St. Martin's Press, 1998.

Lanier, R. O'Hara. "The Problem of Mass Education in Liberia." *Journal of Negro Education* 30 (1961): 251–260.

Latour, Bruno. *The Pasteurization of France.* Cambridge, Mass.: Harvard University Press, 1988.

Lauria-Santiago, Aldo, and Aviva Chomsky, eds. *Identity and Struggle at the Margins of the Nation-State: The Laboring Peoples of Central America and the Hispanic Caribbean.* Durham, N.C.: Duke University Press, 1998.

Léger, Jacques M. *Haiti: Her History and Her Detractors.* 1907; New York: Negro Universities Press, 1970.

LePrince, Joseph A. and A. J. Orenstein. *Mosquito Control in Panama: The Eradication of Malaria and Yellow Fever in Cuba and Panama.* New York: Putnam's, 1916.

Leuchtenberg, William E. *Franklin D. Roosevelt and the New Deal, 1932–1940.* New York: Harper Colophon, 1963.

Levine, B. B., ed. *The Caribbean Exodus.* New York: Praeger, 1987.

Levins, Richard, and Richard C. Lewontin. *The Dialectical Biologist.* Cambridge, Mass.: Harvard University Press, 1985.

Lewis, Arthur W. "Economic Development with Unlimited Supplies of Labour." In *The Economics of Underdevelopment*, ed. A. N. Agarwala and S. P. Singh, 400–449. London: Oxford University Press, 1970.

———. "Unlimited Labour: Further Notes." In *Selected Economic Writings of W. Arthur Lewis*, ed. Mark Gersovitz, 365–390. New York: New York University Press, 1985.

Lewis, Earl. "Expectations, Economic Opportunities, and Life in the Industrial Age: Black Migration to Norfolk, Virginia, 1910–1945." In *The Great Migration in Historical Perspective*, ed. Joe W. Trotte, 23–45. Bloomington: Indiana University Press, 1991.

Lewis, Gordon K. *The Growth of the Modern West Indies.* London: MacGibbon and Kee, 1968.

Lewis, Julian H. *The Biology of the Negro.* Chicago: University of Chicago Press, 1942.

Lhérisson, Camille. "Diseases of the Peasants of Haiti." *American Journal of Public Health* 25 (1935): 924–928.

Lichtenstein, Alex. *Twice the Work of Free Labor: The Political Economy of Convict Labor in the New South.* London: Verso, 1996.

Lieberson, Stanley. *A Piece of the Pie: Blacks and White Immigrants since 1880.* Berkeley: University of California Press, 1980.

Lifton, Robert J. *The Nazi Doctors: Medical Killing and the Psychology of Genocide.* New York: Basic Books, 1986.

Lifton, Robert J., and Eric Makusen. *The Genocidal Mentality: Nazi Holocaust and Nuclear Threat.* New York: Basic Books, 1990.

Link, Arthur S., ed. *Papers of Woodrow Wilson.* Vol. 46. Princeton, N.J.: Princeton University Press, 1966.

Lipset, Seymour M., and Reinhard Bendix. *Social Mobility in Industrial Society.* Berkeley: University of California Press, 1964.

Litoff, Judy B. *American Midwives 1860 to the Present.* Westport, Conn.: Greenwood, 1978.

Litwack, Leon F. *Trouble in Mind: Black Southerners in the Age of Jim Crow.* New York: Alfred A. Knopf, 1998.

Logan, Rayford W. "Education in Haiti." *Journal of Negro History* 15 (1930): 401–460.

———. "Liberia in the Family of Nations." *Phylon* 7 (1946): 5–11.

———. "The United States Mission in Haiti, 1915–1952." *Inter-American Economic Affairs* 6 (spring 1953): 18–28.

———. *Haiti and the Dominican Republic.* London: Oxford University Press/Royal Institute of International Affairs, 1968.

Lowenkopf, Martin. *Politics in Liberia: The Conservative Road to Development.* Stanford, Calif.: Hoover Institution Press, 1976.

Lowrance, William W. *Modern Science and Human Values.* New York: Oxford University Press, 1985.

Mack, Andrew, David Plant, and Ursula Doyle, eds., *Imperialism, Intervention, and Development.* London: Croom Helm, 1979.

Macleod, R., and M. Lewis, eds. *Disease, Medicine, and Empire: Perspectives on Western Medicine and the Experience of European Expansion.* London: Routledge, 1988.

Maddox, James G., et al. *The Advancing South: Manpower Prospects and Problems.* New York: Twentieth Century Fund, 1967.

Mancini, Matthew J. *One Dies, Get Another: Convict Leasing in the American South, 1866–1928.* Columbia: University of South Carolina Press, 1996.

Manigat, L. F. *Haiti of the Sixties, Object of International Concern (A Tentative Global Analysis of the Potentially Explosive Situation of a Crisis Country in the Caribbean).* Washington, D.C.: Washington Center of Foreign Policy Research, [1964].

Manning, Kenneth R. *Black Apollo of Science: The Life of Ernest Everett Just* (New York: Oxford University Press, 1983.

Marland, Hilary, and A. M. Rafferty, eds. *Midwives, Society, and Childbirth: Debates and Controversies in the Modern Period.* New York: Routledge, 1997.

Marshall, Ray F. *The Negro and Organized Labor.* New York: John Wiley & Sons, 1965.

Martin, Sandy D. *Black Baptists and African Missions: The Origins of a Movement, 1880–1915.* Macon, Ga.: Mercer University Press, 1989.

Marx, A. W. *Making Race and Nation: A Comparison of South Africa, the United States, and Brazil.* New York: Cambridge University Press, 1999.

Matthews, J. R. *Quantification and the Quest for Medical Certainty.* Princeton, N.J.: Princeton University Press, 1995.

May, Stacy, and Galo Plaza. *The United Fruit Company in Latin America.* National Planning Association, 1958; New York: Arno, 1976.

McBride, David. "Solomon Porter Hood, 1853–1943: Black Missionary, Educator, and

Minister to Liberia." *Quarterly Publication of the Lancaster Historical Society* 84 (1984): 2–9.

————. *Integrating the City of Medicine: Blacks in Philadelphia Health Care, 1910–1965.* Philadelphia: Temple University Press, 1989.

————. *From TB to AIDS: Epidemics among Urban Blacks since 1900.* Albany: State University of New York Press, 1991.

————. "Medicine and Medical Care." In *Macmillan Encyclopedia of World Slavery*, vol. 1, ed. Paul Finkelman and Joseph C. Miller, 383–387. New York: Macmillan, 1998.

McClintock, Ann. *Imperial Leather: Race, Gender, and Sexuality in the Colonial Contest.* New York: Routledge, 1995.

McDougall, Walter A. *Promised Land, Crusader State: The American Encounter with the World since 1776.* Boston: Houghton Mifflin, 1997.

McFadyean, Andrew Sir, ed. T*he History of Rubber Regulation, 1934–1943.* London: International Rubber Regulation Committee/George Allen & Unwin, 1944.

McKinney, Kitzie. "Costa Rica's Black Body: The Politics and Poetics of Difference in Eulalia Bernard's Poetry." *Afro-Hispanic Review* 15 (1996): 10–16.

McLaughlin, Russell U. *Foreign Investment and Development in Liberia.* New York: Praeger, 1966.

McMurry, Linda O. *George Washington Carver: Scientist and Symbol.* New York: Oxford University Press, 1981.

McNeill, William H. *Plagues and Peoples.* New York: Anchor, 1976.

————. *The Pursuit of Power: Technology, Armed Force, and Society since A.D. 1000.* Chicago: University of Chicago, 1982.

Melhorn, K. C. "Public Health in Haiti: A Resume of 10 Years' Work." *United States Naval Medical Bulletin* (1929): 568–573.

Miles, Toni P., and David McBride. "World War I Origins of the Syphilis Epidemic among Twentieth-Century Black Americans: A Biohistorical Analysis." *Social Science and Medicine* 45 (1997): 61–69.

Miller, Basil. *George Washington Carver: God's Ebony Scientist.* Grand Rapids, Mich.: Zondervan, 1943.

Miller, Jake C. *The Plight of Haitian Refugees.* New York: Praeger, 1984.

Millspaugh, Arthur C. *Haiti under American Control, 1915–1930.* Boston: World Peace Foundation, 1931.

Minority Rights Group, ed. *No Longer Invisible: Afro-Latin Americans Today.* London: Minority Rights Publication, 1995.

Montague, Ludwell L. *Haiti and the United States, 1714–1938.* 1940; New York: Russell & Russell, 1966.

Moore, Barrington, Jr. *Political Power and Social Theory: Six Studies.* Cambridge, Mass.: Harvard University Press, 1958.

————. *Social Origins of Dictatorship and Democracy: Lord and Peasant in the Making of the Modern World.* Boston: Beacon, 1966.

Moore, John H. *The Emergence of the Cotton Kingdom in the Old Southwest: Mississippi, 1770–1860.* Baton Rouge: Louisiana State University, 1988.

Moore, O. E. *Haiti: Its Stagnant Society and Shackled Economy.* New York: Exposition, 1972.

Morais, Herbert M. *The History of the Negro in Medicine.* New York: Publishers Co./ Association for the Study of Negro Life and History, 1968.

Moran, Mary H. *Civilized Women: Gender and Prestige in Southeastern Liberia.* Ithaca, N.Y.: Cornell University Press, 1990.

Morner, Magnus, ed. *Race and Class in Latin America.* New York: Columbia University Press, 1970.

Moses, Wilson J. *Alexander Crummell: A Study in Civilization and Discontents.* New York: Oxford University Press, 1989.

————, ed. *Liberian Dreams: Back-to-Africa Narratives from the 1850s.* University Park: Penn State University Press, 1998.

Mostofi, F. K. "Contributions of the Military to Tropical Medicine." *Bulletin of the New York Academy of Medicine* 44 (1968): 702–720.

Munro, Dana G. *The Five Republics of Central America: Their Political and Economic Development and Their Relations with the United States.* New York: Carnegie Endowment for International Peace, 1918.

Murolo, Priscilla. *The Common Ground of Womanhood: Class, Gender, and Working Girls Clubs, 1884–1928.* Urbana: University of Illinois Press, 1997.

Myrdal, Gunnar. *An American Dilemma: The Negro Problem and Modern Democracy.* 1944; 20th anniversary ed., New York: Harper & Row, 1962.

National Academy of Sciences, National Research Council, Division of Medical Sciences. *Tropical Health: A Report on a Study of Needs and Resources.* Publ. 996. Washington, D.C.: National Academy of Sciences—National Research Council, 1962.

Newman, Peter. "Malaria Control and Population Growth." *Journal of Development Studies* 6 (1970): 133–156.

Nicholls, David. *From Dessalines to Duvalier: Race, Colour, and National Independence in Haiti.* Cambridge, U.K.: Cambridge University Press, 1979.

Ninkovich, F. "Theodore Roosevelt: Civilization as Ideology." *Diplomatic History* 10 (1986): 234.

Northrup, Herbert R. *Negro Employment in Basic Industry: A Study in Racial Policies in Six Industries.* Vol. 1, *Studies of Negro Employment.* Philadelphia: Wharton School of Finance and Commerce, 1970.

Numbers, Ronald L., ed. *Scientific Authority and Twentieth-Century America.* Baltimore: Johns Hopkins University Press, 1997.

Nye, E. R., and M. E. Gibson. *Ronald Ross: Malariologist and Polymath: A Biography.* New York: St. Martin's Press, 1997.

Odum, Howard W. *Southern Regions of the United States.* Chapel Hill: Southern Regional Committee of the National Research Council/University of North Carolina Press, 1936.

Oshinsky, David M. *"Worse Than Slavery": Parchman Farm and the Ordeal of Jim Crow Justice.* New York: Free Press, 1996.

Ott, Katheryn. *Fevered Lives: Tuberculosis in American Culture since 1870.* Cambridge, Mass.: Harvard University Press, 1996.

Owen, Catherine. *U.S. Relations with the World Bank, 1945–1992.* Washington, D.C.: Brookings Institution, 1994.

Pacey, Arnold. *The Maze of Ingenuity: Ideas and Idealism in the Development of Technology.* 1975; Cambridge, Mass.: MIT Press, 1992.

————. *Technology in World Civilization: A Thousand-Year History.* Cambridge, Mass.: MIT Press, 1990.

Padmore, George. *Pan-Africanism or Communism: The Coming Struggle for Africa*. London: Dennis Dobson, 1956.

Paine, Thomas. *The Age of Reason: Being an Investigation of True and Fabulous Theology* [1794]. In *Thomas Paine: Collected Writings*. New York: Library of America, 1995.

Palmer, Frederick. *Central America and Its Problems: An Account of a Journey from the Rio Grande to Panama*. New York: Moffat, Yard, 1913.

Palmer, Paula. *"What Happen": A Folk-History of Costa Rica's Talmanca Coast*. San José, Costa Rica: Ecodesarrollos, 1977.

Parayil, G. "Models of Technological Change: A Critical Review of Current Knowledge." *History and Technology* 10 (1993): 105–126.

Parsons, R. P., and E. R. Stitt. *History of Haitian Medicine*. New York: Paul B. Hoeber, 1930.

Partridge, W. L. "Banana County in the Wake of United Fruit: Social and Economic Linkages." *American Ethnologist* 6 (1979): 491–509.

Pattison, William D. *Beginnings of the American Rectangular Land Survey System, 1784–1800*. Chicago: University of Chicago Press, 1957.

Patton, Adell, Jr., "Howard University and Meharry Medical Schools in the Training of African Physicians, 1868–1978." In *Global Dimensions of the African Diaspora*, ed. Joseph E. Harris, 206–220. Washington, D.C.: Howard University Press, 1982.

Pavitt, Keith, and Michael Worboys. *Science, Technology, and the Modern Industrial State*. London: Butterworths, 1977.

Penfield, J. K. "The Role of the United States in Africa: Our Interests and Operations." *Department of State Bulletin* 40, no. 1041 (June 8, 1959): 841–849.

Pepper, Charles Melville. *Panama to Patagonia: The Isthmian Canal and the West Coast Countries of South America*. Chicago: A. C. McClurg, 1906; New York: Young People's Missionary Movement, n.d.

Perine, P. L., et al. *Handbook of Endemic Treponematoses: Yaws, Endemic Syphilis, and Pinta*. Geneva: World Health Organization 1984.

Perret, Geoffrey. *A Country Made by War: From the Revolution to Vietnam—The Story of America's Rise to Power*. New York: Vintage, 1989.

Petras, Elizabeth M. *Jamaican Labor Migration: White Capital and Black Labor, 1850–1930*. Boulder, Colo.: Westview, 1988.

Phillips, Ulrich B. "The Decadence of the Plantation." *Annals of the American Academy of Political and Social Science* 35 (1910): 37–41.

Platt, D.C.M. *Business Imperialism: 1840–1930*. Oxford: Clarendon Press, 1977.

Platt, Orville H. "Our Relation to the People of Cuba and Porto Rico." *Annals of the American Academy of Political and Social Science* 18 (1901): 145–159.

Plummer, Brenda Gayle. *Haiti and the Great Powers*. Baton Rouge: Louisiana State University Press, 1988.

———. *Rising Wind: Black Americans and U.S. Foreign Policy, 1935–1960*. Chapel Hill: University of North Carolina Press, 1996.

Poindexter, Hildrus A. "Special Health Problems of Negroes in Rural Areas." *Journal of Negro Education* 6 (1939): 399–416.

———. "A Laboratory Epidemiological Study of Certain Infectious Diseases in Liberia." *American Journal of Tropical Medicine* 29 (1949): 435–440.

Polanco, D. Nuñez. "Antecendens del canal de Panamá." *Invenciones y Ensayos* 492 (June 1991): 99–104.

Porter, J. A. "Picturesque Haiti." *Opportunity* 24, no. 4 (October–December 1946): 178–180, 210.

Preston, Samuel H. *Fatal Years: Child Mortality in Late Nineteenth-Century America.* Princeton, N.J.: Princeton University. Press, 1991.

Price, Sir A. Grenfell. *White Settlers in the Tropics.* American Geographical Society special publication no. 23. New York: The Society, 1939.

———. *White Settlers and Native Peoples.* 1950; Westport, Conn.: Greenwood, 1972.

Proctor, Robert. *Racial Hygiene: Medicine under the Nazis.* Cambridge, Mass.: Harvard University Press, 1988.

Prothero, R. Mansell. *Migrants and Malaria.* London: Longmans, Green, 1965.

Rabkin, Yakov M. *Science between the Superpowers.* New York: Priority Press, 1988.

Raper, Arthur F. *Preface to Peasantry: A Tale of Two Black Belt Counties.* 1936; New York: Atheneum, 1968.

Redkey, Edwin S. *Black Exodus: Black Nationalist and Back-to-Africa Movements, 1890–1910.* New Haven, Conn.: Yale University Press, 1969.

Reid, C. F. "Federal Support and Control of Education in the Territories and Outlying Possessions." *Journal of Negro Education,* Yearbook issue, 7 (1938): 409–414.

Reingold, A. Nathan, and I. H. Reingold, eds. *Science in America: A Documentary History, 1900–1939.* Chicago: University of Chicago Press, 1981.

Rescher, Nicolas. *Scientific Progress: A Philosophical Essay on the Economics of Research in Natural Science.* Pittsburgh: University of Pittsburgh Press, 1978.

Richards, Audrey I., Ford Sturrock, and Jean M. Fortt, eds. *Subsistence to Commercial Farming in Present-Day Buganda.* Cambridge, U.K.: Cambridge University Press, 1973.

Richardson, Bonham C. *Panama Money in Barbados, 1900–1920.* Knoxville: University of Tennessee Press, 1985.

Richardson, Joel M. *The Negro in the Reconstruction of Florida, 1865–1877.* Tallahassee: FSU Research Council, 1965.

Robinson, Cedric. "Du Bois and Black Sovereignty: The Case of Liberia." *Race and Class* 32 (1990): 39–50.

Romualdi, S. "Haiti: Forgotten Dictatorship; Duvalier Rules His Little Country with a Ruthless Hand." *New Leader* 43 (July 18–25, 1960): 15–16.

Rondinelli, D. A. *Development Administration and U.S. Foreign Aid Policy.* Boulder, Colo.: Lynne Rienner, 1987.

Ropp, Steve C. *Panamanian Politics: From Guarded Nation to National Guard.* New York: Praeger, 1982.

Rosen, George. *From Medical Police to Social Medicine: Essays on the History of Health Care.* New York: Science History Publications, 1974.

Rosenberg, Charles E. *No Other Gods: On Science and American Social Thought.* 1976; rev. and exp. ed., Baltimore: Johns Hopkins University Press, 1997.

Rosenberg, Emily S. *Spreading the American Dream: American Economic and Cultural Expansion, 1890–1945.* New York: Hill and Wang, 1982.

———. "The Invisible Protectorate: The United States, Liberia, and the Evolution of Neocolonialism, 1909–40." *Diplomatic History* 9 (summer 1985): 191–214.

Rosenman, S. I., comp. *The Public Papers and Addresses of Franklin D. Roosevelt with a Special Introduction and Explanatory Notes by President Roosevelt.* 1938 vol., *The Continuing Struggle for Liberalism.* New York: Macmillan, 1941.

Rosner, David, and Gerald Markowitz. *Deadly Dust: Silicosis and the Politics of Occupational Disease in Twentieth-Century America*. Princeton, N.J.: Princeton University Press, 1991.

Rotberg, Robert I., ed. *Haiti Renewed: Political and Economic Prospects*. Washington, D.C.: Brookings Institution Press/World Peace Foundation, 1997.

Rotberg, Robert I., with Christopher K. Clague. *Haiti: The Politics of Squalor*. Boston: Houghton Mifflin, 1971.

Rusca, R. A., and C. A. Bennett. "Problems of Machine Picking." In *Crops in Peace and War: The 1950–1951 Yearbook of Agriculture*, by the U.S. Department of Agriculture, 441–444. Washington, D.C.: GPO, 1951.

Russell, Paul F. "Malaria and Its Influence on World History." *Bulletin of the New York Academy of Medicine* 9 (1943): 611–623.

———. *Man's Mastery of Malaria*. London: Oxford University Press, 1955.

Sagan, Carl. *The Demon-Haunted World: Science as a Candle in the Dark*. New York: Ballantine, 1996.

Saha, S. C. *A History of Agriculture in Liberia, 1822–1970: Transference of American Values*. Lewiston, N.Y: Mellen, 1990.

Said, Edward W. *Culture and Imperialism*. New York: Vintage, 1993.

Savitt, Todd L., and James H. Young, eds. *Disease and Distinctiveness in the American South*. Knoxville: University of Tennessee Press, 1988.

Sawyer, Amos. *The Emergence of Autocracy in Liberia: Tragedy and Challenge*. San Francisco: Institute for Contemporary Studies Press, 1992.

Scarborough, William K. *The Overseer: Plantation Management in the Old South*. Athens: University of Georgia Press, 1984.

Scheman, L. R., ed. *The Alliance for Progress: A Retrospective*. New York: Praeger, 1988.

Schlesinger, Arthur M., Jr. *The Age of Roosevelt: The Crisis of the Old Order, 1919–1933*. Boston: Houghton Mifflin, 1957.

Schmidt, Hans. *The United States Occupation of Haiti, 1915–1934*. New Brunswick, N.J.: Rutgers University Press, 1971.

———. *Maverick Marine: General Smedley D. Butler and the Contradictions of American Military History*. Lexington: University Press of Kentucky, 1987.

Schurmann, Franz. *The Logic of World Power: An Inquiry into the Origins, Currents, and Contradictions of World Politics*. New York: Pantheon, 1974.

Sellards, A. W. "Bonds of Union between Tropical Medicine and General Medicine." *Science* 66 (July 29, 1927): 93–100.

Sellers, James B. *Slavery in Alabama*. Tuscaloosa: University of Alabama Press, 1950.

Senghor, Leopold S. "William Tubman: The Building of a Nation." Introduction to *The New Liberia: A History and Political Survey,* by L. A. Marinelli. New York: Praeger, 1964.

Sherman, Mary A. B. "Review: *The Politics of Miseducation* by Donald Spivey." *Liberian Studies Journal* 12 (1987): 88–93.

Shocket, E. "The Mellon Mission in Haiti." *Florida Medicine* 55 (December, 1968): 1098–1100.

Shore, Lawrence. *Southern Capitalists: The Ideological Leadership of an Elite, 1832–1885*. Chapel Hill: University of North Carolina Press, 1986.

Shugg, R. W. "Survival of the Plantation System in Louisiana." *Journal of Southern History* 3 (August, 1937): 311–325.

Sibley, James L., and Diedrich Westermann. *Liberia—Old and New: A Study of Its Social and Economic Background and Possibilities of Development*. Garden City, N.Y.: Doubleday, Doran, 1928.

Siddiqi, Javed. *World Health and World Politics: The World Health Organization and the UN System*. Columbia: University of South Carolina, 1995.

Singer, Charles, E. J. Holmyard, A. R. Hall, and Trevor I. Williams, eds. *A History of Technology*. Vol. 5. New York: Oxford University Press, 1958.

Skolnikoff, Eugene B. *The Elusive Transformation: Science, Technology, and the Evolution of International Politics*. Princeton, N.J.: Princeton University Press, 1993.

Smith, B.L.R. *American Science Policy since World War II*. Washington, D.C.: Brookings Institution, 1990.

Smith, Douglas L. *The New Deal in the Urban South*. Baton Rouge: Louisiana State University Press, 1988.

Smith, Merritt R. *Harpers Ferry Armory and the New Technology: The Challenge of Change*. Ithaca, N.Y.: Cornell University Press, 1977.

Smith, Robert A. *William V. S. Tubman: The Life and Work of an African Statesman*. 1966; 2nd ed., Amsterdam, Netherlands: Van Ditmar, 1967.

———. *The American Foreign Policy in Liberia, 1822–1971*. Monrovia, Liberia: Providence Publications, 1972.

Smith, Susan L. *Sick and Tired of Being Sick and Tired: Black Women's Health Activism in America, 1890–1950*. Philadelphia: University of Pennsylvania Press, 1995.

Smucker, G. R. "Peasant Councils and the Politics of Community." In *Politics, Projects, and People: Institutional Development in Haiti*, ed. D. W. Brinkerhoff and Jean-Claud Garcia-Zamor, 93–113. New York: Praeger, 1986.

Sonbol, Amira el Azhary. "The Creation of a Medical Profession in Egypt during the Nineteenth Century: A Study in Modernization." Ph.D. diss., Georgetown University, 1981.

Spivey, Donald. *The Politics of Miseducation: The Booker Washington Institute of Liberia, 1929–1984*. Lexington: University Press of Kentucky, 1986.

Stanley, W. R. "Transport Expansion in Liberia." *Geographical Review* 60 (1970): 529–536.

Starkey, Otis P. *The Economic Geography of Barbados: A Study of the Relationships between Environmental Variations and Economic Development*. 1939; Westport, Conn.: Negro Universities Press, 1971.

Sterner, Richard. *The Negro's Share: A Study of Income, Consumption, Housing, and Public Assistance*. New York: Harper, 1943.

Stoler, Ann L. *Capitalism and Confrontation in Sumatra's Plantation Belt, 1870–1979*. 1985; 2nd ed., Ann Arbor: University of Michigan Press, 1995.

———. "Making Empire Respectable: The Politics of Race and Sexual Morality in Twentieth-Century Colonial Cultures." *American Ethnologist* 16 (August 1989): 634–660.

———. *Race and the Education of Desire: Foucault's History of Sexuality and the Colonial Order of Things*. Durham, N.C.: Duke University Press, 1995.

Stone, Alfred H. *Studies in the American Race Problem*. New York: Doubleday, Page, 1908.

Strasser de Saavedra, Mia, and David Saavedra. *El libro de Oro: Estudio Completto de las posibilidades agricolas y ganaderas que Panama*. Ciudad de Panamá: R. de P. Imprenta Nacional, 1926.

Strong, Richard P., ed. *The African Republic of Liberia and the Belgian Congo Based on the Observations Made and the Material Collected during the Harvard African Expedition, 1926–1927.* 2 vols. Cambridge, Mass.: Harvard University Press, 1930.

Sundiata, I. K. *Black Scandal: America and the Liberian Labor Crisis, 1929–1936.* Philadelphia: Institute for the Studies of Human Issues, 1980.

Susie, Debra A. *In the Way of Our Grandmothers: A Cultural View of Twentieth-Century Midwifery.* Athens: University of Georgia Press, 1988.

Sutherland, L. G. "Panama Gold: The Story of Negroes on the Panama Canal." *Opportunity* 12 (1934): 336–339.

Takaki, Ronald T. *Iron Cages: Race and Culture in Nineteenth-Century America.* New York: Oxford University Press, 1990.

Tandy, Elizabeth C. "Infant and Maternal Mortality." *Journal of Negro Education* 6 (1937): 322–349.

Tannenbaum, Frank. *Darker Phases of the South.* 1924; New York: Negro Universities Press, 1969.

———. *Slave and Citizen.* 1946; Boston: Beacon, 1992.

Tarpeh, James T. "The Liberian-LAMCO Joint Venture Partnerships; The Future of Less Developed Country and Multinational Corporation Collaboration as a National Strategy for Host Country Development." 2 vols. Ph.D. diss., University of Pittsburgh, 1978.

Taylor, W. C. *The Firestone Operations in Liberia.* Washington, D.C.: National Planning Association, 1956.

Teich, Albert H., ed. *Technology and the Future.* New York: St. Martin's Press, 1993.

Terborg-Penn, Rosalyn, Sharon Harley, and A. B. Rushing, eds. *Women in Africa and the African Diaspora.* Washington, D.C.: Howard University Press, 1987.

Thompson, Edgar T. "Comparative Education in Colonial Areas, with Special Reference to Plantation and Mission Frontiers." *American Journal of Sociology* 48 (May 1943): 710–721.

———. *The Plantation: An International Bibliography.* Boston: G. K. Hall, 1983.

Tindall, George B. *The Emergence of the New South.* Baton Rouge: Louisiana State University Press, 1967.

Tuan-Wich Mayson, Dew, and Amos Sawyer. "Labour in Liberia." *Review of African Political Economy* 14 (1979): 3–14.

Tubman, William V. S. *The Official Papers of William V. S. Tubman, President of the Republic of Liberia, Covering Addresses, Messages, Speeches, and Statements, 1960–1967.* Ed. E. Reginald Townsend and Abedou Bowen. London: Longmans Green, 1968.

Tugwell, Rexford G. *Industry's Coming of Age.* New York: Harcourt, Brace and Co., 1927.

Turley, David. "Slavery in the Americas: Resistance, Liberation, Emancipation." *Slavery and Abolition* 14 (August 1991): 109–116.

U.S. Army, Medical Department. *Preventive Medicine in World War II.* Vol. 6, *Communicable Diseases: Malaria.* Washington, D.C.: Office of the Surgeon General, 1963.

———. *Preventive Medicine in World War II.* Vol. 8, *Civil Affairs/Military Government Public Health Activities.* Washington, D.C.: Office of the Surgeon General, 1976.

U.S. Department of Commerce, Bureau of the Census. *Negro Population: 1790–1915.* Washington, D.C.: GPO, 1918.

U.S. Department of Labor, Division of Negro Economics. *Negro Migration in 1916–17, Reports by R. H. Leavell et al.* 1919; New York: Negro Universities Press, 1969.

U.S. Department of State. *Liberia: Documents Relating to the Plan of Assistance Proposed by the League of Nations, U.S. Department of State*. Washington, D.C.: GPO, 1933.

U.S. Office of Scientific Research and Development. *Science, the Endless Frontier: A Report to the President on a Program for Postwar Scientific Research and Development*. Vannevar Bush, Director. Washington, D.C.: GPO, 1945.

U.S. Senate, 61st Congress. *Affairs in Liberia*. Senate Document No. 457, Vol. 60. Washington, D.C.: GPO, 1910.

UNESCO. *The Haiti Pilot Project: Phase One, 1947–1949—Monographs on Fundamental Education*. Paris: UNESCO, 1951.

Vance, Rupert B. *Human Factors in Cotton Culture: A Study in the Social Geography of the American South*. Chapel Hill: University of North Carolina Press, 1929.

Venzon, Anne C., ed. *General Smedley Darlington Butler: The Letters of a Leatherneck, 1898–1931*. New York: Praeger, 1992.

Vera, Rubin, and R. P. Schnaedel, eds. *The Haitian Potential—Research and Resources of Haiti*. New York: Teachers College Press, 1975.

Von Eschan, Penny M. *Race against Empire: Black Americans and Anticolonialism, 1937–1957*. Ithaca, N.Y.: Cornell University Press, 1997.

Wallace, C. E., F. N. Marshall, and C. Robinson. "A Retrospective Review of the Maternal and Child Health Services at the Albert Schweitzer Hospital in Rural Haiti." *Bulletin of the Pan American Health Organization* 16 (1982): 7–16.

Walters, Ronald. *Pan Africanism in the African Diaspora: An Analysis of Modern Afrocentric Political Movements*. Detroit: Wayne State University Press, 1993.

Walton, Anthony. "Technology versus African-Americans." *Atlantic Monthly* 283 (January 1999): 14, 16–18.

Wang, Jessica. *American Science in the Age of Anxiety: Scientists, Anticommunism, and the Cold War*. Chapel Hill: University of North Carolina Press, 1999.

Washington, Booker T. *Working with the Hands: Being a Sequel to "Up from Slavery" Covering the Author's Experiences in Industrial Training at Tuskegee*. New York: Doubleday, Page, 1904.

Wayne, Michael. *The Reshaping of Plantation Society: The Natchez District, 1860–1880*. Baton Rouge: Louisiana State University Press, 1983.

Weschler, H., and C. F. Cahill, Jr. *The Health Horizons*. Cambridge, Mass.: Harvard University Press, 1978.

Wesley, Charles H. "The Employment of Negroes as Soldiers in the Confederate Army." *Journal of Negro History* 4 (1919): 239–253.

West, J. B. "United States Health Missions in Liberia." *Public Health Reports* 63 (October 15, 1948): 1351–1364.

Westerman, G. W. "School Segregation on the Panama Canal Zone." *Phylon* 15 (1954): 276–287.

Wiley, Bell I. *Southern Negroes, 1861–1865*. 1938; New Haven, Conn.: Yale University Press, 1965.

Williams, Walter L. *Black Americans and the Evangelization of Africa, 1877–1900*. Madison: University of Wisconsin Press, 1982.

Winner, L. *Autonomous Technology: Technics-out-of-Control as a Theme in Political Thought*. Cambridge, Mass.: MIT Press, 1992.

Wise, Frederic M., and M. O. Frost. *A Marine Tells It to You*. New York: J. H. Sears, 1929.

Wolff, Richard D. *The Economics of Colonialism: Britain and Kenya, 1870–1930*. New Haven, Conn.: Yale University Press, 1974.

Wood, Charles H., and J. A. Magno de Carvalho. *The Demography of Inequality in Brazil*. Cambridge, U.K.: Cambridge University Press, 1988.

Wood, Leonard. "The Military Government of Cuba." *Annals of the American Academy of Political and Social Science* 21 (1903): 153–182.

Woodruff, Nan E. "Debates over Mechanization, Labor, and Civil Rights in the 1940s." *Journal of Southern History* 60 (May 1994): 263–284.

Woodson, Carter G. *The Rural Negro*. Washington, D.C.: Association for the Study of Negro Life and History, 1930.

Woofter, Thomas J., Jr. *Landlord and Tenant on the Cotton Plantation*. Research monograph 5. Works Progress Administration, Division of Social Research. Washington, D.C.: GPO, 1936.

World Health Organization. *First International Symposium on Yaws Control, Bangkok, 1952*. Geneva: WHO, 1953.

Yearley, Steven. *Science, Technology, and Social Change*. London: Unwin Hyman, 1988.

Zephir, Flore. *Haitian Immigrants in Black America: A Sociological and Sociolinguistic Portrait*. Westport, Conn.: Bergin & Garvey, 1996.

INDEX

About the Author

David McBride is professor of African American studies and African American history at Pennsylvania State University. He is also with the Center for Health Policy Research at that institution. He has received research grants from the Simon Rifkind Center of the City University of New York, the Rockefeller Foundation, and the National Endowment for the Humanities. McBride has been a visiting professor at the Rollins School of Public Health at Emory University. Among his prior books are *Integrating the City of Medicine: Blacks in Philadelphia Health Care, 1910–1965* and *From TB to AIDS: Epidemics among Urban Blacks since 1900.*